PUBLIC PEOPLE PRIVATE PARTNERSHIP
FOR
SUSTAINABLE FOREST DEVELOPMENT

By the same author

—Community Participation as a Management Tool for Sustainable Forest Development

—Ecotourism and Livelihoods: Capacity Building for Local Authorities

—Training manual on people's biodiversity register

—प्रशिक्षण मार्गदर्शिका—ग्रामीण सामुदायिक जैवविविधता पंजी—जल जंगल जमीन बही

—Forestry for the next decade—managing thrust areas

—सात दिन कंधार में

IC814 Hijack Memoirs

About the Editor

A.K. Bhattacharya, an Indian Forest Service officer of 1983 batch, is presently the CEO of MP Ecotourism Development Board, Bhopal.

A Post Doctoral in Environmental Sciences, he has contributed a large number of articles in national and international journals of repute on a range of topics related to forestry and environment. He also has to his credit a book on *Kandhar Plane Hijack,* based on his own experience as a hostage.

Dr. Bhattacharya's main contributions to the field of ecotourism include organising capacity building workshops and training courses for the local authorities of the various concerned departments, communities and NGOs, extending professional support to the State Government for promotion of ecotourism, supervising research works on ecotourism related topics, and preparing Ecotourism Management Plans for few protected areas.

PUBLIC PEOPLE PRIVATE PARTNERSHIP
FOR
SUSTAINABLE FOREST DEVELOPMENT

Proceedings of the Workshop, Contributory Articles and Relevant Acts, Rules and Regulations

Edited by

A.K. Bhattacharya, *IFS*

Editorial Support

Manoj K. Dubey
Research Associate

CONCEPT PUBLISHING COMPANY, NEW DELHI-110059

Cataloging in Publication Data—DK
Courtesy: D.K. Agencies (P) Ltd. <docinfo@dkagencies.com>

Workshop on Sustainable Development of Degraded Forests and Wastelands Through Private Sector Participation with a View to Increasing/Improving Tree Cover : Issues and Policy Options (2003 : Bhopal, India)

Public people private partnership sustainable forest development : proceedings of the workshop, contributory articles and relevant acts, rules and regulations / edited by A.K. Bhattacharya ; editorial support, Manoj K. Dubey.

p. cm.

Includes bibliographical references (p.)

ISBN 8180692531

1. Sustainable forestry—India—Congresses. 2. Community forestry—India—Congresses. 3. Public-private sector cooperation—India—Congresses. I. Bhattacharya, A. K. II. Title.

DDC 634.909 54 21

ISBN 81-8069-253-1

First Published 2007

Published and Printed by
Ashok Kumar Mittal
Concept Publishing Company
A/15-16, Commercial Block, Mohan Garden
New Delhi-110 059

Phones : 25351460, 25351794
Fax : 091-11-25357103
Email : publishing@conceptpub.com

PREFACE

> ***"Forestry by the masses i.e. participatory forestry can be complemented by forestry for the masses through private sector initiatives."*** —Dr. M.E. Chipeta, 2000

Over the last decade private sector involvement in forestry has significantly increased in many nations around the world. However, in India, forestry professionals and government itself are yet to be convinced for the need for private sector involvement for the promotion of forest cover, in particular for the rehabilitation of degraded forests/wastelands to achieve the ultimate goal of sustainable forest development.

To address on the mentioned topic, a workshop on *Sustainable development of degraded forests and wastelands through private sector participation with a view to increasing/improving tree cover: Issues and Policy options* was organised in Bhopal during June 6-7, 2003 by the Indian Institute of Forest Management, Bhopal. The workshop was sponsored by the Ministry of Environment and Forests, Government of India, New Delhi and attended by more than 30 delegates from the Central and State Governments, NGOs, industrial houses, financial institutions and research organisations. The main objectives of the workshop were to examine and analyse the issues to be addressed for public private partnership for the rehabilitation of degraded forests/wastelands and to suggest remedies for resolving the main constraints of the aforesaid partnership. The distinctive feature of the workshop was the focused discussions by clubbing the participants into groups with assigned themes. Finally, the group leaders presented the recommendations of their groups which was followed by open discussion.

In the backdrop of aforesaid workshop, this book contains various articles relevant to public private partnership in the Indian forestry scenario, contributed and presented by eminent Government officials, NGOs, industrialists and researchers. In the end, recommendations made by each group of the workshop are summarised for follow-up action.

We would like to thank our sponsor, authors and delegates who contributed to the success of this workshop. We express sincere thanks to all who have contributed directly and indirectly for bringing out this book for the future discourse.

Bhopal **A.K. Bhattacharya**

CONTENTS

Appendices

THE CONTRIBUTORS

A.K. Bhattacharya, Conservator of Forests, MP Forest Department, Bhopal.

Anil Agarwal, Associate Professor, Yale University, USA.

Bijendra Basnyat, Consultant, NARMA, Kathmandu, Nepal.

Debasis Bhattacharya, Social Development Consultant, CII, Delhi.

J.B. Nugent, Consultant, CII, Delhi.

J.K. Rawat, presently Conservator of Forests, Social Forestry, Rohtak (Haryana).

K.G. Vyas, Adviser, Rajiv Gandhi Watershed Mission, MP, Bhopal.

Kailash Chandra Bebarta, Conservator of Forests, Chhattisgarh Forest Department, Raipur.

M. Chakrabarti, Director, CREATE; Darjeeling, West Bengal.

M.M. Tripathi, *IFS* (Retd.), MP Forest Department, Bhopal.

Manoj Kumar Dubey, M.Sc., University of Wales, Banjor, U.K.

N. Ali, Consultant, CII, Delhi.

N. Kumar, APCCF, MP Forest Department.

Piare Lal, ITC, Bhadrachalam, AP.

R.C. Saxena, *IFS* (Retd.), Executive Director, National Centre for Human Settlements and Environment, Bhopal.

R. Gopal Krishnan, Secretary to Hon. Chief Minister and Coordinator, Rajiv Gandhi Mission, MP.

S.D. Mukherjee, PCCF, Andhra Pradesh Forest Department, Hyderabad.

S.K. Dutta, Consultant, CII, Delhi.

S.N. Mishra, *IFS* (Retd.), MP Forest Department.

S.S. Shrivastava *IFS* (Retd.), MP Forest Department.

Swati Bakshi, Project Associate, IIFM, Bhopal (MP).

V.P. Singh, CCF (Central) GOI, MoEF, Bhopal.

1

BACKGROUND

The ever-increasing pressure on our scarce forest resources to meet the national, social and industrial needs is a cause of concern for everyone. The per capita forest area has reduced from 0.11 ha in 1983 to 0.074 ha in 2001 (FSI, 2001). According to Wasteland Atlas of India (2000) the total area of wastelands in the country is 6.38 lakh sq km, which includes 1.4 lakh sq km of degraded notified forestlands. Since the biotic pressure on the forests continues unabated, putting a halt to the degradation and rehabilitating the degraded forest area is likely to remain a major challenge.

A very practical method of coping with the biotic pressure is to involve the village communities to protect the forest and to share the benefits therefrom. Joint Forest Management is the outcome of this philosophy. JFM programme has helped to mitigate the process of degradation to some extent. Similarly, some degraded areas can be rehabilitated through agro-forestry practices. However, the magnitude of the problem is too large to be solved by these approaches alone. Moreover, the resources required for this quantum of afforestation would run into thousands of crores. The constraint of funding has been a paramount problem in forestry sector right from the First Five Year Plan. The National Forest Policy of 1988 has, therefore, emphasised the necessity of financial support mentioning that "The objectives of this revised policy cannot be achieved without the investment of financial and other resources on a substantial scale." Along with the participation of people, the participation of corporates, specially the forest based industries, could make a impact in mobilising the financial resources required to restore and increase forest productivity.

Just as Government has realized the need to attract private capital (even capital from abroad) to meet the growing demands of the infrastructure sector (energy, highways etc.), it is high time that the need to do the same for raising forest is also realized. It is, therefore, suggested that some of the degraded forests/wastelands be made available to entrepreneurs who wish to invest their money into this sector, in accordance with preset conditions and terms, to put these lands under effective plantations based on the potential of the site.

The workshop provided a common platform for different stakeholders to examine and analyse various issues pertaining to involvement of communities, individuals and corporate in forestry development, specially the stakes of local communities, leasing of forest lands *vis-a-vis* the Forest Conservation Act, sharing of produce and profit, socio-economic and ecological implications etc., in such a way so that a proper strategy may be evolved to involve this sector on a sustainable basis for mutual benefit of all, ensuring the long term interests of sustainable forest development.

The workshop mainly focused on resolving the bottlenecks and exploring remedies to the problems identified, so as to evolve a pragmatic and effective strategy for rehabilitating degraded forests and wastelands. The main objectives and issues of the workshop are as follows:

Main Objectives

- To identify and analyse various issues pertaining to participation of communities, individuals and corporates in rehabilitation of degraded forests, especially in Madhya Pradesh.
- To examine the issues and constraints and suggest remedies to resolve the conflicts through the opinion of various stakeholders.
- To evolve a strategy for participation of various non-government sectors in forestry development acceptable largely to all stakeholders.

Major Issues

1. Historical background of participation of communities, individuals and corporates in forestry sector and identification of forest areas and wastelands for their participation for sustainable forest development

1.1 History of participation

- History of participation of indigenous people/individuals/ corporates in forest management.
- Previous efforts of participation and their results and experience.
- Traditional agro-forestry practices in different parts of the state.
- Social and cultural linkages with the systems of forest use in different parts of the state.
- Evolution of forest policy/legislative environment for rehabilitation of forest/revenue wastelands.
- Impact of forest policy/laws on reclamation of wastelands; in historical perspective.
- History of individual initiatives in forest conservation.
- History of private plantations in the country/state.
- History of private/public sector investment in government forests and wastelands.

1.2 Identification of degraded forests and wastelands

- The criteria for identification of degraded forests and wastelands.
- Categorization of forest and revenue wastelands.
- Criteria for extent of wastelands (forest and non-forest) to be made available to private sector under PPP arrangement.
- Use of state-of-art technology in identification of degraded forests and wastelands.
- Issues of prioritisation between forest and revenue wastelands.
- Issues of ownership/possession.
- Identification of successful/experienced entrepreneurs in the field of plantation.

2. Ecological and Silvicultural issues pertaining to the development of forests and forest produce including NWFP

2.1 Ecological issues

- Application of ecological principles in the rehabilitation of degraded forests.
- Synchronizing bio-diversity and wood production in management practices.
- Prioritisation between monoculture and mixed cropping.
- Environmental impact of forest degradation.
- Impact of private plantations on environmental amelioration.

2.2 Silvicultural issues

- Determination of objective of management in rehabilitation of forest wastelands.
- Issue of harmonizing multiple objectives of management.
- Sivicultural techniques for maximising yield.
- Standardisation of silvicultural thinning in mixed forests.
- Sustainable harvesting techniques for NWFP in post-rehabilitation phase.
- Issues pertaining to high-tech plantations.
- Optimisation of soil resources.
- Selection of site, appropriate species and techniques of rehabilitation.
- Use of site specific soil and water conservation techniques.
- Development of sustainable silvicultural systems for meeting rural community needs.
- Relevance of traditional silvicultural systems and prospects of change in rehabilitation of degraded forests/wastelands.

3. *Legal issues, Government rules and relevant acts with respect to private investment in forestry sector*

- Strengths and weaknesses in the existing legal framework with respect to private investments in afforestation.
- Changes in forest and land laws necessary for the promotion of private investment in rehabilitation of degraded forests and wastelands.
- Compatible fiscal changes for promoting private sector investment Rehabilitation of Degraded Forests and Wastelands (RDFW).
- Incentives for investments in RDFW.
- Issues pertaining to CPR and PPR (Private Property Regimes) and user rights.

4. *Socio-Economic Issues and JFM perspectives in private sector investment*

- Impact on the existing rights of the local communities.
- Developing appropriate mechanism for usufruct sharing from the plantations.
- Compensation for successful afforestation efforts.
- Integrating 'Nistar' requirements of local people with investment patterns.
- Issues of social responsibility of private sector participating in RDFW.
- Financial Issues including pattern of sharing and investment
- Investment sharing *versus* usufruct sharing.
- Constraints in institutional financing.
- Fluctuating rate of interest.
- Direct lending by NABARD.
- Long term fiscal policy.
- Creation of separate forestry bank for promoting investment in forestry, especially for RDFW.
- Priority for lending in forestry sector, especially in RDFW.

- Tax concessions for promoting tree cultivation in degraded forests/wasteland.
- Creating a separate corpus for forestry sector by raising public money from green thinkers and environment friendly corporates.

5. *Administrative and Management Issues with respect to Institutional Framework*

- Creation of a separate financial institution like Grameen Vaniki Bank to promote and facilitate financial services in the rural sector.
- Developing a practice of 'chartered forester' in the private sector, for preparing management plan for RDWF.
- Decentralization of process of sanctioning management plan.
- Dispute resolution/arbitration.
- Direct investment *versus* investment through community based organizations.
- Creating data bank for (1) status and potentiality of wasteland, (2) monitoring changes in vegetation and productivity of the land.
- Strength and weaknesses of the existing institutions responsible for development of RDWF.

2

PUBLIC PEOPLE PRIVATE PARTNERSHIP FOR SUSTAINABLE FOREST DEVELOPMENT

AN INTEGRATED HOLISTIC APPROACH

A.K. BHATTACHARYA

Backdrop

Forests provide goods and services to society and constitute the life support system to the planet. The survival of mankind depends on the continuance of forest systems. Over time, anthropogenic pressure on our forests increased manifold and transcended the threshold limit. As a result, depletion of forest resources has begun and more and more forests in India are being increasingly degraded day by day.

The phenomenal increase in forest degradation in the country has assumed an alarming proportion over the last two decades. Approximately, 57 per cent forest area of the country is in degraded state. As per NRSA survey report, the estimated wastelands in the country account for about 75.5 million ha, which includes 16.3 million ha of degraded forests. Degraded non-forest land available for plantations is about 33 million ha.

Madhya Pradesh having the largest forest area has to tackle the challenge of rehabilitating 72 lakh ha degraded forest area, which constitutes 46 per cent of State's recorded forest area. Since the anthropogenic pressure on the forests continues unabated, putting a halt to the degradation and rehabilitating the degraded forest area remains a major challenge. In order to retain and subsequently enhance the green cover of Madhya Pradesh as well as of the country, it is essential that both degraded forests and cultivable wastelands be put under effective rehabilitation programmes based on the potential of the respective sites.

The pace of degradation of forests is faster than the present human endeavour to rehabilitate. Therefore, there is an imperative need to arrest the process of further degradation on one hand and rehabilitate the already degraded forests on the other. This phenomenon is witnessed all over the country. In order to achieve this two-pronged objective, we need to embark on Sustainable Forest Development.

Sustainable forest development consists of the management and conservation of forest resource base and the orientation of technological and institutional change in such manner as to ensure the attainment and continued satisfaction of human needs for the present and future generation. Such sustainable development should be environmentally non-degrading, technologically appropriate, economically viable and socially acceptable.

In the case of degraded forestlands, sustainable development necessitates immediate amelioration and augmentation of depleted forest resources, and after developing the depleted resources, the sustainable use of the resources by the people for the present without impairing the potentials of the future generation begins. The process of amelioration and augmentation of the resource base requires people's participation on the one hand and financial and technological support on the other.

Involving communities through JFM has helped to mitigate the process of degradation to some extent. Similarly some degraded areas have been rehabilitated through agro-forestry practices. But the rehabilitation of degraded forests and the goal of sustainable forest development cannot be achieved without assured sustained financial support. Therefore, for the integrated ecological development of forests and promotion of the multiple forest products for use by the communities, it becomes inevitable to explore all possibilities of *participation of communities, individuals and corporates in forest development programmes.*

Non-wood forest products (NWFP) originating from diverse sources ranging from large plants to micro-flora, consisting of heterogeneous products constitute a critical lifeline for the poor forest dwellers by providing family sustenance and livelihood. Rural income generation for the poor forest dwellers can be augmented through value addition and village level processing of the forest produce. Conservation and propagation of NWFP is equally important to ensure non-destructive harvesting on sustainable basis.

In the last decade, the country heralded in an era of market reforms in the country. Apart from this, the nation embarked on a massive programme of people's participation in forest management. These two developments have significant bearing on the future of forestry in India.

Involvement of people in forest management practices has improved the prospects of rehabilitating degraded forests in many parts of the country, most particularly in Madhya Pradesh. The beginning of an era of liberalization in the year 1991, unfolded enormous opportunities for the private sector to invest in activities which were presently undertaken by State. However, after a decade, the forestry sector has failed to attract significant private participation. It is also a fact that without sufficient financial support, it will be impossible to rehabilitate all the degraded forests.

History of Private Participation in Forestry in Madhya Pradesh

The efforts to encourage private participation in M.P. Forestry dates back to 1983-84 when Government of Madhya Pradesh decided to launch companies in Joint Sector. Towards this end, M.P. Forest Development Corporation collaborated with private sector companies to form such joint sector companies which were aimed at increasing private participation in rehabilitation of degraded forests. But, this scheme did not bear fruit because of the Government of India's clarification that such efforts were not in common with the Forest Conservation Act. Again, in the year 1997, the Government of Madhya Pradesh made an effort to lease out the degraded forests for rehabilitation with a system of a tie-up with private sector companies who were to take a part of the harvest. But again the effort did not get the clearance from Central Government.

Presently, the Government of India has issued guidelines for participation of private sector through involvement of NGO's and Forest Departments in afforestation/rehabilitation of degraded forests. The Ministry of Environment and Forests, Government of India has shown great concern over the rate of slow rehabilitation owing to financial and other constraints. It is also an accepted fact that no government is in a position to restore the existing degraded forests from the available resources. Therefore, Government of India advised the States to adopt innovative policies with regard to rehabilitation of degraded forests. It has been found by the Government of India that a large number of environment friendly industrial houses are willing to join hands in this movement through NGOs and under the supervision and guidance of the forest departments voluntarily without any rights whatsoever on the forest land or the usufructs. Towards this end, the Government of India has issued guidelines for participation of private sector through involvement of NGOs and forest departments in afforestation/rehabilitation of degraded forests.

The Governments of these States have also initiated a process of creating a congenial environment for increasing private participation in sustainable forestry with respect to non-forest revenue waste land. The Government of M.P. has already issued directions/guidelines for giving such lands to private sector on lease for afforestation and rehabilitation purpose. Apart from this, the Government of M.P. has embarked on "Lok Vaniki Mission", a programme to promote private sector in sustainable forestry. Government of M.P. is also actively considering to improve the private propriety regime with respect to tree and other forest products; so that the State can emerge as the biggest tree grower in the private land, after having optimized productivity of the land resources, and developing the socio-economic conditions of the people. Towards this end, Government of M.P. has already de-liberalized the forest produce transit system. At present, forest department issues transit passes for six tree species and the Panchayat issues transit passes for another ten species and there is no need of transit passes for the rest of the species.

Community participation has emerged dominantly in the forest management scenario all over the country. Madhya Pradesh Forest Department remains the harbinger in its march towards people's participation in forest management till today.

PPP and Sustainable Development

During the 1970s and 1980s, the corporate responsibility agenda was closely associated with the impact of business activities on the natural environment. The International Chamber of Commerce introduced the ICC Environmental guidelines for Industry in 1974. Although initially quite general, the guidelines were subsequently supplemented by sector-specific codes developed by various industry associations. Good corporate citizenship is closely associated with the idea of sustainability. Sustainable Development (SD) is one objective, for which corporate social responsibility activity has been targeted. One way in which sustainable development has been defined is as the means of meeting the needs of present generations without compromising the ability of future generations to meet their own needs (Brundtland Commission, 1987). Sustainability has four elements: (i) Social Sustainability, which is achieved when social exclusion is minimised and social equity is maximized. (ii) Environmental Sustainability, which is achieved when the productivity of key natural resources is conserved or enhanced for future generations. (iii) Institutional Sustainability, which is achieved when prevailing, structures and processes have capacity to continue to perform. (iv) Economic sustainability, which is achieved if a baseline level of economic welfare can be achieved and sustained (SCL, 2000).

Many financial markets are also waking up and are increasingly interested not only in eco-efficiency, but in business ethics, corporate social responsibility, human rights, which all are potentially part of the Sustainable Development Agenda. The environment, ethics, human rights are key issues at play. The ability of business leaders and corporate boards to pick up and amplify "weak signals" in their business environment will be an increasingly crucial factor determining which corporations are and can remain globally competitive.

The proper social responsibility of a business is to turn a social problem into economic opportunity, economic benefit into productive capacity, into human competence, into well paid jobs and into wealth (Anon, 2002). Two major issues that are being felt to be the foundations of future global CSR are poverty and education. It is felt that the vicious circle – poverty, generator of poor quality of life, depriver of education, generator of poverty – must somehow be broken if people are ever to hope for a better way of life for all.

The solutions to the world's most urgent problems will require the private sector to embrace, support and enact a set of core values in the areas of human rights, labour standards and environmental practices, as proposed by UN.

Challenges and the Opportunities through PPP

A government circular focus on participatory and farm forestry was reaffirmed by both the Ministry of Environment and Forests expert committee on 'Review of afforestation policies and rehabilitation of wastelands' (1997) and the Planning Commission's working group on leasing of degraded forest lands (1998). Both committees considered the potential extension of the principle to participation in company ventures to rehabilitate wastelands. The idea is to encourage private investors to invest in afforestation in degraded areas by forming partnerships with local farmers. (However the Planning Commission, in particular, did not endorse the involvement of large industrial investors).

While increased private sector participation, either by individual farmers or forestry companies, is recognised as a key component of forestry reform in India, less attention has been given on how to create an environment favourable to private investment and how the forestry administrative structure should be changed to reflect its changing role. The first of these gaps in policy appears to have been tentatively addressed by the expert committee. It proposes three sets of market-based instruments to attract private finance. These include the liberalisation of the timber market, which has been controlled by felling and transport restrictions, the introduction of a new tenure arrangement to permit private leases of large forest areas, and the introduction of financial incentives such as subsidies and tax breaks. The committee's report makes no mention of the need for forestry authority restructuring. However, it appears that none of these proposals have yet been acted upon to any great extent.

Private forest ownership amounts to only 4 per cent of the total forest area, as compared with the state's 85 per cent and communities 11 per cent (FSI, 1998). The private sector is not permitted to own natural forest, but its ownership of planted forests is limited by the Private Forests (Acquisition) Act of the 1950s and the Land Ceiling Act of the 1960s. The former provided for the nationalisation of private forests and led to much felling, while the latter specifically limits the area private enterprises can own for tree plantations (except in the case of plantation crops such as rubber or tea). Land ceilings vary by state, but are normally set so low that they effectively bar commercial investment in plantation forestry.

Given that a lot of developmental challenges face the nation, the PPP agenda for India is vast. Opportunities are immense to convert the poverty issues into business opportunities. Environmental issues and empowerment of poor through employment generation, increasing their earnings and purchasing power can upgrade their living standards and also harness rural energies for business advantage. By focusing on the conditions and reasons for poverty, businesses can create assets by creating proper strategy and using established rural markets.

Therefore, there is an imperative need to integrate people's participation and private sector participation in the task of rehabilitating the vast chunk of degraded forests. Without harmonized synchronization between the two aforesaid factors, this task shall not be accomplished.

3

RE-GREENING INDIA
ISSUES AND CHALLENGES TO ENCOURAGE PRIVATE SECTOR PARTICIPATION

DEBASIS BHATTACHARYA

Introduction

In spite of increase in private sector participation in forest management and development in many countries, currently in India there are two opposing views being debated on the role of the private corporate sector in raising captive plantations on degraded forestlands. One view supports it based on the following rationale: (i) resources required for large scale technology based high-yielding plantations are considerable and may not be easily available in the public budget; (ii) involving the industry owners in establishing high-yielding fast growing plantations in a partnership arrangement on degraded forest lands may result in accelerated rehabilitation of these lands and thereby help re-greening of India; (iii) making industry self-reliant in their own raw material supply is cost-effective as more public resources could then be diverted for better protection and conservation of natural forests; and (iv) making private sector responsible for wood production will intensify joint forest management (JFM) efforts of the forest departments.

Another view vehemently opposes any partnership arrangement with the private sector industries for establishing technology-based plantations on public forest land. They call it a sell-out of forestlands. Their opposition centre around the following: (i) private sector in

India particularly those in the wood-based industry has traditionally been exploiters of forests and therefore, cannot be trusted in their new role as the developer of forest resources; (ii) leasing or outright sale of degraded forestlands may result in diversion of forestlands for other uses; (iii) forestlands should as far as possible be managed for promotion of biodiversity and meeting community's needs and should not be used for monoculture plantations; (iv) land for such plantations should be procured from government non-forest lands and private lands in the market place; (v) allowing use of degraded forest lands may deprive the community of their rightful requirements of small timber, fuel wood and leaf fodder.

Opportunities for India

Except in the case of Joint Forest Management (which is also not as transparent as it should be) until now, in India forestlands are directly managed by the government, or given in usufruct to communities under various arrangements. Private trees are grown on farms or community lands. Harvesting is usually carried out by the State Forest Development Corporations (SFDCs) whereas transport and processing are the responsibilities mainly of the private sector. The management of natural forests and direct sale of timber provide substantial revenue to the State Governments for financing major development activities. Forests also provide direct employment to a large section of rural people. Forests in India also help preserve biodiversity and regulate global climate including air pollution.

There is a consensus that degradation of forest cover in the country has been caused by population growth and illicit removal and other development pressure. Thus the Joint Forest Management (JFM), which involves instituting sustainable forest management options in degraded natural forests through participatory processes and collaborative arrangements with the local communities, has become one of the corner stones of future forest management options in India. It will be mainly a government-sponsored programme and may cover about 10-15 million ha. of degraded forest lands. The restoration of these lands will be time consuming and will cater to the subsistence demand of the local population. The pace of the JFM programme will be constrained by government budgetary support.

The national demand for timber has to be still met from India's forestlands.

In the coming decades the natural high forest management as discussed in earlier chapters will decline as a source of industrial timber supply and instead preserve management of parks and protected areas through eco-system based approaches (in which local communities including indigenous peoples are recognised as an integral part of the eco-system) for biodiversity conservation and providing non-timber values will become a more important management option.

Wood production and timber trade trends indicate that the contribution of natural forests in the production process for the timber-based industries will continue to decline and technology based plantation forestry which permit substitution of capital and technology for land will be the main source for meeting the industries requirements for timber. *Should then large areas of degraded forest lands which are currently producing very little continue to remain under the managment of the public agencies particularly since the state governments do not have enough resources to invest to make these lands productive? Would it not be better for the Government to divest this resource as a non-performing asset in the market place just like any other non-performing public sector agencies as is being done in respect of the PSUs.*

This policy option will open new opportunities for investment in forestry plantations as well as in carbon sequestration with a scope for carbon trading in the international market given that forestlands are available to the private sector at reasonable prices in the market place or in some sort of long-term tenurial arrangements to the private sector.

Option 1 : Divesting Degraded Forest Lands in the Market

Option 1 explores the possibility of divesting in the market place about 5 million ha. of degraded forest lands in India which can be put to technology based high yielding plantations (at 0.25 million ha./year) for the next 20 years at an investment cost of US $130 million per annum by the private corporate sector. This may at the end of the period yield on a sustained basis enough timber to meet the demands of the timber-based industries on a sustained basis in the market place. Further it may enable the private sector to earn an income through carbon trading at the same time providing rural employment (each ha. of plantation generates about 875 man-days work per ha. Therefore, 0.25 million ha. will provide about 220 million days of work annually).

The rationale for this disinvestment option of these non-performing forest lands are the following: (i) the investment costs for restoring the productivity of these forests are enormous and the public agencies will never be able to access this quantum of resources from the public budget in the near future to make them productive; (ii) due to inadequacy in enforcement measures these forest lands are getting further degraded and may reach such an ecologically degraded state from where restoration of these lands for productive purposes may be difficult; and (iii) considering the costs involved and the difficulty to manage these under JFM option (due to already a current ambitious JFM programme) the proposed policy option may be a viable alternative.

It may be appropriate and logical for the government to *spin-off to the private sector,* who have necessary resources through outright sale (after proper valuation), these 5 million ha. of lands which could be suitable for intensive management through plantations. This may provide a total quantum of resources from Rs. 2500 to 5000 crores (US$ 1 to 2 billion) to the Government Funds received from this sale can be re-deployed either in a *trust fund* to support non-timber and other aspects of development on the remaining forests (e.g. non-timber forest production) that provide substantial public values. This will reduce the problem of enforcing protection measures also. This will also be in line with the government's current *disinvestments* policy of the public sector agencies. This option to generate funds for the forestry sector may be a good option as more funds can then be injected to the sector through "Trust Fund" mechanism (Binkley *et al.*, 1999).

Option 2: Joint Sector Plantation Companies with Private Corporate Sector

This option is proposed as the second best solution in view of the political difficulties that may constrain selling off forestland in the market place and the difficulty of dedicating the sale proceeds specifically to the forstry sector due to budgetary process. The option considers the establishment of a joint venture what state forest corporations (SFDCs) with private corporate sector (forest based and other green face companies) in collaboration with local communities. The SFDCs will provide equity in terms of the forest lands which will be valued and the private sector company will bring the technology, financial support and management know-how. The choice of species and sharing of products specially the wood may be regulated through a management plan acceptable to the three stakeholders along with a memorandum of understanding (MOU) for the sharing of produce. The technology based plantations by the FDCs or in collaboration with industry and the local people should follow the draft guidelines developed in this regards.

New Policy Orientation and Policy Research

The required new policy orientation needed to promote private sector participation has been grouped as issues under the following categories for further deliberation in the workshop (i) what should be the appropriate incentive framework to promote PSP; (ii) what types of tenurial reforms should be adopted to promote PSP on public forest land; (iii) trade policies; (iv) redefining the role of the public forest agencies; (v) institutional sensitisation and main streaming information; (vi) policy research; and (vii) building partnership initiatives and cooperation with global and regional initiatives.

Tenurial Reforms *Versus* Forest Land Ownership

There is no doubt in the context of India this is the most important issue. There are three options available. (i) Outright sale disinvestments of public forestland; (ii) leasing of forestland on renewable basis; and (iii) tripartite joint sector venture with the public agencies—private sector and the local community.

Sale/Disinvestments of Public Forest Lands in the Market: This is probably the ideal solution as the private ownership of forest lands provides the opportunity to the private sector for long term investment commitment. In order to dispose of/sell the public forest land an appropriate valuation of the land will be necessary. This may require consultancy from outside, as the government must get a fair price for the land. The required policy decision to divest the public asset will be the first prerequisite before venturing on divesting.

Long Term Renewable Lease: In case outright sale of public forest land is politically unacceptable, a long term renewable leasing arrangements as has been carried out in the Philippines may be a second best solution. Here also valuation for the purposes of determining lease rental needs to be initiated preferably with the help of outside consultants. In this case the stake of the communities living in and around these forest lands have to be taken into consideration by incorporating in the lease deeds their right to the usufructs once the plantations are established.

Tripartite Joint Sector Venture : This alternative of a tripartite public-private community joint venture arrangements may be an ideal solution which will enable the public agencies to keep the land as equity, the private sector providing the financial resources and the technical know-how and the community providing the protection and labour within the framework of some cost-sharing arrangements. In the context of a country that is reluctant to give away forestlands and yet wishes to involve the private sector, this could be an ideal solution.

Redefining the Role of the Public Agencies: If the production functions of the forests are going to be privatised through policy changes and tenurial reforms then the role of the public sector agencies have to be redefined to undertake (i) preserve management of natural forests; (ii) management of parks and protected areas; (iii) promoting forestry as an alternate land-use; (iv) management of refractory sites; and (iv) promoting private sector participation. Is it necessary for the Public agencies to change their roles?

Changed emphasis in sector management will require a management process to continuously define the roles and responsibilities of the public agencies consistent with the envisaged changes. It may be necessary to prepare a code of *Forest Practice* and strategic environmental overview and a monitoring and evaluation of performance built around (i) monitoring of key performance indicators in line with the new roles; (ii) programme management information particularly if the discreet production responsibilities are privatised;

and (iii) independent evaluation studies including beneficiary assessments. Public agencies will need to be stakeholder orientated, adaptable and sensitive to the fact that there will be a change to a decentralised decision making process. The responsibilities of the foresters in the public agencies will change, as a result there will a need for improved skills and skill up-gradation.

Institutional Sensitisation and Main Streaming Information: It is necessary for the public sector agencies to modify the unclear institutional arrangements in order to raise its efficiency to enhance PSP. The weak linkages between various organs of the agencies should be streamlined and division of responsibilities should be made more explicit. Domestic pressures from cause-oriented pressure groups who oppose PSP should be resisted by mainstreaming information. The institutional culture should be modified to integrate broader reforms in the forestry sector along with the other sectors of the economy. The institution should also align itself with the directional shifts being adopted by the global community. It will also be necessary for the institution to abide by the global commitments on PSP.

It may be prudent to examine the six process-focused IT modules that are in use in the forestry sector elsewhere for improved institutional capacities and mainstreaming PSP: (i) finance and administration module; (ii) forestry operation module; (iii) forest management plan preparation module; (iv) biodiversity information module, (v) planning information module; and (vi) monitoring and evaluation module. The prospect for introducing *local area network* (LAN) as a communication link is essential for improved institutional arrangements.

Policy Research

Forest policy related research can play a crucial role in promoting PSP. The challenge is to make it demand-driven and an interactive process involving the private sector. Due to smaller economic interest compared to, for example agricultural research, financial investment in forest policy related research has traditionally been low, resulting in inadequate infrastructure and facilities, a smaller number of research personnel and a low level of skill.

Research capacities (infrastructure and human capital) need to be improved. While the planning and co-ordination of research and development activities and the co-operation of people and institutions need to be strengthened. In this regard, changes are required in the planning and management of forestry policy related research, with a view to involving *users* of research in identifying problems and opportunities and to providing appropriate field orientation. Promotion of *user-managed* research is an essential step in the right direction. Also a more rigorous approach to the preparation and implementation of research is needed in order to raise the status of forestry-related research to increase its relevance to real-life situations.

Forest policy related research activities in line with the management options outlined in this paper have to be embedded in: (i) the identification of relevant research topics, in a demand-driven process; (ii) the generation, compilation and dissemination of information to interested stakeholders in a consultative and participatory way; (iii) the generation and preparation of and making available of research results to stakeholders such as policy decision makers, development planners etc.; (iv) the setting up and running of local, national and international network communication system to improve the exchange of information among researchers, extensionists and users.

A list of worthwhile policy based research topic is detailed for further discussion: (i) identification, listing and analysis of a set of conditions that will promote PSP in plantation development on public forest lands; (ii) an inquiry into the nature and need for specific incentives that may encourage PSP in plantation as well carbon sequestration on public forest land; (iii) analysis of costs and benefits of various land tenure options—open market sale, long term lease, joint sector ventures for involving PSP on public forest lands; (iv) analysis and synthesis of conditions for promoting PSP on public forest lands; (v) analysis of benefits and costs of alternate models of private—public-community partnership—contractual dimension, equity in partnership, outgrower's scheme, partnership in processing etc.; and (vi) criteria for selection of the private sector for participation on public forest lands. These policy researches may help in focusing attention of policy and decision makers to make informed decisions.

Building Partnership Initiatives and Cooperation with Global and Regional Initiatives

The complementarity and close relationship of the private corporate sector and the public agencies in forests resources management in the pursuit of achieving environmentally sustainable development have resulted in many international and regional initiatives in support of PSP. It is necessary to be proactive to reinforce and/or establish partnerships with these initiatives to derive maximum benefits in the coming decades. Forest management in developing countries particularly in India will greatly benefit from maintaining and improving its regular consultation on forest issues with such international initiatives.

Special attention should be given to building partnership with the new funding organisations like the *Sylvan Fund* and the Geneva based *World Business Council* for Sustainable Development, a body of 150 Corporate Groups who have decided to follow the sustainable development path as their corporate goal, multinational corporations with a "green face" and international transfers aimed at PSP and SFM. In this regards the following three groups are important: (i) participation of companies and associations interested in undertaking projects for joint implementation of forestry programmes; (ii) agri-business companies interested in the conservation of biodiversity in order to save genetic information in wild species that may prove valuable in the production of drugs or other substance; (iii) portfolio or other investors interested in acquiring stakes in environmentally friendly companies who should be encouraged to take up commercial seedling production.

Conclusions and Consequences of Choice

Forestry in the developing countries particularly in India is heading towards a new process change. The forestry will find it difficult to function in the new environment if they follow the traditional forest management practices that espouse multiple use forestry but basically focus on timber management. Developing countries foresters have long worried about the image of forestry and sought to "educate" the public in the hope that people would take an interest in forestry. Now people are interested and in fact probably more interested than foresters. Foresters are being asked by them to be managers and producers of forest products but also the curators of the rare, the fragile and the beautiful. Social pressures are pushing Indian foresters to take a wider landscape or eco-system approach to management. Thus forestry particularly in India is at the crossroads where consequences of choice will have a great bearing on the way forests will be managed in the new millennium. Foresters can choose to give the responsibility of producing timber, fuelwood and others to the stakeholders and manage the wider eco-system by themselves. This will make them more efficient in preserve management and enable them to serve the society better.

4

SUSTAINABLE DEVELOPMENT OF DEGRADED FOREST AND WASTE LANDS

ISSUES AND POLICY OPTIONS

K.G. Vyas

It is strange that even after 50 years of Independence, we are still without an effective national policy for sustainable development of degraded forests and waste lands. No nation can allow this situation to persist. The continued degradation of our degraded forests and waste lands must be arrested. It is now being felt and persistently realised that poverty and degradation go hand in hand and implementation of appropriate programmes to save degraded forests and waste lands must address poverty alleviation to reverse the situation and achieve sustainable development.

The problem of deteriorating natural forests and land resources is complex and exceedingly difficult, not only because of its sheer size, but also because of various internal and external factors influencing the process. We cannot ignore the problem yet the challege ahead is extremely difficult to address without associating human beings in real sense. We must accept the reality that in spite of our concentrated efforts we have not been able to get the desired success. Before we consider what would be the best way of meeting it, it would be necessary to understand and examine, in some detail, the major problems which need solution.

Education and Involvement of the People

The foremost problem as conceived by all for appropriate activities is the education and the facilitation of the people in all matters related with the issue. Till now, we have, through legal and penal interventions tried to reverse the cycle of degradation of forests and waste lands. Our thrust has been on the rules, regulations and more authority for managers. In many cases this has not worked or has not given desired results. What seems to be more reasonable and appropriate is to address the problems, needs and issues of community in improving their income generating capabilities through better management of forests and waste lands. It is necessary to convince ourselves rather than to convince community. Efforts made by Madhya Pradesh in empowering and involving rural community has unmistakenly demonstrated that natural resources can best be managed and regulated by community but lot more has to be done in favour of community to ensure success. This is not simple because managers feel otherwise. Madhya Pradesh experiment has yet to gain strength and widespread coverage and acceptability by the community. Lot more has to be done in not only empowering people but also through appropriate policies with a swing in favour of community and community controlled management and implementation of natural resource management.

Resource Assessment

The second major problem in the field of degraded forests and wastelands concerns our vision about these resources in the light of the past experiences. We have been insisting since long for proper nation-wide status survey. In insisting this, we forget that till now wastelands have been our last priority. We have ignored the complexities and variations associated with the degraded forests and wastelands. In search of biomass, community has mercilessly exploited them and our rules, regulations, provisions and vision has not been able to provide adequate strategy to combat the situation.

Surveys reflecting true status of waste lands in degraded forests have been carried out in many areas but in many cases have suffered in details and accuracy. Probably we have to look for a new approach and work culture to address the problem of resource assessment. We also need to have an accurate idea of our soil resources in these areas and to locate and identify the areas which need priority attention. Many people, now argue that in the process of resource assessment, it is the primary stakeholder who should be given the key role and the external agencies to provide facilitation along with appropriate legal cover.

Erosion by Water

The national scene of erosion by running water is poorly known. It is estimated that erosion by water has affected around 90 million hectares out of a total geographical area of around 320 mh. Erosion by water constitutes the most serious menace to the land resources of the country not only because it denudes the soil directly but also because it leads to the siltation of reservoirs and tanks, to the occurrence of floods and to the loss to the sea of a great deal of sweet water. This referred water is generally contributed by higher reaches which are mostly forest area. This contribution of water is enhanced every year due to land degradation and decreasing forest cover. It can safely be understood that erosion by water in waste lands and degraded forests affect the area most and these are the areas which receive least attention and remedial measures. The situation in Madhya Pradesh is not different. The situation in the state is peculiar in the sense that most of the rivers originate from mountain ranges located in various parts of the State and flow on higher slopes with faster velocity which in turn augment erosion and further harm our degraded forests and waste lands.

Maintenance of Soil Fertility

Maintenance of soil fertility is essential but is a high-tech area. Analysis of soil samples involves intensive and indepth analysis of thousands and thousands of soil samples. This is a Himalayan task and at times narrated as something which meagre institutional arrangement cannot meet. Due to above narrated reason, it has not been possible to even analyse soil samples in requisite number in arable areas. In this context it is a dream to think of analysis of soil samples and maintenance of soil fertility of soils occurring in waste lands and degraded forests. The changes in outlook is urgently needed to improve the situation to achieve sustainability of natural resources.

Urbanisation

Lot of revenue wastelands are being lost to urbanisation. This diversion of degraded revenue land is to provide for the growing needs of urbanisation. Town and Country Planning Organisations equipped with regulatory authority will have to be sensitised to the pros and cons of the problem and needs of the country thereof.

Degraded Forests

Area under degraded forests has a significant role in influencing the land use policy. A sizable area is under degraded forests cover and it not only influences the climate but also influences the human being residing in and around forest. The existing laws are still not free from the pre-independent thinking. It is a welcome step that participatory concept is taking roots in the forest department and joint forest management is becoming popular and developing roots but lot more has to be done to refine the JFM concept to make it more user friendly. Another issue relating to the development of degraded forests cover, is re-thinking on the situation and a refined policy intervention along with selection of species suggested to be grown in the said area. A new look and vision is necessary with support of suitable policy environment and its enforcement.

Suggestions

The following suggestions are considered appropriate for sustainable development of degraded forests and waste lands :

1. View waste lands and degraded forests as resource whose health directly relates to rural poverty. The policy intervention should view an address the improvement of natural resources as a means of providing basic needs of rural population. This has

to be a necessary component of sustainable land use policy for these areas. Separate policies are required for wastelands and degraded forests.

2. View lands and degraded forests as resource management problem of the community which should be made the peoples agenda for their livelihood. Policy intervention should provide suitable legal instruments to the community. At the same time, community should be provided with adequate funds. We may think of providing GRAM KOSH—*community controlled investment fund* for environment building through natural resource conservation and management. This will lead to a sound rural economy for the people residing in the above areas of the State.
3. View rural community as the primary stakeholder in planning, managing and controlling land management. Community should restore environmental equilibrium, promote biomass protection along with its distribution on the acceptable norms of local community and Panchayat Raj Institutions.
4. View lands and degraded forests as resource management issue for the people and should be made a regeneration movement for reversing the cycle of degradation. Everybody involved in the above movement should work in a mission mode. The policy intervention is expected to reverse the cycle of natural degradation.
5. View waste lands and degraded forests as an income generating resource for resource poor families, women, SC, ST and other BPL without entitlement.
6. Promote forest land with more community friendly interventions. People should be brought to play an environment friendly key role in addressing poverty and developing need based green forests.
7. View waste lands and degraded forests, water and people as interconnected, inseparable and interdependent essentials of successful policy.
8. Natural Resource Management should be addressed in a holistic manner and monitored by a high power empowered committee. It should be supported by a strong political will.

The management of land and soil resources is also linked with the management of water resources in these areas and therefore be attempted by subjecting the latter to community discipline. We all know that water is not an end in itself but is an essential life saving input or means. It is needless to stress that water supports greater productivity if used in appropriate quantity at appropriate time. It therefore, needs desired attention. A sound policy taking into consideration all natural resources, is the need of the day.

Sustainable development of degraded forests and waste lands is a goal. It is possible to achieve the goal if sustainable development of degraded forests and wastelands is attempted by introducing the component of water in the said area. It is needless to stress or point out that degraded areas can only be improved if moisture is provided to support activities of greening them. The moisture augmentation along with improving soil health and providing deficit nutrients has to be brought on the agenda of programme managers and policy makers. It is in this context, suggested that nearly 15 per cent area be reserved for water harvesting structures on forests lands and waste lands. Without this, sustainable development of degraded forests shall remain a dream.

The community involvement is a second most important issue. We must give them the key role in planning and implementation of activities and make them accountable. The programme managers will have to switch over and accept the responsibility of the facilitator. The action plan of degraded forests or waste lands will be conceived and implemented by the primary stakeholders and the promoters/ managers will provide legal and financial support. Past experience, existing rules and status of these areas undoubtedly suggest that it is the opportune time in the new millennium to adopt an appropriate strategy and a effective policy environment.

(The views expressed in the article are the views of the Author. These views have nothing to do with his official position.)

5

STRATEGY TO INVOLVE RURAL PRIVATE SECTOR TO CREATE MORE STOCK AND NEW FOREST

R.C. SAXENA

Summary

Our forest areas and resources are shrinking and degrading continuously. Demands and pressure of the increasing population of both human-being and cattle would result in further and irrepairable degradation. Madhya Pradesh's rural population, about five crores, depend for their needs mainly on the existing forests. Unless degraded forests are developed and an alternative source is created, the future seems dark. Land in the private sector (with farmers) seems to be a suitable alternative to create more stock and new forests. Possible funding agencies will have to be explored to generate additional funds.

1. Degradation of forests is of various nature and intensity. It should, therefore, be clearly defined. Strictly speaking, all such forest areas which, at present, have stock less than its natural potential should be considered as degraded. Its intensity has to be determined to ascertain as to which areas should be given priority for sustainable development. The general view is that those potentially productive areas which, at present have density 0.4 and below, should be given priority.

2. Two main causes of degradation of forests are: (i) Diminishing forest areas on account of various kinds of demand over them made by various Socio Economic Development activities, e.g. diversion for cultivation, industrialisation, irrigation and power projects, rehabilitation, etc., and (ii) Unbearable demand of increasing population both of human being and cattle.
3. Regarding the first, there could be possibility of further demands in future causing reduction of forest areas. Thus, only the residual areas will be available for meeting all social demands of forest products including provision of suitable environmental conditions for the population, which will continue to grow.
4. Madhya Pradesh having largest forest area has 72 lakh ha. of degraded forests which is 46 per cent of the States recorded forest area. The rural population of the State is 76.8 per cent of the total population, *i.e.* about five crores, which depends for their needs mainly on the existing forests. There is always a limiting factor regarding the potentiality of each of the degraded areas, depending on a number of natural factors and these areas could be developed to the extent of their natural potentiality. Unless all degraded forests are developed and brought under operation, five crores of the rural population will have to depend on the use of 54 per cent of the better stocked forests, most of which may be occurring under the Reserved Forests. If this happens, these forest areas (54%) will not be able to bear such a heavy demand with the result that sooner or later these will also be degraded.
5. Demands and pressures over the forests mainly consist of need to meet the *bona fide* requirement for healthy survival and for reasonable standard of life. Another demand, perhaps, more responsible for degradation is the demand of greed for monetary gain. The group which is involved in this activity consists of influencial people who work through rural poorer class.

The above situation could be averted only by diverting the demands to some other areas by creating new forests and/or by sustained development of already degraded areas or by voluntarily reducing demand or meeting them through other alternate sources, and if possible by creating their new stock in their own holdings. As per the available information (based on 1997-98 statistics) M.P. State has 96.03 lakh land holdings occupying total land area of 218.90 lakh ha, the average being about 2.3 ha. only. Out of this an area of 199.0 lakh ha. is under plough leaving an area of 19.5 lakh ha., available for creating new forests. Farmers could also be motivated to plant trees on bunds to augment the stock.

6. In order to correctly and fully appreciate all limiting and operating factors/reasons, it would be necessary to identify and locate the degraded areas occurring in various classes of forests, viz. R.F., P.F. unclassed and wastelands. The location becomes relevant and important, because it would determine its legal status and all circumstances influencing directly and indirectly, for the present degradation. In R.F. everything is prohibited, except what is permitted; in P.F. everything is permitted except what is prohibited, in unclassed forests there is hardly any law. Wastelands are governed by the Land Revenue Code under the control of the Revenue Department.

7. In the above circumstances developmental process is likely to consist of :—

    (i) Assessment of present and prospective *bona fide* demands.
    (ii) Areas likely to be available to meet these demands, both in government sector and private sector.
    (iii) Difficulties in getting all these areas for necessary development and methods to tide over them.
    (iv) Preparation of suitable and workable schemes to rehabilitate and/or afforest these areas, indicating financial requirements, possible sources to obtain them and proper implementing agencies.
    (v) Public education by propaganda, publicity, involvement of NGOs, identification of motivated village leaders, etc.

8. Areas likely to be available will be—

    (i) Government land, where all activities will be undertaken according to the prepared project report, and

(ii) Private holdings, where it will depend on the owner (farmer), and his willingness to take up the responsibility and how he is motivated to get interested in the activity.

9. In the past government was able to protect the Government Forests with the help of law and authority, when these were respected by the public. This was possible during pre-independence under the colonial rule. Government employees were servants of the crown or of rulers with more loyalty towards their employer and showed very little concern about the public. There was a change after independence when the public under our 'Constitution' became supreme. Government servants are now required to work as public servants for public service and for public welfare. Gradually the respect for law and authority got eroded and there is a positive change in public attitude. The public now strongly feels that forests are meant for their welfare and their *bona fide* demands should be met at any cost. It is the responsibility of the government to decide, how they do it? The public so far has not realised that their demands cannot be met without the protection, conservation and development of forests and enforcement of regulated use—for which they have to cooperate and play a definite partnership role. The public, thus, became conscious of their rights, but has conveniently forgotten their obligations necessary to ensure their rights.

10. Under these circumstances the Forest Department has not been receiving the desired necessary co-operation and help in protection, conservation and development of forests. This has led to over exploitation and resulted in increasing degradation.

11. It is now realised and established that it is impossible to protect and rehabilitate forests without active cooperation, involvement and participation of the users under persuasive technical guidance. The concept and experiment of JFM is a positive step in this direction. JFMs are in operation for the last decade and it is sufficient period to assess their success and/or failure and suitably modify its contents to make it more effective.

12. In order to compensate for the loss of government forest areas, it has almost become imperative to find out alternative suitable areas in the private sector, with the farmers, where required stocks of forest produce could be raised, used and surplus marketed. This activity in the private sector will ensure sizable diversion of *bona fide* demand and pressure from already degraded adjoining Government Forests enabling them to naturally develop. The experience of social forestry schemes had not been encouraging. Farmers have not shown desired interest nor have they expressed their anxiety and keenness to protect what was planted at great cost for their future benefit. There is apprehension that suitable private areas for planting may not be available unless some special efforts are made to motivate and persuade farmers for voluntary involvement. Closer and trustworthy association and contacts with package of suitable and attractive schemes assuring positive and definite socio-economic benefits may be able to create interest and ensure better success. Still better may be to create a suitable village level leadership amongst the local social worker or NGO like that of village Kulahar in Vidisha District. Sustainable Development of degraded forests should be tackled as a socio-economic problem with cooperative partnership attitude and approach, and not as an administrative/executive scheme. It would also require attitudal change among the government staff, particularly in relation to the rural population.
13. *Arrangement of funds:* It may not be possible that such an extensive programme could be tackled by the Forest Department alone. Coordination of connected sister departments like Panchayat, Revenue, Agriculture, etc. would be necessary. It may also be pointed out that many schemes dealing with natural resources have either failed or have not given desired results, on account of paucity or continuity of funds or improper formulation of the scheme. In a forestry/tree planting programme, period of establishment is important and cannot be short. It requires positive thinking on a long term basis and liberal funding agencies. There are cases where desired success was achieved when successful plantations have been raised, at a sufficient

cost and were nursed till their establishment. As against this in certain short-sited schemes, the plantations ahve disappeared in absence of proper watch and ward, and shortage of funds.

Funding Agencies, if properly approached, would come forward to promote this big venture, in national interest. There are agencies in, government as well as in private sector (National and Foreign Agencies) like NAEB, NWDB, CAPART, CASA, NTGFC and large industrial houses, who are keen to support this noble environment friendly cause. There are good NGOs who have laboured hard to achieve people's cooperation and have given successful results in afforestation schemes, who could also be involved in such a venture.

A definite approach, deep thinking and sincere effort in this direction will positively give desired results.

6

PEOPLE'S PARTICIPATION FOR SUSTAINABLE DEVELOPMENT

S.N. Mishra

India is a country of 850 million people afflicted with numerous major problems, such as abysmal levels of poverty, illiteracy, low rate of economic growth, galloping population, trade deficit, inflation, etc. The government in shouldering the constitutionally ordained responsibilities of development, has launched massive welfare and sectoral development programmes, such as integrated rural development, irrigation, agriculture, electrification, animal husbandry, urban development, housing and many poverty alleviation programmes.

These development programmes constitute the base of the national policy approach since the First Five-Year Plan. While the benefit or the positive impact of these programmes cannot be denied, they are not free from dysfunctional impact too. Most of these programmes have also left an unintended adverse impact on the biological equilibrium in the environment. The damage to the environment is both tangible as well as intangible. Therefore, it is not amenable to lay evaluation. In fact, it has been found that the rate of environmental degradation very often offsets the rate of development. It implies that there is a net negative development *albeit* the massive development investment.

The Focus of Sustainable Development

Sustainable development is development that meets the needs of the present without compromising the ability of future generations to meet their emergent needs. It embraces two key concepts: (i) the concept of 'needs', in particular the essential needs of the poor, to which overriding priority should be given; and (ii) the idea of limitations imposed by the state of technology and social organisation on the environment's ability to meet the present and future needs.

Development involves a progressive transformation of economy and society. A development path that is sustainable in a physical sense could prectically be pursued even in a rigid social and political setting. But physical sustainability cannot be secured unless development policies tackle the averages in access to resources and in the distribution of costs and benefits. In the narrow notion of physical sustainability implies a concern for social society between generations, a concern that must logically be extended to society within each generation.

Sustainable Development—Defined

The satisfaction of human needs and aspirations is the major objective of development. The essential needs of vast numbers of people in developing countries—food, clothing, shelter, jobs—are not satiated beyond basic levels and the people develop legitimate aspirations for an improved quality of life. In the world, in which poverty and inequity are endemic, will always be prone to ecological and other crises. Sustainable development requires meeting the basic needs of all extending to all the opportunity to satisfy their aspirations for a better life.

Living standards that go beyond the basic minimum can be sustained if only consumption standards are conceived on a long-term basis. Yet, there is an in-built contradiction in the use of world's ecological means, for instance the patterns of energy use, irrigation, etc. The need identification must be socially and culturally anchored, and sustainable development requires, promotion of values that encourage consumption standards within the bounds of the ecological possibilities and to which all can reasonably aspire.

Meeting the essential needs depends in part on achieving full growth potential, and sustaining the development which clearly requires consistence with economic growth, provided the content of growth reflects the broad principles of sustainability and non-exploitation of weaker sections of the population. High levels of productive activity should be conterminus with sustainable development.

An expansion in numbers can increase the pressure on resources and slacken the pace of rise in living standards in areas where deprivation is widespread. Though, the issue is not merely one of population size but of the distribution of resources, and sustainable development can only be pursued if demographic developments are in harmony with the changing productive potential of the eco-system.

A society may in many ways compromise its ability to meet the essential needs of its future generations by overexploiting resources. The direction of technological developments may solve some immediate problems but lead to even greater ones in future. Large sections of the population may be marginalised by ill-conceived development priorities.

Growth has no set limits in terms of population or resource use beyond which lies ecological disaster. Different limits hold for the use of energy, materials, water, and land. Many of these will manifest themselves in the form of rising costs and diminishing returns, rather than in the form of any sudden loss of a resource base. The accumulation of knowledge and the development of technology can enhance the carrying capacity of the resource base. But there are ultimate limits, and sustainability requires that long before these are reached, equitable access to the constrained resources and reorienting of technological efforts to relieve the pressure have to be ensured.

In essence, sustainable development is a process of change in which the exploitation of resources, the direction of investments, the orientation of technological development, and institutional change are all in harmony and enhance both current and future potential to meet human needs and aspirations.

The concept of sustainable development is not only development which will conserve our resources and not destory and damage them, as our present development models are doing. Sustainable development is equally important because it is development for all and not for a few in society. Therefore, the practical application of sustainable development is, first, that we must choose projects which are not resource consuming. It is because once we begin destroying resources, little can be done about it. So the first thing to do is to see whether we can agree upon going in for projects which are not resource consuming, which are not resource-damaging and resource-destroying. And, secondly, to build into every project the means of conserving the resources which have to be used, renewing resources which have to be consumed and countering the damage and pollution to environment that emerges as a result of development, whether it is in agriculture or in manufacture.

Sustainable development is wide-ranging in its application reaching out to: (a) Land use, including forestry, (b) Water, (c) Air and Pollution Control, (d) Industry, (e) Energy, and (f) Human resources. The doctrine of sustainable development points to the possibilities of turning our development around, not only to repair the damage caused but also to lay the basis for holistic development of the economy.

People's Participation

Development, being a multi-dimensional phenomenon, requires effective involvement of the community in achieving its goals. In a developing country like ours, the resources for development at the command of the government are very limited and scarce. As such, in order to carry out local sustainable development measures effectively, community needs to be mobilised—physically, psychologically, financially and materially—to supplement the resources provided by the government and other extra-community sources. In the overall strategy of sustainable development policies and implementational work plans, community input in the form of a systematic arrangement for participation of the people of the community in policy making and for a regular feedback in regard to sustainable development programmes may help in decision-making process at different levels. The community could raise substantial financial and human resources from the community for sustainable development which otherwise might remain unused, such as the enthusiasm and energy of the youth of the local community and women of the area. Instead of relying upon voluntary and individual contributions, the community with institutional structures—such as local self-government, cooperative societies, etc.,—having a control over the local community, could mobilise resources for community social welfare purposes more easily. Therefore, the community participation is most widely visualised as a means to mobilise resources for sustainable development which otherwise might not be available. In spite of its sincere and best efforts, the government cannot promote sustainable development entirely from its own resources. It is, therefore, desirable and imperative that the intended beneficiaries themselves contribute to the extension of sustainable development just to minimise the cost of the development projects.

When we talk about community participation in sustainable development schemes, we mean their participation in planning, implementation and monitoring of such activities. Thus, instead of being asked to either contribute or to take advantage of the services offered under the programmes, the community should be fully involved in it. In other words, community participation is the process in which individuals and families assume responsibility for their own development. It is because they better know their own situations and are motivated to solve their common problems. Instead of being passive beneficiaries of development assistance, this strategy may enable the members of the community to become agents of their own development. Therefore, they need to realize their own obligations, not to accept the conventional solutions that were not suitable and to acquire and develop capacity to appraise a situation, examine the various possibilities and estimate what their own contribution could be.

Purpose of Participation

Based on our above statement, it becomes imperative to examine the purposes which participation could achieve. They are as follows:

(i) The most important purpose of participation is to create will and determination among the members of the community to improve upon their present and future life.

(ii) To identify and develop the local resources to generate self-reliance among the members of the community.

(iii) To achieve integrated area development by coordinating among various agencies interested in sustainable development activities.

(iv) To mobilise available human resources for productive and meaningful activities.

(v) To intimate community about the development in the area.

(vi) To help the community to understand the new technology by way of arranging functional literacy programmes.

(vii) To create centres for discussion and development by organising various clubs of youths and women.

(viii) To provide an open forum to the members of the community where they could discuss their problems and find indigenous solutions which may prove efficient and economical.

(ix) For continuous education and mobilisation of the people in the area to develop local leadership.

(x) For accelerating socio-economic development of the community to encourage people to adopt modern changes.

(xi) To arrange extra-curricular activities for generating social awareness by way of well-designed publicity.

(xii) To encourage people to develop their own confidence instead of depending upon the government, and become self-reliant which is key to development.

Etinne Berthet has rightly said that all human behaviour aims at satisfying a need or an aspiration, and people will only agree to change their habits when they are convinced that it will be to their advantage. So it is through informing, motivating and encouraging the people that we can best hope to improve their living conditions.

To improve the efficiency and efficacy of sustainable development programmes, the hand in glove relationships between the people and development organisations may be beneficial for both. The largest number of people who come in contact with development administration need a sense of sympathetic urgency which could be made meaningful if administration intended to seek their cooperation. According to Brook Adams, "administration is the capacity of coordinating many, and often conflicting, social energies in a single organism, so adjointly that they may operate as a unity". In nutshell the sense of identification with the people should be the most essential quality which the new development administration must acquire. In winning the confidence and affection of the people whom they are required to serve, it was this identity which had been responsible for the success of the eminent development administrators and reformers of the past. As such, there must be harmonious and intimate relationship between the development administrators and the people.

The administration exists for the well-being of the citizens and not for its own well-being. While delivering his inaugural address at IIPA, New Delhi, Pt. Jawaharlal Nehru said, "Administration like most things is, in the final analysis, a human problem to deal with human beings, not with some statistical data...there is the danger that pure administrators at the top (not to much at the bottom, because they come into contact with human beings) may come to regard human beings as mere abstractions...and come to conclusions which are justifiable apparently, but which miss the human element. After all whatever department of government you deal with, it is ultimately a problem of human beings and the moment we forget them, we are driven away from the reality".

The true test of public administration is the welfare of the people and as such the development administration should realise the ground realities and it should not exist in some kind of ivory towers. The (All India) Administrative Reforms Commission (1966) in this connection has also observed that if in the prosperity of the people lies the strength of a government, in their contentment lies the security and stability of democracy. The knowledge of the people is more valuable and useful than the wide general knowledge. The young administrator should mix up more with all kinds and conditions of the people than to retire into his cell when he leaves his office. He may learn more things from them in comparison to what are there in his files.

The Fifth Five-Year Plan recognised it well when it stated "the involvement of the people and their elected representatives is a pre-requisite for effective planning. A plan which does not take into account their aspirations and preferences can have no operational validity, especially since its successful implementation can be ensured only if the majority of citizens functioning as entrepreneurs and decision-makers in relation to consumption, savings, investment, etc., endorse the envisaged policies and programmes by that whole-hearted participation". More emphatically this philosophy was retained in the Seventh Plan document.

The question at this juncture arises as to how the community could participate in sustainable development programmes? A model is being provided here to examine and experiment with community participation.

Process of Participation

Since the community consists of many heterogeneous groups, the understanding about the community itself becomes the first pre-requisite of participative management. It helps in understanding the positive as well as negative attitude of the community towards a particular set of activities and ultimately helps in accelerating the process of participation. However, this exercise must be carried on scientific lines and attempt should be made to locate informal leaders within the community through the process of sociometry. If such informal leaders are located, they may help in changing the heterogeneous character of the community in homogeneous one and may help in avoiding the wastage that may result from the friction among different groups.

The identification of the priority needs of the community may be treated as second prerequisite and the development workers in the area must design the development programme to meet the identified and realistic needs of the community. At the stage of identification of community needs, it is desirable that the opinion leaders—both formal and informal—are involved in such exercise. It may help in obtaining such information which might not have been collected during the formal survey. It may also help in minimising the clash of interests.

After identification of needs, when the programmes of sustainable development are designed and got approved by the headquarters, the development workers should educate the people both through formal and informal channels about the usefulness and potentialities of the programmes. At this stage, development workers are also required to assess the resources which could be tapped locally, involving no cost or very little cost, so that people could

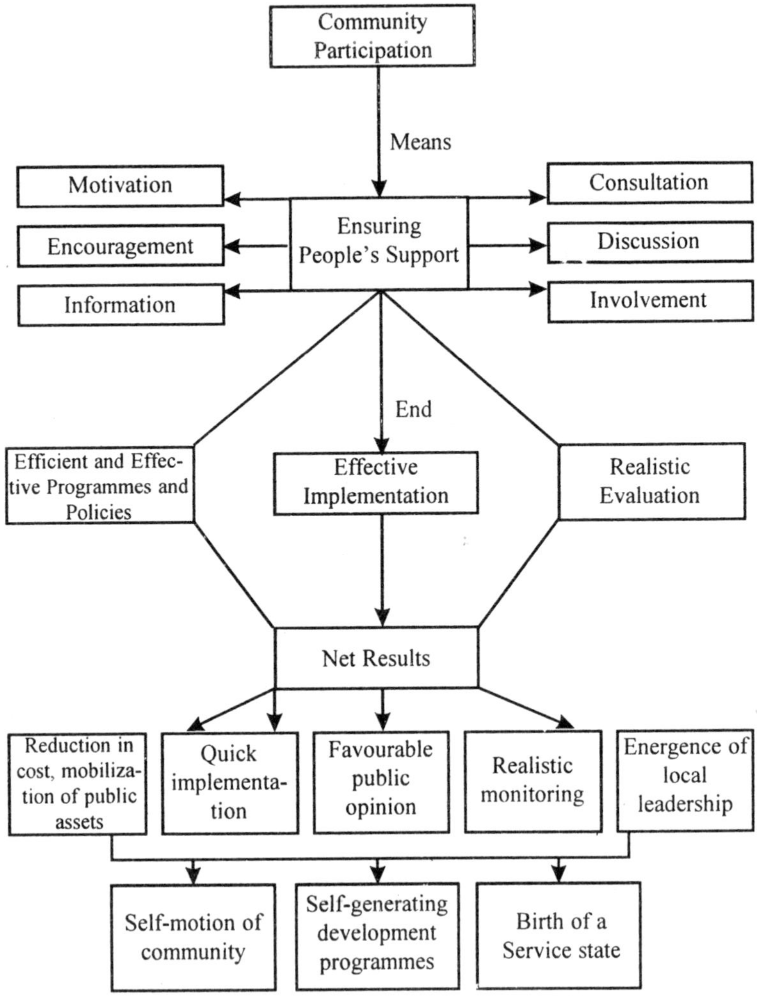

manage the development programmes on their own at some later stage.

The leadership role of the development workers is one of the most important steps in the process of participative management. However, the development workers can play their crucial role in participative management only when they are adequately and properly trained in science and arts of management. Thus, the success or failure of participative management would largely depend on administrative capability, and leadership and motivation development workers. It becomes even more significant because of the fact that administrative capability is an important means of converting programme's inputs into outputs with economy and efficiency. V.L. Gabriel has rightly remarked that "what makes the leadership variable so crucial in the implementation process is its dynamic, not passive quality, *i.e.*, its capability to act and react on these critical inputs. It is this administrative and transforming quality of leadership that could significantly determine the administrative capability of implementing organisation". In other words, it may be said that administrative capability is the capacity to obtain intended results through organisations. In this connection, Saul M. Kutz remarks "administrative capability for development involves the ability to mobilise, allocate and combine the actions that are technically needed to achieve development objectives".

To achieve the operational goals with dedication and perseverence the development workers should be the persons of vision, initiative and desired in the words of Pt. Jawaharlal Nehru, "No administrator, I suppose, or any one else for that matter, can really do first-class work without a sense of formation, without some measure of crusading spirit. I am doing this, I have to achieve this, as a part of great movement in big cause. That gives a sense of function not the sense of individual, narrow approach of doing a job in the office for a salary or wage, something connected with your life's outlook of anything perhaps being interested, as people inevitably are, on one's personal preferment in that particular work."

When the sustainable development programmes take-off with aided guidance of bureaucracy, it becomes important for development workers to encourage the people to run the programmes independently without their help. In other words, the development workers should intentionally becomes passive and monitor the interest of the community. This may repose confidence among the members of the community and bring a sense of confidence and responsibility to take up the development measures independently, which is key to the participative management. However, it is expected that members of the community may face some problems in initial stages and as such the development workers are required to occasionally monitor the running of the programme and if needed, solve the problems faced by

Chart

Process of Participative Management

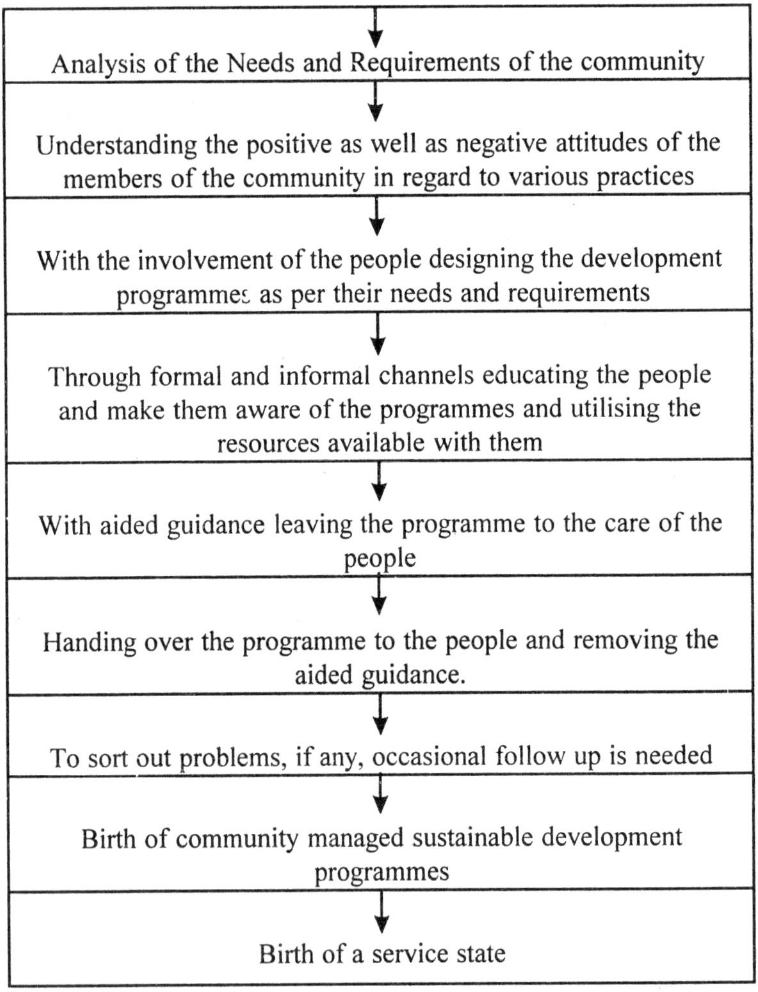

the community. Such types of problems may be varied and many, *i.e.*, political, religious, social, economic, pychological, administrative and so on. The development workers must provide their seasoned guidance to sort out such problems which may look petty in the initial stages but can lead to disastrous results in the long-run. They should be alive to this fact and must intervene at the opportune moment. If we follow this strategy, it is supposed that the sustainable development programmes may take deep roots in the soil of the community and may continue till there is a need for such programmes by the community even for the future generations.

Whatever, narration of participative management has been given is preceding pages, they could be summarised in above Chart. We hope that the chart shown below may help in understanding the mechanism of participative managment and goal achievement. This may further help us in applying our mind to find out an even better framework for participative sustainable development management.

Summing Up

Sustainable development has above all wide-ranging application to the human resources factor. It demands a reduction in fertility rates by increasing the literacy and education of women and girls. It demands drastic reduction in the inefficiency and wastage in our schooling system expressed in the massive dropout rates and educated unemployment. It demands such a policy which is based on a dispersed pattern of human settlements, low-cost water supply and housing, and the recycling of wastes. It demands optimal use of the medical infrastructure that we have built and will further build particularly in rural areas. Above all, it demands a positive, as against the present normative, approach to the major socio-economic problems of unemployment poverty and inequality, which can never be addressed, unless there is a turn-around in society, where all power is concentrated in a thin crust at the top and where a start can be made by helping and allowing the poor and the dispossessed to organise themselves to fight for their rights.

And so to return to the starting point, the major question posed is whether our growth centred development plans, including the Eighth Plan that we have now launched visualising a six per cent growth rate, can be replaced by a plan aiming at sustainable development, which addresses itself simultaneously to safeguard and enriching the environment as well as to meeting some of the basic needs of the neediest of society.

In sum and substance, when we talk of sustainable development, the easiest approach is that we, the present generation, have inherited a certain amount of ecology and environmental surrounding in terms of land, water and air, when we leave it to the next generation, we should leave it at least int he same condition. When we talk in these terms, we talk of rules and regulations and enforcement, in terms of being able to do something really positive to stop environmental degradation and encourage the community to play its positive role in preserving it.

7

PEOPLE'S FORESTRY "LOK VANIKI" IN INDIA

AN INNOVATIVE AND DECENTRALIZED APPROACH IN MANAGING PRIVATE FORESTRY

A.K. BHATTACHARYA AND BIJENDRA BASNYAT

Rapid population growth has created a demand and supply gap for the forestry products which is increasing day by day. The demand and supply gap of fuelwood was 9.75 million cubic metres in 1999 (MP Forestry Action Plan, 1999). Apart from this, the productivity of the forest resources is decreasing day by day due to over dependency on forest products. This has created tremendous pressure on government forests, which leads to their degradation, and number of forest-based industries has closed or started to close down. Singh 1998, estimated that 14 forest-based industries in the state have been closed due to shortage of raw materials and many others are operating below their installed capacity. In addition to above, the verdict of Honorable Supreme Court on 12-12-1996 states that felling of trees in all forests (including private forests) is banned except in accordance with approved working (management) plans. Hence, it has been felt that there is an imperative need to shift the wood production functions from natural forests under forest development to private forests and wastelands in order to make the government forest play the ecological role exclusively. As a result, the government of M.P. conducted a thorough study and discussion with farmers and officials in order to hammer out a programme to promote tree growing in the revenue area and private holdings, which generally fall in the category of wasteland.

The Government of M.P. launched "the Lok Vaniki Mission" a programme to promote tree growing in private holding and revenue area with people's participation in April 1999. Initially, the programme was launched in four districts on a pilot basis, but now it has been extended to ten districts. The chronological event of evolution of Lok Vaniki is summarized in Table 7.1 below.

Table 7.1 : Evolution of *Lok Vaniki* in MP

Year	*Chronological events*
1996	Supreme Court decision on banning of green felling of trees
1999	Launch of the private forestry concept in Madhya Pradesh and opening of *Lok Vaniki* Cell under *Lok Vaniki* with Additional PCCF (Production as the mission leader).
	Decision to implement in four districts of MP.
	A high-powered committee composed of senior officials from the *Panchayat,* Tribal, Forest and Revenue departments was set up to look into legal requirements of the *Lok Vaniki* scheme.
2000	Scheme was added to three more districts.
2001	Formulation of *Lok Vaniki* Act.
2002	Formation of *Lok Vaniki* Rule and its implementation in 10 districts of MP.

Lok Vaniki is an innovative programme launched in Madhya Pradesh in April 1999. The concept contemplates capturing the inter-relationship between economic growth, environmental preservation and poverty alleviation through development of forestry in the private sector. The major focus of the scheme is on promoting multi-tier scientific management of neglected and degrading forests on private holdings and also inculcating a culture of tree cultivation by considerably spreading green cover on areas unsuitable for good and profitable agriculture. Most of these areas are lying barren and degraded and are creating environment problems. The vision statement, envisaged as "Lok Vaniki : Vision 2020", states—

'To transform the forestry Sector in MP, so as to enable it to fulfil its dual role of maintaining ecological balance and environmental stability while simultaneously meeting not only

the domestic demand but also a share of the international market of forest products through people's participation thereby contributing to the socio-economic development of the State'

Lok Vaniki aims to:

1. Increase production of wood and non-wood forest products in the private sector by managing and harvesting standing forests on private holdings;
2. Rehabilitating degraded forests on private holdings and on lands under the control of government departments (other than Forest Department) mainly Revenue Department;
3. Plantations on private and Revenue Department waste lands.
4. Empowering and strengthening of Panchayats and
5. creation of other necessary institutions like 'Kisan Sangh' and 'Chartered Foresters' to manage, supervise and monitor forestry and forest based activities as also tax collection in Non-Government sector.

Hence, *Lok Vaniki* is initiated for the scientific management of private forests and promotion of tree cultivation on marginal lands. This would be achieved by providing suitable legal, institutional and market environment. The *Lok Vaniki* management process can be summarized as below in Box 7.1.

Box 7.1 : Procedural steps for Managing Private Forestry under "Lok Vaniki"

- Making the farmers aware of the concept, benefits to them through implementation of the scheme and their responsibilities.
- They are clearly told that the land cannot be diverted for any other use.
- Land has to be clearly demarcated and certificates obtained from Forest and Revenue departments that no forest or government land has been included in their holding.
- A management plan has got to be prepared and got approved by a competent authority.
- Trees to be felled are marked strictly in accordance with the approved management plan.
- Regeneration of forest has to be ensured.
- Regular monitoring will take place and implementation of the management plan will be suspended if working is not found to be in accordance with management plan. All illicitly felled trees will be confiscated.

The State Government has taken adequate measures to provide sufficient legal support to the whole initiative. *Lok Vaniki* Act, 1999 and *Lok Vaniki* Rule, 2001 has been formulated which provided an enabling legal framework to people willing to manage forests and tree clad areas on scientific lines. The objectives of the Act are to regulate and facilitate management of tree clad private and revenue areas in the state of Madhya Pradesh. *Lok Vaniki,* which emerged after the historic decision of the Hon'ble Supreme Court has now assumed a shape that is based on sound legal grounds. Key features of *Lok Vaniki* legislation are summarized in Box 7.2 below:

Box 7.2 : Key Features of Lok Vaniki Legislation

- Fairly free from Colonial Shadows.
- *An Enabling Law-voluntary in application:* Private forest owners voluntarily prepared plan for managing their wood lots.
- *Preparation of management plan:* Management plan is prepared to ensure continuity and improvement of the forest so that it fulfils its environmental and economic roles optimally. In preparing the management plan of provate forests, the same silvicultural principles are applied as in the case of government forests.
- *Chartered Foresters :* Availability of Technical Forestry Services to people on commercial basis. These are the private independent parties who will prepare the management plan for the farmer for sustainable management of private forest.
- *Putting People First :* Self-assessment by the owner of Private Forests the Act itself is very regressive, the process for framing rules under the Act was also participatory.
- *Single window* deals with the issue of management of such private holdings for which a management plan is prepared under the provisions of the Act. Such lands shall remain outside the purview of the Madhya Pradesh Land Act.
- *Recognition of role of Local government in Sustainable :* Gram panchayat after having received the approved management plan from the competent authority should implement the management plan according to prescribed time schedule.
- *Institutional Framework :* State level coordination Committee, Forestry Boards at district and block level, 'Kisan Samiti' and Sangh and Chartered foresters to manage, supervise and monitor forestry and forest based activities.

The *Lok Vaniki* was initiated in four districts but now it covers ten districts, namely Dewas, Hoshangabad, Narsinghpur, Seoni, Damoh, Jabalpur, Katni, Mandla, Dindori and Sidhi. Until now

1196 farmers owning 4211 hectares of land from five districts have got involved in the scheme (Saigel *et al.*, 2002). The State Government has made elaborate arrangements for training and capacity building of these peoples and institutions to evolve a transparent and convenient mechanism for private peoples active participation in management of their own tree clad areas. More than 150 training programmes are organized in ten districts and 20,000 farmers are trained on various aspects of forest management with the introductory module of *Lok Vaniki* legislation. 50 management plans of *Lok Vaniki* have already been prepared of which four are being sanctioned and three have been forwarded to Government of India for approval as the plans cover more than 10 ha of land. In anticipation of the income that would be generated, many farmers invested in raising of plantations to further increase their assets. As per available information, approximately 3,97,000 seedlings were planted in 1999 planting season in Dewas, Sidhi and Raigarh districts only. Three farmers have started harvesting and their name and annual income is summarized in Table 7.2 below.

Tale 2 : Income from *Lok Vaniki*

District	*Name of farmer*	*Annual income*	*Total Trees*
Hosangabad	Kamta Prasad Mahoto	Rs. 97,705	12
Devas	Shohadra Bai	Rs. 21,905	219
Damoh	Kodulal Dubey	Rs. 18,758	—

Lok Vaniki is a decentralized approach adopted by Madhya Pradesh Forest Department where the role of Panchayat has been duly acknowledged. The concept of chartered forester is not only innovative but could also contribute a lot in reducing the unemployment rate of the country. There are immense potentialities of managing the private forests in Madhya Pradesh. There are about 200,000 private forest owners in the state owning forest patches of above 0.4-hectare (Singh, 1998). It is estimated that out of a total of 8.4 million farmers in the state, 200,000 farmers have a total of about 400,000 hectares of standing forests, which can be 'unlocked' immediately. Further, about 600,000 farmers each have about one hectare of degraded forest. The farmers can make this productive without much investment (Saigal *et al.*, 2002). This would not only help for conservation and development of private forestry but also help in socio-economic upliftment of people. In spite of greater political commitments and adequate legal and policy framework, implementation of the *Lok Vaniki* scheme is not encouraging as it is being implemented in 10 districts out of 45 districts of State. The process of handing over is very slow due to long administrative procedures, as it requires approval from Central Government if the areas in more than 10 ha. of land. Hence, *Lok Vaniki* is not a panacea but it can contribute a lot towards sustainable management of private forests if properly implemented.

References

GOMP, 2002, *Madhya Pradesh Lok Vaniki Act, 2001.* Madhya Pradesh Government, Bhopal, MP, India.

GOMP, 2002. *Madhya Pradesh Lok Vaniki Rule, 2002.* Madhya Pradesh Government, Bhopal, MP, India.

MPFD, 1999. Madhya Pradesh Forestry Action Plan, Madhya Pradesh Forest Department, Bhopal, India.

Saigal, S., Arora, H. and Rizvi, S.S., 2002. *The new foresters: the role of private enterprise in the Indian forestry sector. Instruments for sustainable private sector forestry series.* Ecotech Services, New Delhi and International Institute for Environment and Development, London.

Singh, D.P., 1998. *Socio-economic Development of Madhya Pradesh through Forestry by the people, for the people.* Government of Madhya Pradesh, Bhopal, India.

8

ROLE FOR INDUSTRY IN THE DEVELOPMENT OF DEGRADED FORESTS

A NEED FOR CHANGE IN POLICY

J.K. Rawat

Introduction

Degraded forests constitute nearly one-fourth area of the total wastelands identified in the country. Several agencies such as Forest Department, panchayats or village societies, voluntary groups and non-government organisation have been involved in the development of wastelands. One agency that has been kept out deliberately in this activity is "Industry". National Forest Policy of 1988 specifically lays down that forest based industries should raise the raw material outside the forest area by establishing a direct relationship between the factory and the individuals who can grow the raw material. The National Forest Policy thus gives a very casual or step-motherly treatment to forest based industries.

The forest based industries consist of pulp and paper, news-print, plywood, match, rayon, panel products and charcoal manufacturing firms. These make a sizable contribution to our domestic economy as well as to the exports. The above mentioned industries can play a much greater role in the national economy but many of their units are working below their respective installed capacities. This is largely due to inadequate or uncertain supply of raw material, *i.e.* wood. The most practical way to overcome this problem and to strengthen these industries is by giving them long term lease over degraded forest lands for raising plantations so as to meet their requirement of raw material. This will necessitate a change in the existing forest policy.

The most compelling reason for suggesting change in forest policy is that industry has resources that can be utilized for wasteland development. The technology for afforestation and reclamation of degraded forests is costly and requires heavy investment. The government does not have adequate resources to develop these wastelands at the desired speed. Thus resources available with the industry can be harnessed for this purpose. Government will get rent or royalty for the area leased out to the industry and the latter, in return, will get an assured supply of raw material.

Discussion

What has been proposed above is not new. Therefore, it is important to analyse why such a step has not been taken so far. The industry has been kept out of forest land because of certain objections and problems that are anticipated to arise. Before contemplating any change in policy, it is necessary to examine, with an open mind, these objections and fears and to suggest how the interests of society and the nation can be safeguarded. These are discussed in the following paragraphs.

1. The fulfilling of basic needs for fuel and fodder of the local population will be adversely affected!

The industry will be allowed to exercise rights over the forest lands under well thought out terms and conditions. The terms would be laid out in such a way that access of local people into the forest for grasses and other minor forest products is not restricted. Though grazing would not be permitted, the need of people for firewood would, however, be met from the forest produce grown in the leased area. This will be done by inducing high productivity in that area since forest soils are quite fertile but need proper management. The productivity of degraded forest if left to natural regeneration, is very low, viz. about 1-2 m^3 per ha. per year. Average annual productivity for the country as a whole is about 0.7 m^3/ha. However, with improved afforestation techniques, additional inputs of fertilizers and irrigation, and intensive managment, productivity within 10-20 m^3 per ha. per year can be achieved. It has been recorded that over 20 per cent volume of a natural tree consists of lops and tops which can be preserved for the use of local people as firewood. This amounts to an annual availability of 2-4 m^3 per ha. which is far more than the produce available from the undeveloped forest land. A committee consisting of representatives from Forest Department, industry and local people can be formed to oversee the distribution of forest produce.

2. *Rights on forest land will further enhance economic and political power of the industry!*

Economic and political power is derived from the extent of asset and manpower controlled. Employing labour for afforestation work and owning an inventory of tree wealth will certainly increase power of industry to a certain extent. This is what is meant by strengthening of an industry so that it is on a surer footing than before. However, to presume that this will be to the detriment of the society may not be correct. Moreover, certain safeguards will be kept to protect the interests of the society and the country. For instance, industry will be given right only to produce grown by them but will have no right to land. It will raise only those trees which supply raw material for its factories. The land will be subject to Forest Conservation Act of 1980 whereby land cannot be diverted for non-forestry purposes. Hence, there is no fear of industry using the land for speculative purpose or for gaining any political mileage out of it. Moreover, only the degraded and open forest lands will be leased to the industry. The right of an industry over forest land will be subject to several terms and conditions controlled by the State Forest Departments. The industry will have to submit annual and long term management plans which will be scrutinised and approved by the Forest Department before any harvesting is permitted in the area. In order to safeguard interst of the tree growers and private suppliers, it is suggested that insutry be leased an area adequate for producting not more than 50 per cent of its raw material needs.

3. *Industry will exploit land by mining resources on unscientific basis and raise monoculture plantations creating adverse ecological impact!*

Mining of resources will take place only if lease period is short, or the extension of lease is uncertain and the industry has to recover its investment in a short period. In order to ensure that the industry uses the land on sustainable basis and not as a mine of wood raw material, the period of lease should be long enough to cover at least 2-3 rotations of plantation crop. In case of fast growing species this should be anywhere between 15-20 years. The lease should be on an extendable basis and there should be a provision to give extension for, say 5 years after the expiry of every 5 year period, subject to satisfactory compliance of terms of lease by the industry. This will help industry in having a long term perspective as far as their raw material needs are concerned. In addition, industry can invest in research and development and in improving the forest productivity further if the rent or royalty is charged on per ha. basis rather than on output basis. This will give an incentive to the industry to increase productivity by applying tree improvement and advanced afforestation technology. The rate of royalty should, however, be linked with the price index of the finished goods manufactured by the industry. Regarding question of scientific management of plantations, it is noted that forestry is being taught in several universities now and there should be no dearth of trained personnel who can be employed by the industry to manage these plantations.

Demand of industries such as paper and pulp, rayon, match, fibre board, charcoal, etc. using small sized trees is of the order of 20 million m^3 per year. If 50 per cent of this volume is to be grown from captive plantations, with an induced productivity of say 10 m^3/ha./year, a total area of just 1 million ha. will be needed. It is only a small fraction of the country's forest land and monoculture over it is not likely to have any adverse ecological impact.

4. *Industry will control the price of wood to the detriment of private suppliers!*

Wood is a bulky product and is difficult to transport to far off places. Therefore, by its very nature it is difficult to have a competitive market for industrial wood. Hence, it is inevitable that a wood based industry, whether having lease over forest land or not, will have a say in deciding price of raw material. The wood market will either be a monopsony market where a large number of private wood growers supply to a single industrial unit, or it will be a case of bilateral monopoly where an industrial unit faces a single supplier such as Forest Department. In the first case the prices is decided by the industry while in the second case it is negotiated between the buyer and the supplier. Since it has been suggested that extent of forest area leased to the industry should be adequate only to produce upto 50 per cent of its wood requirement, the industry will have to depend on private suppliers for the other half. Whether private suppliers, mostly tree growing farmers, are exploited or not exploited by the industry depends upon the tree growers themselves. If they take initiative and form tree growers cooperatives, they will be in a stronger positions to negotiate wood price and other terms of supply with the industry.

Conclusions

From the discussion above we can see that there are no logical grounds for prohibiting involvement of industry in afforestation activities on degraded forest lands. The objections and problems generally put forward are not insurmountable. We have to overcome our prejudice against industry first and examine each case on its merit and, through terms of contract, safeguard the interest of the nation.

At present, in several states, the industry is procuring forest produce from the government forests on payment of royalty. The rates of royalty are generally low and thus, in other words, we are subsidising the supply of wood to the industry. Why should the State Forest Department grow wood for industry when the latter has means and know how to grow its own raw material. Instead, we only have to allow them long term lease for afforestation over otherwise unproductive degraded forest lands. Although, implementation of this policy will not be easy and will need a thorough discussion before acceptance, it is necessary to at least not to have a closed mind towards such a move.

The forest based industries play an important role in the economic development of the country. It is important to modify forest policy so as to involve them in the development of degraded forest lands and thereby the helping the industry as well as the nation.

9

PRIVATE SECTOR FORESTRY RESEARCH
A SUCCESS STORY FROM INDIA

PIARE LAL

Introduction

Almost the entire 76 million ha recorded forest area is owned and managed by the State Governments in India. Unfortunately, the forests are under intense biotic pressures leading to degradation of forest resources and 31 million ha of forest area has less than 40 per cent crown cover density (Anon, 1996). This is a great paradox and tragedy of immense proportions for an overpopulated country. And yet, forest-based industries or corporate sector are denied any role in the reforestation of degraded forest lands.

There are statutory ceilings limiting the agricultural landholdings both for individual farmers and corporate sector to extremely low levels in all States, e.g. upto 54 acres of worst category of land in Andhra Pradesh (Reddy and Reddy, 1995). It is, therefore, not possible for any wood based industrial unit or private sector company to own adequate land for raising intensively managed technology based captive plantations.

Forestry research by the State Forest Departments and various Forest Research Institutes under the Indian Council of Forestry Research and Education (ICFRE) is funded by the Government. Legislation for *sui generis* protection for breeder's rights is still pending, and there is no mechanism for certification of seed of forestry species or registration of clones. Research in agro-forestry is also carried out by the universities and Indian Council for Agricultural Research (ICAR). There is no wonder, therefore, that there is very little incentive for private sector investments in forestry research.

A very limited number of Non-Governmental Organisations, certain research institutes supported with donations like Tata Energy Research Institute (TERI), New Delhi, are engaged in research in limited areas related to forestry. Some integrated pulp and paper mills, of late have started promoting farm forestry plantations with limited research and extension support. The only major exceptions of significant forestry research and promotion of technology based plantations by the private sector being applied research in case of Poplars by Wimco Seedlings Ltd., and clonal technology research and technology based plantations of Eucalyptus and Casuarina by ITC Bhadrachalam Paperboards Limited (ITC Bhadrachalam), (Nair *et al.*, 1996)

This paper discusses the success story of development and commercial scale deployment of high yielding, fast growing and disease resistant clones of Eucalyptus and Casuarina, and the promotion of technology based clonal plantations through the pioneering efforts of ITC Bhadrachalam.

Farm Forestry Plantations

An integrated pulp and paper mill with 40,000 tonnes/year capacity was established by ITC Bhadrachalam, at Sarapaka, near the temple town of Bhadrachalam in a remote tribal belt of Khammam District of Andhra Pradesh State during 1979. The mill at present has a capacity of 65,700 tonnes pulp and 210,000 tonnes high quality paperboards and paper annually. For pulp production, 40,000 tonnes bamboo from government owned forests and 124,000 tonnes hardwoods (bone dry weight) primarily from the farm forestry plantations, are procured annually. Company has been very keen to invest in captive industrial plantations for ensuring future supplies of high quality pulpwood in a cost effective and sustainable basis (Anon, 1997). However, because of statutory ceilings on the agricultural landholdings and retrograde government policy not to involve industries in reforestation of degraded forest lands owned by the government, it has not been possible to develop captive industrial plantations.

The visionary management of the company, therefore, opted for the only alternative to promote farm forestry plantations on marginal agricultural lands by providing high quality seedlings, technical extension services and buy-back guarantees at remunerative prices to farmers. 6185 farmers in 1138 villages were assisted to promote 7441 ha of eucalyptus plantations with 17.4 million seedlings during 1987-95. Unfortunately, genetically improved planting stock of short rotation pulpwood species like eucalyptus, leucaena and Casuarina, popular with the farmers, was not available. Therefore, the productivity and profitability of seed route plantations has been very low ranging between 6-10 cum/ha/yr. (Lal 1993, 1995).

Clonal Technology Research and Development

With a view to improving the productivity and profitability of plantations and making farm forestry an attractive land use option, the company has been implementing a major research and development project since 1989 with main thrust on genetic improvement of planting stock and improvement of package of practices. Major gains in productivity of eucalyptus have been achieved in a short time span through applications of vegetative propagation and cloning techniques with gainful exploitation of existing useful variation and development/ deployment of locality specific, high-yielding, fast growing and disease resistant clones (Lal *et al.*, 1993, 1994).

Methodology adopted has been selection of candidate plus trees with most desirable qualities, cloning of the Candidate Plus Trees (CPTs) through rooting of juvenile coppice shoots under controlled environment in the greenhouses. Evaluation of comparative genetic superiority of clones through replicated field trials, genotype X environmental interaction studies to identify a group of superior clones most adaptable to specific sites and soil types and commercial scale multiplication and field planting of selected genetically superior clones.

Starting with cloning of 64 CPTs of *Eucalyptus tereticornis* and Mysore Gum during 1989, more than 500 CPTs have been cloned so far. Progeny of all clones has been planted in the gene banks as well as replicated field trials for evaluation. 49 clonal testing areas covering 16 ha, 6 clonal demonstration plots covering 14 ha, 2 clonal seed orchards covering 1 ha and extensive gene banks covering 17 ha have been planted.

Based on the performance of individual clones in the field trials, 72 promising, fast growing and disease resistant clones have been identified. These clones are known as 'Bhadrachalam' clones with productivity ranging between12-44 cum/ha/yr. under rainfed conditions compared to 6-10 cum/ha/yr. productivity of normal seed route plantations. Further, research work for identification of still better clones is continuing.

Current Research and Development Focus

Vegetative propagation and cloning techniques exploit full productivity potential of superior geno types. However, no further enhancement of productivity or improvement of genetic qualities of selected clones is possible. In order to develop still better clones than the best available presently and for widening the genetic base of clonal plantations, following research and development priorities have been identified and the same are under implementation:

— Selection of additional candidate plus trees for cloning and field evaluation continues as a regular feature of the strategy for development and deployment of new clones.

— Field testing and selection of highly productive site specific clones for refractory sites including saline and alkaline soils through genotype X environment interaction studies is a high priority research focus. 'Bhadrachalam' clones 1, 71, 130, 272, 404, 405 and 406 have already demonstrated good tolerance to calcareous/alkaline soils with fairly high productivity.

— Development of intra-specific hybrids through control pollination between best 'Bhadrachalam' clones of *E.tereticornis* and inter-specific hybrids of these clones crossed with *E.urophylla* and *E.torelliana.* F1 hybrids showing good heterosis will be cloned.

— Development of clonal seed orchards for production of improved genetically superior seed for future seed route plantations.

— Further improvement of technical package of practices for field plantations and clonal nurseries.

Several intra-specific hybrids have been developed through controlled pollination between selected best 'Bhadrachalam' clones of *E.tereticornis.* Hybrid progeny is under evaluation and many hybrids are showing extremely good promise at one-year age. Development of inter-specific hybrids between *E.tereticornis* and *E.urophylla/E.torelliana* is progressing well (Lal *et al.,* 1996, 1997).

Encouraged with the unique success story of 'Bhadrachalam' clones of Eucalyptus; similar research work for development and deployment of genetically superior clones of Casuarina has been making rapid progress since 1994. 129 CPTs of Casuarina have been successfully cloned and gene banks have been established. Field trials are in progress for evaluation of comparative genetic superiority of these clones, with a view to selecting locality specific, disease resistant clones with high productivity (Lal *et al.,* 1996 a).

Improved Package of Practices

Thrust areas in respect of improvement of package of practices have been thorough preparation of site through deep ploughing, effective preventive and control measures against termites during the critical establishment stage after transplanting, cultural practices including timely weeding and soil working, protection against damage by insect pests and cattle, maintenance of soil fertility through raising leguminous crops in between the 3M wide planting rows for green manuring and application of fertilisers to supplement deficient plant nutrients.

Improvement of Nursery Practices

The entire clonal planting stock of eucalyptus and Casuarina is being raised in modern root trainer nurseries using vermiculite as the growing medium. Improved package of practices for ensuring best possible results in respect of rooting of cuttings in the green houses under controlled environment and subsequent maintenance and nursing of plants in the shade houses and open nurseries have been standardised through intensive research. Schedules for fertiliser and micro-nutrients applications have been refined and perfected for large-scale commercial production of clonal planting stock. Prophylactic and control measures against common leaf spot disease of eucalyptus at nursery stage have been identified, e.g. Bavistin has been found to be most effective to control *Cylindrocladium* and Dithane M-45 as well as Blitox are quite effective against *Alternaria* leaf blight.

Plantations Department—Organisation

The Plantations Department of the Company headed by the Vice President (Plantations), has complete freedom to decide research priorities, technical issues and implement research and clonal production/marketing plans subject to the approval of financial expenses by the Managing Director. Vice President (Plantations) is assisted by Deputy Chief Manager (Plantations Research and Clonal Plant Production), supported by eight managers. Likewise, Deputy Chief Manager (Plantations) supported by eleven managers assist the Vice President (Plantations) in respect of farm forestry extension, promotion of technology based plantations and marketing of clonal planting stock. Four junior managers provide secretarial and office support services.

Extremely well defined research objectives, competent and committed scientific personnel, continuity of research plans and personnel, adequate delegation with accountability and wholehearted support at the topmost level of the Company have been key factors leading to emergence of ITC Bhadrachalam as respected pioneers in the field of clonal technology and technology based plantations in India. Company's efforts have been recognised with three prestigious awards and technical consultancy assignment for World Bank aided Forestry Development Project in Maharashtra State (Lal *et al.*, 1997).

Application and Transfer of Research Findings

Planting stock of selected 'Bhadrachalam' clones has been supplied to Forest Development Corporations/Forest Departments of various States and Forest Research Institutes to enable them plant clonal seed orchards and maintain their own gene banks for future multiplication of these clones. Andhra Pradesh Forest Development Corporation (APFDC) is implementing a World Bank funded Industrial Plantations

Project to plant 17500 ha degraded forest lands with 'Bhadrachalam' clones of eucalyptus.

Clonal planting stock of 'Bhadrachalam' clones of eucalyptus has been supplied to Forest Departments/Forest Development Corporations of Tamil Nadu, Kerala, Karnataka, Andhra Pradesh, Maharashtra, Gujarat, Madhya Pradesh, Orissa, West Bengal and Haryana. Likewise, 'Bhadrachalam' clones have been supplied to Institute of Forest Genetics and Tree Breeding, Coimbatore, Kerala Forest Research Institute, Peechi and many wood-based industrial units.

Clonal saplings of 11 best clones with productivity of 20-44 cum/ha/yr under rainfed conditions are being supplied to farmers for raising clonal farm forestry plantations with technical extension services and buy back guarantees provided by the Company. Two Clonal Demonstration Plantations (CDPs) were established during August 1991 based on five shortlisted clones with outstanding performance in CTA I at two years age. Additional CDPs have been established in three districts of Andhra Pradesh covering 15 ha area.

Clonal planting stock of the most promising 'Bhadrachalam' clones was released to farmers on selective basis from 1992 onwards. As seeing is believing for the farmers, these demonstration plantations and successful extension campaigns by the field staff of ITC Bhadrachalam have been instrumental in large scale acceptance and vast popularity of 'Bhadrachalam' clones.

Between 1992-95, 0.78 million clonal saplings were supplied. Supply of clonal plants exceeded one million saplings during 1996. 1.3 million clonal saplings were supplied during 1997 and supplies during the current year 1998-99 will be around 2 million saplings. A uniform has been charged from all clients including the farmers, Forest Departments and Forest Development Corporations for supply of clonal planting stock. Selling price was Rs.7 or roughly 19 cents per sapling ex-Clonal Plant Production Centre at Bhadrachalam for the years 1996 to1998. Total area planted with 'Bhadrachalam' clones of Eucalyptus up to 1998 will be more than 3000 ha involving nearly 2000 farmers.

Findings of the clonal technology research and experiences gained in the promotion of technology based plantations have been published by the company's scientists in the form of research papers in the *Indian Forester,* and papers presented in various national level and international seminars/conferences (Lal, 1991, 1992, 1993, 1994, 1995 1996; Lal *et al.*, 1992, 1993, 1993a, 1994, 1996, 1997). Printed copies of the improved package of practices for clonal plantations have also been supplied to the Forest Departments and farmers raising clonal plantations.

Ever growing demand for planting stock of 'Bhadrachalam' clones is a testimony of their popularity and genetic superiority. Visionary approach and pioneering achievements of the company in clonal technology research and promotion of large-scale clonal plantations have been recognised and rewarded with three prestigious awards.

Funding of Research and Extention Costs

The entire clonal technology research and farm forestry extension programmes are funded internally by the Company. The sale proceeds of clonal planting stock at the current level meet only part of the production, research and development, staff and administration, depreciation and the interest costs. Yet, the company continues to support this activity which is vital for securing future pulpwood supplies.

During the period 1986-87 to 1997-98, ITC Bhadrachalam spent Rs. 3.19 crores, on promotion of farm forestry plantations and providing technical extension services. Likewise, during the period 1989-90 to 1997-98 the company spent Rs. 2.88 crores towards operating expenses and administration costs in respect of clonal technology research/development and production of clonal planting stock. Capital expenditure incurred during the period 1989-90 to 1997-98, mainly for establishment of greenhouses, shade houses, modern root trainer nurseries, lab equipment and vehicles, etc., has been Rs. 1.50 crores. A sum of Rs. 2.36 crores has been recovered towards sale proceeds of seedlings and clonal planting stock up to 1997-98.

However, the Company now wants the Plantation Department to be self-reliant and self-supporting for funding future expenses. As the demand for clonal planting stock is growing very fast and production capacity has been expanded to 2.3 million clonal saplings per year during 1997-98. Research and Development expenses and fixed overhead costs will be spread over larger volumes. Plantations Department hopes to be self-sufficient in funds through sale proceeds of clonal planting stock within 1-2 years and simultaneously achieving the primary objective of securing future pulpwood supplies through clonal farm forestry plantations.

Co-ordination of Research and Linkages

After the visit of top executives of ITC Bhadrachalam to Aracruz in Brazil during 1987, the company solicited transfer of technology and consultancy services from Aracruz Florestal, S.A. and some renowned international consultants. However, the arrangements could not be finalised because of the exorbitant price tag, which ITC Bhadrachalam just could not afford. Therefore, it was resolved to be self-reliant and make earnest efforts to succeed through in-house efforts. A separate Plantations department was created and Vice President (Plantations) with supporting research and extension managers were recruited during 1989.

Collaborative arrangements initially for two years were finalised with Tata Energy Research Institute (TERI), which were extended for another four years term. It was a learning process for both the organisations. The vast experience of Shri A.N. Chaturvedi, a well-known forester and Shri Sujan Singh, plant pathologist of TERI, was of immense help. Initial selection of candidate plus trees was carried out jointly by ITC Bhadrachalam scientists and Shri A.N. Chaturvedi. Shri Sujan Singh provided very useful help in identification of fungal pathogens and finding appropriate preventive and control measures against the same at the greenhouse and open nursery stages, as well as field plantations.

The pioneering and pain-staking efforts of ITC Bhadrachalam for the development and deployment of genetically superior clones of eucalyptus would not have been successful, but for the generous help provided by APFDC and the farmers of Andhra Pradesh who permitted the company scientists to select CPTs for cloning from their seed

successful results in a short time span with minimum costs. However, lack of availability of sophisticated equipment, inadequate genetic resources, absence of breeding populations, non-availability of land for field trials, limited staff and shoe-string budgets for research and development and publicity for promoting sale of clonal planting stock are some of the major constraints under which ITC Bhadrachalam had to operate. But for these constraints, the achievements would have been far more commendable and faster.

Contribution of Clonal Technology Research

Extremely fast growing, high yielding and disease resistant clones of eucalyptus have been developed for the first time in India by ITC-Bhadrachalam. Productivity of 'Bhadrachalam' clones of eucalyptus is 2-3 times higher compared to normal seed route plantations. Clonal eucalyptus plantations in India promoted by ITC Bhadrachalam are the first successful example of commercial scale clonal plantations of any forestry species traditionally propagated through seedlings. More than 2000 farmers and a large number of Forest Departments/Forest Development Corporations in many States have already planted these clones.

Clonal Plantations covering 1.25 million ha or 33 per cent of the degraded forest areas in Andhra Pradesh alone can yield 25 million tonnes of pulpwood annually, sufficient for meeting our country's entire pulp and paper requirements projected at 8.5 million tonnes by 2010-11, based on 70 per cent wood fibre furnish. Likewise, high yielding short rotation clonal plantations can meet country's firewood requirements thereby minimising biotic pressures on remaining natural forests and conserve their rich bio-diversity.

ITC-Bhadrachalam will be able to release genetically improved clones of Casuarina for commercial scale farm forestry and reforestation projects in the near future. Immense gains in productivity and improvement of quality of produce in a short time span are possible through applications of similar cloning techniques for many important indigenous species amenable to vegetative propagation, e.g. *Tectona grandis, Gmelina arborea, Pinus roxburghii, Dalbergia sissoo, Anthocephalus chinensis* and *Dipterocarps,* etc. (Lal, 1995,

M. Pal, 1993). Unfortunately, forestry research organisations of the State Forest Departments and ICFRE institutions have not accorded richly deserved high research priority for development and large scale deployment of genetically improved clones of important indigenous species. No wonder, the country pays very high price for continuing large scale use of un-improved seed for most tree species in the farm forestry and reforestation programmes.

Acknowledgements

Author conveys sincere and grateful thanks to his dedicated research and extension team for their untiring efforts, and the present and past Managing Directors and Chairmen of the company for their support and encouragement, without which this success story would not have been possible. I also express my sincere gratitude to all the farmers and State Forest Departments/Forest Development Corporations who have adopted 'Bhadrachalam' clones in a big way. Grateful thanks to Ms. Lorraine White for her competent secretarial support in preparing and revising the text of this paper.

10

SUSTAINABLE DEVELOPMENT OF DEGRADED FORESTS BY SEMI-PRIVATE PUBLIC CORPORATION/COMPANY

M.M. TRIPATHI

Introduction

Madhya Pradesh has approximately 1,55,400 sq. kms. of forest area, which is under the control of Forest Department. There is some forest in Revenue areas and private holdings. The government forests are classified as Reserved, Protected and Unclassed forest which are governed by Forest Act/Forest Laws and Rules made thereunder. All the Reserved Forests are properly demarcated and Working Plans have been prepared for them. Some of the Protected Forests are demarcated, and Working Schemes have been prepared for all of them.

It is estimated that about 72 million ha. which form about 46 per cent of recorded area is presently degraded. The reasons for this degradation are increase in population at fast rate, pressure of demand of forest produce, demand of forest land for cultivation, habitation, power and irrigation projects etc.

The Forest Department had been in existence for over 100 years. After independence, the department was required to manage the forests under great stress and strain with the result that 46 per cent of forest areas have been rendered degraded for which schemes of sustainable development are under study.

It may be stated that the Forest Department has got an independent wing Madhya Pradesh Rajya Van Vikas Nigam (MPRVVN), which was given responsibility of raising plantation and managing the industrial units, MPRVVN has got a number of restraints and difficulties including funding on account of which they have not been able to give satisfactory results.

Looking to the magnitude of Sustainable Development of Degraded Forests it appears that only Forest Department or only one wing of Forest Department, *i.e.* MPRVVN eventually will not succeed in meeting this challenge. It is also realised that such activities for Development of Degraded Forests cannot succeed unless there is required participation/involvement and accountability of the public who ultimately will be major beneficiary.

Keeping in view the prevailing socio-economic, techno-political situation, it appears that some kind of *via media* has to be evolved which can take care of all the adverse situations and achieve desired success in raising plantations and protecting them till they are fully established.

Most important restraints in such developmental activities appears to be:—

(i) requirement of funds for such development for a long term,

(ii) requirement of suitable sites areas,

(iii) participation of government and public particularly in the neighbourhood of areas of development,

(iv) involvement, coordination and participation of other concerned beneficiaries like Industries, Environmentalists, Financial and Educational institutions etc.,

(v) a non-Governmental Semi-Private Public Corporation/ Company can only take care of all the above constraints.

The main constituents of the Semi Private Public Corporation/ Company will be:—

(i) Government who will be a major partner of the Corporation/ Company holding 51 per cent equity which can be in the form of lease of land and provision of funds,

(ii) MPRVVN and other Corporations of the Government having equity according to their financial capacity,

(iii) Financial Instiutions,

(iv) Panchayats who may be partners in decision making and/or in equity participation according to their financial capacity, and

(v) Industries and such other organisations.

The advantages of this proposal are:—

(i) the Corporation/Company will arrange to meet requirement of funds.

(ii) it will evolve suitable methods for ensuring public cooperation.

(iii) implementation can be within shorter period with lesser cost, better efficiency and more assured success.

The government will always remain the watchdog because these areas will always remain as government forests. If any such forest development is required to be done in private forests/tenancy holdings this proposed organisation can act a consultant/adviser and, if necessary, executor.

If the proposal is accepted in principle the details of its implementation could always be worked out.

11

POSSIBILITY OF SUSTAINABLE DEVELOPMENT OF DEGRADED RESERVED FORESTS

A JOINT SECTOR PARTICIPATORY APPROACH

S.S. SHRIVASTAVA

Introduction

On the 11th May 2000 India crossed the mark of 100 million in population. This is an event on which we have to ponder whether it is a matter of rejoicing or holding our heads in shame. Against this progress, availability of all life giving and life sustaining natural resources like forests are fast deteriorating and have reached a stage of alarm in many places requiring immediate remedial action. To add to this, natural calamities of higher intensity are becoming a normal feature, rendering all kinds of socio-economic developments doubtful and uncertain. It is now time to honestly realize and boldly admit that the present situation is man-made and is the result of mankinds past misdeeds and of arrogance against nature and its laws. Nature is known for kindness and benevolence (like God) but is equally cruel and revengeful against all who knowingly and deliberately cross the limit, ignore or disobey laws of nature. Land, water, climate, environment in particular, are natural resources which suffered most and are creating problems. Increasing degradation of forest is an instance, now, under serious study.

Basic features and issues involved in the scope of the subject are (i) what will constitute "degraded forest" and "waste lands"? (ii) where these actually occur? (iii) why and how they are rendered so? (iv) what circumstances and who are responsible for it? and (v) why it could not be prevented? It is also to be considered as to what is actually meant and understood by "sustainable development" in the above context. Strictly speaking, all such forest areas which, at present, have stocks less than their natural potential, should be considered degraded. But, if this standard is honestly applied, perhaps, 90 per cent of recorded forest area in Madhya Pradesh will fall in the category of degraded forest. For the present and for purposes of fixing priority for taking up areas for sustainable development of them, "degrades forest" should include only such potentially productive forest lands which, for reasons of past working or natural calamity, presently certain stock of 0.4 density and below. They should not include naturally unproductive lands. Such lands, it there be any, which contain stock over 0.4 density consisting of un-utilizable species may also be included in this category. "Wastelands" should mean such village lands as are rendered unproductive under agriculture but, still, are reasonably productive under tree planting.

It is relevant to state that 90 per cent of lands classified as "forest" are owned and controlled by the State Forest Department and are governed under the Forest Acts and laws. Laws are favourable for protection, conservation and development of forests. Under the law, there is prohibition in the Reserved forest and regulation in the Protected forest. Forest under the management of Forest Department have the status of the "property" of Government but does not receive the protection given to the property envisaged under the Constitution. During the Pre-Independence Colonial British rule, Government used Government forest as their property, appropriated profits and benefits, permitted concessions to the adjoining public according to the reciprocal services rendered by them. Public was allotted separate forest lands for "Nistar" which was managed by the Revenue Department through a large number of ex-proprietors. Necessary authority was given to the Forest Department and its officials and full freedom to exercise it. There was, therefore, reasonable acountability and natural loyalty towards forest. Public was co-operative, and appreciation of value of forests, respect for law, and sense of reciprocal obligation to protect forests for receiving benefits from there in perpetuity. Situation and circumstances changed after Independence. Under the philosophy of democracy which is the foundation of our Indian Constitution and under the present form of governance which is of the people, by the people and for the people, public is now supreme, and public interest and public welfare are the main considerations in all socio-economic activities and developments. Since majority of Indian population lives in villages, it is natural that rural development should receive priority. Sustainable development of degraded forests has, therefore, to be considered within the above parameters.

Exact location of degraded forests is important as it has direct bearing on the intensity of and reasons for the extent of degradation. Elimination of all factors which have caused degradation can only guarantee success of sustainable development. These fctors could be social, economic, political, personal, for fulfilment of certain objective, or even the result of connivance of forest officials. Main reasons of degradation are increasing demands of forest products lack of appreciation of protective, productive environmental value of forests, selfish short-term interest to make best use of the situation, lack of required loyalty of public and public servants towards forest, hostile attitude of the public to press and claim their rights over forest without fulfilling their necessary reciprocal obligation to protect and conserve adjoining government and community forests and many times positive political encouragement as a vote catching strategy. Unfortunately there is widening gap at mutual distrust and increasing loss of mutual confidence between the forest officials and the public for which politicians are more responsible. In some cases and to some extent even non-technical bureaucrats and professional rivalry among technocrats are responsible.

Under the circumstances stated above it appears that the success of sustainable development of degraded forest can be achieved only if it is dealt with as socio-economic-technical undertaking with necessary flexibility and not as administrative/Executive project with rigid statutory guidelines giving no scope even for practical deviations. It can be implemented better either by such agency which has necessary competence, experience, executive powers motivated by expectation of positive reward; monetary or of any other kind of by an agency wholly dedicated to public service living as a respected leader among those whom they are serving for long time. There should be no constraints of required funds, and, their accountability is so linked that if they shirk or escape their responsibility, they should heavily suffer socially and/or financially.

Human nature has certain weaknesses and also certain strengths. People generally, are not motivated unless there is sense of belonging, some kind of owner's interest and/or assurance of some definite personal benefit, gain or advantage in terms of money, kind, recognition of reward. Forest Department is begging co-operation of involvement and participation of the public in schemes for protection and development of forests with package of benefits but required response is not forthcoming. It is a common knowledge that private sector gets willing co-operation of all kinds and of all concerned with much lesser efforts. For this reason, it is now well established that private sector is more efficient and records better success at more reasonable cost within the lesser time with more profit (where profit earning is the objective). This is a peculiar example of this kind in Forestry Sector. Large number of plantation companies—ENBEE Plantations, Golden Forest, to name a few, purchase small lands of degraded quality or even marginal, prepare economically viable project reports, obtain required loans from financial institutions, promise to raise successful fast growing plantations (it is another thing whether they actually raise plantations or not) and confidently receive funds in advance from the public by issuing bonds guaranteeing attractive monetary benefits in futrue. How they were able to gain trust of the public after extracting money from them, would be an interesting matter for investigation. Forest Department is holding much better productive areas, presently degraded where success of plantations are best assured within shorter period at lesser cost than projected by plantation companies. This is also true that success of plantations will depend on continuing willing co-operation of neighbouring public. Government can now have good reasons to ask the Forest Department as to why they could not give such handsome returns in plantations when excellent areas were taken up for plantations in good old days.

The above being the actual situation of facts and about the people and their attitude, good and bad points of the public and private sectors, it will be positive wisdom to be more practical and evolve a suitable model of an organization, pure or in joint sector more competent, efficient, accountable to undertake sustainable development of degraded forest ensuring better success. This organization should work under strict vigilance and control of the Forest Department and the public in mutual interest. If the experiment succeeds and success of plantations are more assured, it will be of advantage to link it with industrialization of rural areas with forest based small and medium size industrial units to provide more employment to rural population and for boosting rural economy. Details could be worked out and safeguards provided if the proposal meets the approval of all concerned. I am purposely keeping back my personal views, lest I am misunderstood being a retired Officer looking at the proposal from outside the Department. However, I will have no hestitation to discuss it.

12

PROTECTING INDUSTRIES, NOT TREES

ANIL AGARWAL

Forest Convention

In India, for instance, the key forestry issue is the reduction of the over-centralised power of New Delhi and various state capitals and to transfer power to communities. The convention would be an exercise in super-centralisation—transffering power from national capitals to global capitals and the World Bank and FAO.

The people who stand to lose the most if the forest convention goes through are the poor forest dependent communities in the South.

The world's highest award for pigheadedness should definitely go to those who dreamt up the forest convention—the idea proposed by a few Northern countries to develop and legal framework for managing the world's forests and opposed by most developing countries for the last 10 years. The recently concluded session of the Intergovernmental Forum on forests (IFF) has proposed in sufficiently obtuse terms to please both the opponents and the proponents of the convention that in the next five years they will consider meeting with a view to recommend the parameters of a mandate for developing a forest convention.

The idea was first mooted by US President George Bush at the 1990 G-7 summit when he was under pressure from his German counterpart, Helmut Kohl, to do something on climate change. He did not want to commit himself to something that would, in his words, 'cost Americans their jobs' and suggested that the world should instead focus on another urgent green issue, saving the tropical forests. This led to the beginning of negotiations to formulate a legally binding global forest instrument.

But in 1992 at Rio, opposition to the 'C' word meant that instead of legally binding convention the world endorsed an important statement of principles for forest management. Within months, the pressure built up again. And in 1995, at the meeting of the Commission on sustainable development the tension was evident. This time the agreement was to set up the Intergovernmental Panel on Forests for a two-year term. With squabbling unabated, in 1997 the IFF was set up, this time for three years. Now with the term of the IFF nearly over and no consensus reached on the need for a convention, a new configuration has been worked out—a permanent panel on forests and this time for a five-year period for breaking down the opposition to the idea of a convention.

It is easy to understand the desperate insistence of the proponents. One only has to remove the smokescreen that shrouds the idea behind the convention to find that the objective is not to protect the forests but to protect the industry that logs the forests. This also explain why the two keenest promoters of the convention are Canada and Finland who are heavily involved in the business of forest produce.

The reasons are simple. Northern industry, that has always controlled the world wood and wood product trade, wants to ensure it keeps churning out profits. It faces new threats; from environmentalists who want control over forestry operations and from Southern companies who are making inroads with an advantage of their own—a climate in which trees grow faster. The strategy, therefore, is to get a level playing field with the South so that there are environmental controls on forestry in the Tropical world. The way to do this is through a legally mandated set of indicators and criterion that would bind world forestry in a common definition of what is "green" wood.

As the southern wood exporters—Malaysia and Indonesia—have understood the business strategy behind the move, they have taken the smart way out—joined their opponents to play the profit game. These countries have developed their own criteria and indicators that define what is sustainable wood and in this process are able to legitimise their industry in the face of green opposition.

There is another reason why the convention is so important to these wood traders. The World Trade Organisation (WTO) rules provide their process and production methods (PPM)—in other words, on the basis of how the wood was grown and how sustainable is its method of production. The only way WTO will accept this marketing strategy is if there is a multilateral agreement that sanctions a globally harmonised definition of what is sustainable forestry. The convention is the only way. It is also for this reason that the opposition to the convention is from another big player in the world forest product business—the US, which is today the most strident critic of the idea it has floated because it sees its own economic interests being hurt in this convention game.

But in this game of forest competitors, the people who stand to lose the most if the convention goes through are the poor forest dependent communities in the south. The convention would create supernational rules for forest management that would, in fact, strengthen the hold of the forest bureaucracy in most developing countries. In India, for instance, the key forest issue is the reduction of the over-centralised power of New Delhi and various state capitals and to transfer power to communities. The convention would be an exercise in super-centralisation—transferring power from national capitals to global capitals and the World Bank and FAO.

It is time that the world stopped playing this charade. When the Commission on Sustainable Development meets to consider this proposal, it sould take a clear stand to say that the world has had enough time—and meetings—to discuss this idea and that this debate is sterile, one-track and dangerous. It must stop this fraud now. (Down To Earth Features.)

13

ECOLOGICAL AND SILVICULTURAL ISSUES PERTAINING TO DEVELOPMENT OF FORESTS AND FOREST PRODUCE INCLUDING NWFP

S.D. MUKHERJEE

Forest management in India is one of the oldest in the tropical world. It was looked upon as a model for other countries in the region not long ago. Guidelines for the preparation of working plans for a wide variety of high and coppice forest, yield regulations, preparation of yield and volume tables, thinning regimes, natural regeneration technologies and the like were among India's outstanding contributions.

The object of Forest management was sustained production of preferred timber species. Forests of India contained a large number of timber species, of which only a handful were marketable. The rest were treated as weeds (even though they were useful to the forest dwellers). They were, therefore, removed to the advantage of the preferred species. Thus forest management was designed to convert mixed forest crops into almost pure even-aged artificial entities consisting largely of the preferred species. Such converted forests yielded higher volumes of desired timber and hence greater revenue.

Mistakes made in forest management, unlike in other land uses such as agriculture, manifest only after many years giving a false impression in the mean time that all is well. This is exactly what happened. It is only about 20 years ago to that doubts were raised about the appropriateness of the availability of forest management systems. It is only recently that it has become clear that these systems have diluted ecological diversity, jeopardized watershed functions and undermined the livelihood and culture of about 50 million forest dwellers! As indicated earlier, other factors have also contributed and accelerated the process. However, it is difficult to apportion the responsibility at the present state of our knowledge and understanding. Even now, we know precious little about tropical forest ecology. Is it a surprise that, we do not know why certain plants in the neighbourhood of certain specific plants, grow better? It seems that there are friendly plants, enemy plants and neutral plants about which our knowledge is nil. Understanding nature is an extremely difficult task which may take decades of research. However, there is a silver lining. Forest dwellers who lived for generations in harmony with forets, possess a lot of knowledge and wisdom. Unfortunately this vast and precious knowledge is not documented. It is this invaluable asset that is being lost at a faster rate with the progressive destruction of forests.

Forests have been treated primarily as reservoirs of timber. They have been rarely looked upon as a system consisting of several sub-systems, chief among which are :

1. Bio-diversity and Environmental sub-system
2. Tribal sub-system
3. Wildlife sub-system
4. Social and Cultural sub-system
5. Pharmaceutical sub-system
6. Industrial sub-system
7. Rural development sub-system
8. Agricultural sub-system
9. Water sub-system

Such multiple uses of forests have created several challenges in its management. Environmental awareness was the international response to the widespread destruction of forests, dilution of biological diversity and pollution of air, water and land. The National Forest Policy was revised in 1988 prohibiting clear cutting of well stocked natural forests and banning plantations of exotics.

The shift in forest management, from man-made forests to natural forests caught the foresters unawares. This shift came at a time when the art and science of managing natural forests had almost been forgotten; it came at a time when there was little research support. This is perhaps the greatest challenge ever faced by foresters.

There is no reason whatsoever to be alarmed. Our forests have the resilience and capacity to regenerate themselves and provide all our basic needs. What our forests need is a little ecological regard and some sacrifice from us.

Ecological Implications

The most important commercial forests of India which have been intensively worked in the past are Moist Deciduous covering an area of about 23m ha. and Dry Deciduous occupying about 29m ha, totalling about 52m ha. This represents 70 per cent of the total forest are of the country and 87 per cent of the total exploitable area. Treatment of ecological implications is, therefore, mainly confined to these two types of forests.

The basic classification of forests into timber forests and miscellaneous forests is ecologically wrong. So is the classification of trees into Superior and Miscellaneous. Each forest community has its own place in nature. All different species are interdependent and perform complementary and supplementary ecological functions which together constitute an eco-system. If you harvest one, you disturb all the rest. Such a diverse system had the maximum stability and resistance against pests, diseases and drought. It is, therefore, obvious that any intervention in the existing relationship among plants (also animals, birds, insects and forest dwellers) is bound to have an ecological impact. The degree of impact, however, depends upon the degree of intervention and the sensitivity of the affected species. Of all types of interventions, clear cutting is the worst, and it is this which is largely employed in the bulk of our forests.

To appreciate the ecological impact, it is necessary to know the intrinsic character of tropical forests. Firstly, they are multi-layered mixed-species of different ages. Nature designed them this way to ensure optimum utilization of incident solar energy and soil productivity at different depths. The alignment and the architecture is such that there is minimum competition above and below ground. In short, a tropical forest is a self-fertilizing, self-sufficient and self-perpetuating machine. Its chemistry and mechanism are little understood even today. However it is clear that in such a design, the total production is the highest. It is incorrect to say that man-made forests are more productive than natural forests.

Implication on Soil

Clear cutting exposes soil to the tropical sun and rain storms. This changes the soil composition and structure so much so that the capacity of the soil to support vegetation deteriorates. In other words, retrogression sets in. This is what has happened in our forests. The soils have become poor in nutrients and water-holding capacity.

The ground water level in many areas have gone down and quite a few wells have gone dry. Most of the natural springs in the hills run dry during summer. Most of the rainwater reaches the sea as run-off in the absence of any ground cover. The present situation is that there is too much water during the rainy season and too little during all other seasons. The erosion is so great that siltation rate has more than doubled in most of the dams and reservoirs of the country. Thus, one of the most important ecological functions of forests is being undermined.

Wildlife habitat is substantially altered. The animals' food and shelter are being reduced at an alarming rate. Thus one integral and inseparable component of ecology is being decimated.

Impact on Agriculture

In India which enjoys tropical climate, agriculture is highly dependent on forests for maintaining soil fertility and for protection from pests and diseases. Nature maintains checks and balances among various pests to the extent that the damage is within tolerable limits. About 50 years ago, pesticides were unknown in India. Their role was performed by forests. Modern forest management has interfered with this balance. To aggravate the situation, native species are being replaced by high-yielding exotics which are vulnerable to diseases. We have, therefore, taken recourse to chemical insecticides which upset the balance further, necessitating stronger chemicals in larges doses. Thus a vicious circle is set up.

Impact on Vegetation

Some ecosystems have suffered more than others on account of our modern forest management. For example, mangrove eco-systems and desert eco-systems have suffered most severely. The desert eco-systems are facing a similar threat. The difference in this case is the introduction of an exotic—*Prosopis juliflora*—over extensive areas. The native grasses, which were very hardy and nutritious, are ousted by this very aggressive plant. In the absence of a baseline survey, it is hard to pinpoint specific localities where some of the grass and tree species have become extinct.

Few attempts have been made to evaluate the full ecological, social and economic values of non-timber species. For instance felling of local trees in some area killed tree sericulture which was till recently a flourishing industry. Similarly, the virtual extinction of *Givotia rottleriformis* has created unemployment among the Toy making artisans in Andhra Pradesh. Yet another cultural loss has resulted from the scarcity of shrubs and herbs yielding natural dyes.

Teak forests were considered the most valuable forests. They, therefore, received the best attention, meaning the greatest intervention. Wholesale clear cutting followed by controlled burning created the picutre of a clean shaven head. The planting is usually done after a rain or two. The maximum soil wash takes place during these early showers which are usually very intense. It is relevant to mention here that, a moist forest possesses a rich diversity of trees, shrubs, herbs, climbers and grasses.

In multi-canopied forests the shade bearers which occur in lower canopies have suffered the most. It is only the strong light demanders such as Teak and Sal, that have prospered at the cost of the rest. As far as teak is concerned, its percentage in a natural forest is never more than 40, whereas, it is planted pure in man-made forests.

As mentioned earlier, focus on teak, sal and eucalyptus has resulted in some unsuitable sites being selected for planting. Such areas have completely failed, they are now colonized by obnoxious weeds like as *Lantana*.

The over-emphasis on commercial trees has created another very serious problem: healthy and vigorously growing trees of non-commercial species are difficult to come by. How are we, then, to restore the mixed composition of the forests? There are few seed banks to provide the required quantities of quality seed. What a colossal ecological loss!

New Silvicultural Practices Suggested

(a) Site-Specific Planning : Just as the diversity and complexity of social systems have been recognized, so silvicultural responses must be equally site-specific and responsive to local conditions. Recent work under JFM programme indicates that it is possible to create management-objective zones within a community-protected forest. However, a number of questions have to be considered when working with local people to decide how a site may best be managed, and to develop silvicultural responses appropriate to different conditions.

(b) Encouraging natural regeneration : To ensure that the growing space is maximized it is essential that all levels of the forest architecture should be used. This should include use of shade-tolerant understorey trees and shrubs; management of shrub and herb layers; introduction of herbaceous medicinal plants; management of forest floor to enrich soil and encourage natural regeneration and the production of edible trubers. If JFM is to be replicated widely, low-cost options based on natural regeneration are more viable than strategies based solely on plantations. However, to assess whether a site is capable of supporting natural regeneration, a series of nested questions need to be asked to decide what level of intervention will be required to ensure the success of the natural regeneration option. The following options require assessment.

- Is there adequate root stock or advanced growth to ensure natural regeneration without any inputs other than protection?
- If not, is there sufficient root stock density and/or advanced regeneration for a minimum of soil working and site improvement with some enrichment planting or gap filling to be enough to supplement and enhance the natural regeneration?
- If plantation is necessary because root stocks and advance

regeneration are rare, is it possible to plant to ensure that existing root stocks are encouraged to grow as part of the mixed stand? For example, in a number of areas where fully stocked plantations were undertaken on supposedly barren lands, teak coppice growth from semi-buried and quiescent stumps emerged after several years of protection. Experience indicates that deferring the planting activity for 3-4 years on barren lands, while the closed area is treated with soil/moisture conservation inputs, allows the regeneration of grasses and assessment of the viability of root stocks.

Favouring natural regeneration does not mean that the plantation option has been rejected, but just that there is a continuum of silvicultural options. The continuum aproach means that the focus shifts to enhancing the growth potential of existing plants and emphasising local diversity. There will always be a series of interventions that run from total dependence on natural regeneration to total dependence on plantations. Innovative approaches can extend the boundaries of the use of natural regeneration and allow experimentation in areas of overlap.

If we accept the importance of relying more heavily on natural regeneration because it meets the multiple objectives of users and is more cost-effective, it underlines the argument that in general natural forests are more appropriate for meeting local people's needs. However, where forests have become so degraded that it is not possible to rely on natural regeneration plantations will have to be undertaken. Plantation strategies have already moved way from the monocultures of the social forestry era to more complex polycultures with a variety of different species. In Andhra Pradesh, plantations raised under TGP included as many as 63 species.

Once again, the interface between technical and social innovation must be carefully managed, and the composition and management of these plantations should be negotiated on a site-specific basis with the users of the land. Simply applying more complicated models and responding to the site's ecological potential does not mean that joint management has been fully embraced. As the variety of tree, shrub and grass species increases, innovative planting practices will also need to be evolved. For example, many species propagate well by direct seeding and may not need to be raised to the sapling stage in nurseries before planting out.

(c) Multiple Time Horizons : Many forest products are seasonal in nature like fruits, fresh leaves and flowers—while others, like certain fodder or fibre grasses, can be harvested periodically. Fuelwood often consists of twigs and branches which can be harvested on a periodic basis, every few years, while poles are of medium duration and timber requires very long temporal rotations. People depend on a steady supply of different products from the forests over a multiple time horizon. The new **silvicultural approaches** will also need to accommodate these temporal sequences and events with a more complex series of **silvicultural activities** over both time and space than is currently advocated through occasional thinning regimes under most standard **silvicultural prescriptions.** They will also require an understanding of the impacts over time of different activities on different livelihood-dependent users. The value of economic flows from the forests, aggregated over seasons and years, should not be underestimated. Trading some of the delayed return from timbers in favour of more frequent short-term returns may make good financial sense and fulfil the needs of a larger number of people.

(d) Individual Plant Manipulation : Indigenous practices of arboriculture and vegetation management include a wide variety of individual plant manipulation methods including pruning, lopping, pollarding, rotooning, weeding and protecting stems and fruits. In the same way, modern plantation management includes conservation at an individual seedling level. How can these cultural practices be applied more intensively within forests themselves to help meet multiple objectives? Several strategies are needed for drastic thinning to open space for fruit tree crows to develop, with the intention of creating a forest 'orchard' of about 250 NWFP-yielding trees per hectare. While contesting the theory that all trees in JFM systems should be managed purely for crown-based products there is much to be said for increased attention to managing the produtivity of foliage, small stems, fruits, etc., through more frequent pruning and pollarding, as one important component of the silvicultural mix. If such individual manipulation for one type of product is carried out, the implications for other products must be considered, since such manipulation may reduce stem and timber production.

Ecological and Silvicultural Issues Pertaining to Development of Forests and Forest Produce Including NWFP with Specific Reference to Andhra Pradesh

Forestry Resources of Andhra Pradesh

The forests in Andhra Pradesh extend over an area of 63,814 sq. km accounting for 23.2 per cent of the total geographic area of the state. The per capita forest area is about 0.10 ha. Over 94 per cent of forests in Andhra Pradesh fall under Southern Tropical Forest type as per Champion and Seth's classification. Bulk of this, (over 44%), is of the Southern Tropical Dry Deciduous category. With respect to legal status, over 79 per cent of forest area is classified as Reserved Forest.

The classification of forest in the state as per different parameters is given as follows :

Classification of Forest Area

Sl.No.	*Particulars*	*Forest Area (Sq. km)*	*%*
I.	**Classification by legal status**		
	(a) Reserved	50,478.63	79.1
	(b) Protected	12,365.34	19.4
	(c) Unclassified	969.76	1.5
II.	**Classification by Composition**		
	(a) Sal	47.16	0.1
	(b) Teak	9,145.00	14.3
	(c) Others	54,621.57	85.7
III.	**Classification as per Champion and Seth Norms**		
	(a) Southern Tropical	16,110.00	25.3
	(b) Soutehrn Tropical Moist Deciduous	16,100.00	25.2
	(c) Southern Tropical Dry Deciduous	28,431.00	44.6
	(d) Littoral Forest	2,856.00	4.5
	(e) Tidal Swamp Mangrove Forests	317.00	0.4
IV.	**Functional Classification**		
	(a) Protection Forests	38,449.47	60.3
	(b) Production Forests	25,364.26	39.7

The forests of Andhra Pradesh host a wide variety of species. While forests are managed mainly for timber and firewood, the proportion of Non-Wood Forestry Products is given below :

Sl. No.	*Year*	*Revenue from Timber and Fuelwood*	*Revenue from NWFP*	*Miscellaneous Rvenue*	*Total*
1.	1994-95	4682.26	5545.05	60.00	10286.31
2.	1995-96	4984.33	9349.74	142.49	14476.56
3.	1996-97	3628.21	6364.67	1596.12	11589.00
4.	1997-98	3652.84	5112.44	1334.15	10100.43
5.	1998-99	3135.72	4490.22	1223.61	8849.55

The NWFP which includes all forestry products other than timber and fuelwood is seen to account for a major proportion of revenue realized in the state.

Andhra Pradesh is endowed with a very rich inventory of NWFPs. The contribution of NWFP to A.P. Forest Department's income is highest compared to other states in the country.

NWFP plays a very important role in the rural economy, especially of villages located within and in close proximity of forest areas. Collection of NWFP is a major subsistence occupation of most of the villagers. NWFP species are collected for sale and/or for domestic consumption. *The species that are usually collected are Bauhinia vahilli, Semicarpus anacardium, Strychnos potatorum, Sapindus emarginatus, Sterculia urens, Boswellia serrata, Anogeissus latifolia,* etc.

Rapid deforestation, excessive exploitation of existing natural forests and growing population have created a tremendous pressure on the forests in Andhra Pradesh.

The population residing in and near the forests are traditionally dependent on forests for their subsistence. Several social practices of these villagers are intrinsically linked with forest resources. During lean periods of employment, the inhabitants depend on forests even for their food. Several forestry products are collected for self-consumption and/or for sale in the local markets.

In a situation as above, the dwindling forest resources which definitely pose ecological problems create severe hardships to the dependent population. Earlier efforts to re-forest degraded forest lands have often fallen short of expected achievement, since the relationship between man and forests was not appreciated in social perspective and was more policing in approach.

In the recent years, the important of symbiotic relationship between forests and dependent popultion area has been appreciated and appropriate provisions are being made. The *JFM* concept has been fully appreciated and taken up on a large scale covering almost 25 per cent of the total forest area in the state. 6585 *Vana Samrakshana Samithis (VSS),* i.e. Forest Protection Committees are actively engaged in protecting and developing the degraded forests in the state.

Level of Dependence on Forestry Resources

Majority of the households in and around the forests are engaged in farming activity, either as cultivators or agricultural labourers. Agriculture is very poorly developed and cultivation is mostly for subsistence. Agicultural labourers migrate to other villages. Cultivators within the village often employ household labour and opportunities for hired labour is very marginal. Collection of *NWFP* as an economic activity is common among all these households which, however, is not perceived to be a major occupation and is undertaken as a subsidiary income generation, especially during lean periods of employment.

The income-wise ranking of the occupation based on a sample survey by Operations Research Group (ORG). Baroda, a consultant appointed under AP Forestry Project is given below:

Income	*% Households Reporting Rank*			
	I (Major)	*II*	*III*	*IV*
Agriculture	31.1	11.1	6.5	2.6
Agriculture Labour	29.8	33.7	9.1	1.8
Non-Agriculture Labour	18.2	14.9	12.2	1.2
Business	0.5	0.3	0.1	0.0
Service	0.5	0.5	0.1	0.1
NWFP Collection	19.5	36.7	40.0	10.7
Others	0.4	1.0	4.3	3.2
No Response	—	1.8	27.6	80.4
Total	100.0	100.0	100.0	100.0

It can be seen from the above table that the proportion of households reporting agriculture as second rank and upwards show a sharp decrease. This is accompanied by a relative rise in responses for agriculture labour. This is due to the fact that most of the households engaged in cultivation are undertaking agriculture labour activity also. This is a reflection of poor development of agriculture in the forest areas.

It can also be seen from the above table that Service class households earn the maximum average income, for the activity performed. However, the weighted average shows this activity to be among the lowest, accounting to one per cent of the aggregate weighted average income.

On weighted average, *income from labour is the highest (43.8%) followed by returns from sale of NWFP (34.3%).*

The above analysis shows that while NWFP may be treated as a subsidiary income generating actvitity, in effect, *the return from NWFP account for sizable proportion of the annual household income.*

Due to lack of modern medical facilities, use of several locally occurring medicinal plants is high. Several medicinal plants collected by households have laxative and antacid properties and are used to treat gastronomic problems. NWFPs with properties to cure skin diseases, rheumatism, pains, sprains, etc. also are known to many villagers. *With the increasing demand for herbal medicines in urban and semi-urban centres, collection of such NWFP has gone up during the past few years.*

Generally collection of NWFP and firewood is done by women and children, who colllect the same on their way to and from work and while looking after cattle grazing in the forest.

Dynamics in Forest Resource Depletion

While it is established that the local population depend on forestry for social-cultural economic as well as for self-consumption, it is natural to anticipate a system which would rationalise the dependence for sustainable production. However, due to various reasons, the forests started decreasing both in area and density. This automatically reflects in the decrease of the production of NWFP. The dwindling of

NWFP resources is due to a composite set of factors each adding to beyond of Sum of Totals.

Another reason for depletion of NWFPs is acute land hunger. Social pressures are very high to shift the land use from forest to agriculture. While forest lands are in effect shrinking, the demand for NWFP is on the rise in urban markets as well.

Thus, NWFP management requires complete reorientation to protect the interests at micro (local villages) and macro (environment amelioration) points of view, while appreciating the fact that the level of dependence of local population on social, economic and cultural considerations is much higher than as perceived by the villagers themselves or by the administration.

NWFP Coverage

As mentioned above, several species of domestic importance and/or commercial value are being collected. Several varieties of tubers, wild fruits and other parts etc. are collected for domestic consumption and they do not command significant market. The survey has essentially focused on those species which are commercially important and the local villagers get economic returns.

List of Important NWFP collected from the forests of Andhra Pradesh.

Sl. No.	*Local Name*	*Botanical Name*	*Use*
1	*2*	*3*	*4*
1.	Tapasi gum	*Sterculia urens*	Edible gum
2.	Tirumanu gum	*Anogiessus latifolia*	Edible, calico printing, paper industry etc.
3.	Tamarind	*Tamarindus indica*	Edible fruit
4.	Amla	*Emblica officinalis*	Edible fruit, medicines
5.	Pongam fruits and	*Pongamia glabra*	Oil used in soap making
	seeds		and has medicinal use.
6.	Rella	*Cassia fistula*	Tannin
7.	Tangedu	*Cassia auriculata*	Tannin
8.	Adda	*Bauhinia vahilli*	Leaf cups plants, rope making bark and Tannin

(Contd.)

Sl. No.	*Local Name*	*Botanical Name*	*Use*
1	*2*	*3*	*4*
9.	Hill brooms	*Thysanolaena maxima*	Brooms
10.	Wild brooms		Brooms
11.	Naramamidi	*Litsea glutinosa*	Bark for in incense sticks, mosquito repellant, medicinal use
12.	Cleaning nut	*Strychnos potatorum*	Cleaning water-coagulant
13.	Soap nut	*Sapindus emarginatus*	Soap
14.	Chironji	*Buchanania lanzan*	Kernel edible as dry fruit
15.	Bamboo	*Dendrocalamus strictus*	Bamboo articles for household use
16.	Shikaki	*Acacia concinna*	Soap
17.	Musni, Musthti	*Stryuchnos nuxvomica*	Medicinal use, rat and fish poison
18.	Honey	—	Edible sweetner, medicinal use

Critical Review on Methods of Harvesting of NWFP and their Impact

Most of the NWFP, being collected have local importance. Except Mohwa, Tendu, Bamboo and Gum karaya, the extent of collection is not yet to commercial scale. The pressure on forest resources stem more from demand for fuelwood, timber and land for agriculture purposes.

The harvesting of *Bauhinia vahilli* is in fact beneficial to forest management. Leaves, seeds and bark of this climber have local importance. Debarking of *Bauhinia* is fatal to the plant. However, since, this species strangulates the host plant such as teak, removal of *Bauhinia* and other such climbers are important forest management practices for better timber production. Farther, this plant is a good coppicer and hence it regenerates naturally.

Collection of Amla and Tamarind which involves an entire fruit bearing brach being felled is beneficial for better vegetative regeneration. Further, these trees are also cultivated in homesteads, farm bunds etc. the cut branches yeild good firewood. The matter of concern is that due to high pressures, fruits are plucked even before maturing and excessive cutting of branches stunt the growth of the trees.

The barks of *Cassia fistula* and *Litsea glutinosa* are in fact incidental to collection of firewood. Before collecting firewood from these trees, they are debarked and ignited.

Improper harvesting is more distinct in case of collection of Gum karaya. The blaze depth, spacing, freshening, etc., do not confirm to recommended norms. The method of hacking blazes instead of making incisions with sharp instruments often injure the tree to fatal extent. Excessive tapping often results in the tree "bleeding" to death.

While it is true that the pressure on forest is increasing due to social reasons, it is not appropriate to place the entire responsibility on poor harvesting methods. Since the vast majority of population is dependent on forests, it will be a herculean task to train the villagers in better harvesting methods and monitor their work. Further, traditionally forests were managed for timber. Fuelwood was the only social consideration. Plantations undertaken were mainly timber yielding species.

It is only since recent past particularly after the advent of JFM, that the social dimension to forest management is expanded and the role of NWFP is appreciated in the contest of regional socio-economy.

Plantations raised were usually timber-species and NWFP species were not adequantly replenished. With the trees diminishing in number and dependent population increasing, it is natural that the disappearance of the trees increases at increasing rate. It is, therefore, pertinent to conclude that pressure on forest resources is due to diminishing number of trees and a sure measure to reverse the trend is to undertake extensive afforestation and reforestation activities.

Ecological and Silviculture Impact of JFM in Andhra Pradesh

During the course of an international World Bank/WWF workshop on Performance Indicators for Sustainable Forest Management conducted at Hyderabad a comprehensive list of measurable indicators was finalised to measure the effectiveness of forest management by the VSS. For the first time, a set of 14 indicators were deployed in 5 per cent of the total VSS formed in the state prior to April 1998. Data from over 120 VSS was collected and analysed. All the 14 indicators may not be relevant in all the VSS villages and hence the appropriate indicators for each VSS were considered for the analysis.

The analysis was done to assess changes in growing stock, production of NWFP, regeneration status, species diversity, water Chart and agricultural yield.

Growing Stock

The change in growing stock as a result of introduction of new silvicultural practices in JFM areas generally varies around 50 per cent except in Anantapur district, which is the driest in the state with heavy incidence of grazing.

Impact of New Silvicultural Practices followed in JFM Areas on NWFP Production

The traditional use of forest by the local people had been based on non-timber forest product (NWFP) that does not involve felling of trees and provides regular income to the local people. The scientific forestry practiced in the past 130 years brought timber based silviculture that neglected the NWFP. In many places high yielding varieties of plant species such as eucalyptus were introduced by uprooting all existing miscellaneous tree species. As the villagers never felt the sense of ownership for the forest, NWFP was collected unscientifically and some time killing the trees by over-exploitation. There were instances where felling the tree itself was restricted to collect the fruits, leaves etc. In Andhra Pradesh, Girijan Co-operative Corporation has the monopoly on the procurement of NWFP. According to their data, the yield of NWFP has been coming down year after year.

The situation has changed after introduction of JFM : People's interest in NWFP has revived. In many degraded forests NWFP species are coming back naturally. Planting of NWFP species have been taken up in most of the VSS. A survey on the availability of

NWFP in the JFM areas was conducted recently. The result from the analysis of data on NWFP availability from representative sample of 62 plots spread over 7 divisions covering two distinctly varied geographical area of Telangana and Coastal regions shows an increasing trend in availability of NWFP.

This change can be attributed to direct result of protection and better management of forest by the people. These trends, although quite encouraging, are only initial trends which are required to be observed for a longer period of time.

Regeneration and Species Diversity

The increase in regeneration and species diversity promises better forest with the needed bio-diversity for sustainable development.

It is seen that in addition to the increase in forest cover, the diversity of species is going up. There were several species that had disappeared in the past but are found to be reappearing in the JFM areas. The maximum diversity is noticed in Paderu Division, a predominantly tribal area.

Increase in Water Table and Agicultural Yield

The indirect ecological benefit from the regeneration of forests is reflected in a data reported which shows water table increase ranging from a minimum of 0.13 metres to a maximum of 13.92 metres with an average increase of 3.6 metres, which is quite significant especially considering the fact that there was poor rainfall during this year compared with the previous year.

An important change noticed is the improvement in agricultural production. The results have been compiled from 79 samples in 8 divisions spread over the state. The percentage increase has been found to be ranging from 17 per cent to 131 per cent.

The increase in water has improved the availability of sub-soil moisture and water for irrigation. This has resulted in increase in agricultural yield, ranging from 16.88 per cent to 131 per cent. The average increase is 51.71 per cent. Further evidence of the improvement in sub-soil moisture availability is provided by the more diversified cropping pattern as reported below:

	Impact on Cropping Pattern in Tribal Areas			
	Before VSS formation		*After VSS formation*	
	Kharif	*Rabi*	*Kharif*	*Rabi*
Mono-crop	*Jowar* Paddy	***Chillie***	• Jowar • Paddy • Groundnunt • Redgram • Pulses • Maize	✔ Paddy ✔ Sesaumu ✔ Chilly ✔ Sunflower ✔ Cotton ✔ Tobacco
Inter-Cropping	*Jowar + Redgram* Groundnut + Redgram	—	• Jowar + Redgram • Groundnut + Redgram	✔ Cotton + Chillie

Reclothing Encroached Lands

Encroachment on forest lands was a major problem for the Forest Department right from the day forest lands were notified as reserved forest and as reported in Chapters VI and VII. The encroachment on forest land was on the increase year after year. From 1980 to 1990 about 11,000 ha. of forest land was encroached annually. The forest land under encroachment is estimated to be 3,50,000 ha. One of the major gains of JFM is the stoppage of further encroachment in VSS areas. No new encroachments in forest land has been reported in the forest land put under VSS because of the people's participation. The significant development is that in many of the VSS, especially in the districts of Visakhapatnam and Adilabad, nearly 24,000 ha of land under *Podu* cultivation has come back to forest because the people have realized the mistake of cultivating steep hill slopes, (where the agricultural yield was diminishing year after year and sometimes exposing the rocks due to severe erosion) and the advantage of growing forest. Forest Department has taken care to ensure that such reversal from *Podu* cultivation to forestry is done with the full knowledge and willingness of VSS members. Forest Department has also raised high yielding varieties of NWFP species such as Tamarind, Amla, Neradu, Seethaphal etc. and also some high yielding Eucalyptus clones as demonstration plots so that people can see the economics of growing the plantations in place of cultivating the forest land. Similarly, growing of medicinal plants has been introduced on an experimental basis with a tie up with Pharmaceutical Companies for certain identified species like Aswagandha. The initial response from the people is encouraging.

14

TECHNIQUE FOR OPTIMISING USE OF DEGRADED FOREST LAND

KAILASH CHANDRA BEBARTA

Introduction

Soils are vital natural resources. Sustainable use of soil resources determines the life-supporting system and socio-economic development of any country. Deterioration of forest soil is a matter of great concern to the country. Degradation of forest soil resources, *inter alias,* is a cause as well as an effect of deforestation. Presently forest soil resources are under different degrees of degradation and are getting further deteriorated at an alarming rate. Rapid degradation of soil resources jeopardises the ecological security of the country. Therefore, there is a great need to maintain the soil health and quality of degraded forest land. This can be best possible by developing technologies and strategies for optimising use of degraded forest land.

Making effective use of available soil and water resources shall imprOve the forest cover and productivity in India. This requires a systematic and scientific appraisal of soils, which forms an important base for planning sustainable forest development. Soil resource maps are prepared during working plan exercises. But, such maps are sketchy and are based on maps prepared by state soil survey and other organisation after having adopted inferior technology and method since 1960s. Apart from this, no uniform technology and sound criterion are used to generate soil resource map on a scale, which can be used for optimisting degraded forest land use.

In view of the aforesaid problem, the National Bureau of Soil Survey and Land Use Planning (Indian Council of Agricultural Research), Nagpur, in co-operation with Soil Survey staff of the Department of Agriculture, Government of Madhya Pradesh, took up soil resource mapping of the state on the scale of 1 : 250,000 during the period from 1989 to 1994. It adopted a three-tier approach in mapping, viz. satellite image interpretation, field surveys and lab analysis, and cartography and printing in order to generate soil resource maps.

The soil resource maps/inventory prepared and published by NBSS&LUP indicates the kind, extent and distribution of soils on the landscape. It specifies different characteristics of soil of different sites. It unfolds the erosion status of the soil. Therefore, the soil resource map can be used for a variety of purposes. Few of such uses are illustrated below:

- To generate forest soil resource map indicating the kinds, extent and distribution of different soils in a forest division.
- To delineate problematic/degraded/eroded for their amelioration.
- To generate several thematic maps as per the requirement of forest management practices.
- To promote forest research on mapped and defined soils for generating soil-based technology and its analogous transfer to other areas having comprable soil site characteristics.
- To generate database for storage and retrieval as and when desired for future interpretation and use.
- To predict forest productivity based on soil site suitability models.
- To monitor the soil health and to forewarn the emerging constraints.
- To prepare a perspective land use plan for the area, including degraded forest land, in order to optimize land use for sustainable forest management.

Geographical Setting of Madhya Pradesh

The Madhya Pradesh is a largest state of India and located between 18°05′ to 26°01′ N latitude and 72°05′ to 84°30′ E longitude. It covers an area of 44.34 m hectare, which is 13.5 per cent of total geographical area of the country. According to 1991 census, the total population of M.P. was 66.2 million and the population density was 149 per sq. km. The distribution of population is uneven throughout the state.

The state has been divided into three major physiographic regions depending upon elevation, slope and ruggedness of terrain. These regions are: Northern Deccan plateau, Central Highlands and Eastern plateau. The state is mostly drained by Narmada, Son, Chambal, Betwa, Ken, Sind, Kali Sind, Parvati, Dhasan, Mahanadi, Indravati, and Sabari etc. The geological formations are diverse in origin. It comprises the archaean, metamorphic to sub-recent and recent alluvial deposits, while the remaining part encompasses a wide variety of rock formations like sedimentary, crystalline and Deccan trap and basic flows.

The state has a wide variations in climate ranging from semi-arid (dry moist, dry and moist), sub-humid to humid tropical. The mean annual rainfall varies from 500 to 3500 mm covering 20 to 80 per cent of the mean annual potential evapo-transpiration (PE). It receives rainfall in the monsoon period from June to September.

How Soil Resource Mapping was Done and How it Can be Used

The project of soil resource mapping was undertaken by National Bureau of Soil Survey and Land Use Planning (Indian Council of Agricultural Research) in collaboration with Department of Agriculture, Government of M.P. It adopted a 3-tier approach, comprising image interpretation, field surveys and laboratory investigation and cartography. The LANDSAT imageries on 1:250,000 scale were interpreted after developing a detailed legend for generating a landform analysis map of the area. The physiographic delineation were then transferred into the Survey of India topographical sheet (1:250,000 scale).

The soil samples were collected horizontal stripwise and analysed for various physical and chemical characteristics, such as soil texture, pH value, organic carbon, calcium carbonate, CEC and base saturation, etc. by following the standard procedures (Sarma *et al.,* 1987).

The state has been divided into different major landforms, which were again divided into regions (Table 14.1). Such regions constitute a group of homogenous mapping units and are elucidated with their descriptions, soil use, soil constraints, soil potentialities and the strategy required to minimise the constraints. Based on landform analysis, field surveys, laboratory investigations and field reviews, the soil resource map of the state was prepared on 1:250,000 scale comprising 64 sheets of the state. The same was printed and published on 1:500,000 scale (9 sheets) without any generalisation. Soil family association with dominant phases formed the soil mapping unit.

Table 14.1 : Districts and Mapping Units in Different Landforms of M.P.

S. No.	*Name of the Major Landforms or Region*	*Districts*	*Mapping Units*
1.	North Deccan Plateau Satpura Range (Basalt)	Jhabua, Khargone, Betul, Hosangabad, Seoni, Jabalpur, Narsinghpur, Mandla, Shahdol.	01 to 88
2.	North Deccan Plateau Satpura Range (Granite)	Betul, Chhindwara, Seoni, Balaghat, Durg	89 to 122
3.	Northern Deccan Plateau Satpura Range (Sedimentary)	Khandwa, Hosangabad, Betul, Chhindwara, Narsinghpur	123 to 135
4.	Northern Deccan Plateau Satpura Range (Laterite)	Chhindwara, Betul, Seoni	136 to 139
5.	Central High Land, Narmada Valley (Basalt)	Raipur, Raigarh, Ambicapur, Sidhi	140 to 166
6.	Central High Land, Narmada Velley (Alluvium)	Jabalpur, Raisen, Hasangabad, Narsinghpur, Khandwa, Dewas, Indore, Sehore, Khargone, Dhar, Jhabua	167 to 199
7.	Central High Land, Vindhyan Ranges (Sedimentary)	Raisen, Bhopal, Sehore, Damoh, Sagar, Narsinghpur, Panna, Satna, Rewa, Sidhi	200 to 277
8.	Central High Land, Vindhyan Range (Alluvium)	Sagar, Damoh, Satna, Panna, Rewa	278 to 298
9.	Central High Land, Vindhyan Range (Granite/Basalt)	Vidisha, Raisen, Sagar, Chhatarpur	299 to 317
10.	Central High Land, Malwa Plateau (Basalt)	Dhar, Jhabua, Indore, Dewas, Sehore, Bhopal, Ratlam, Ujjain, Shajapur, Mandsaur, Rijgoch, Vidisha, Raisen, Sagar, Guna and Shivpuri	318 to 419
11.	Central High Land, Malwa Plateau (Granite Genius)	Jhabua	420 to 423

(Contd.)

S. No.	*Name of the Major Landforms or Region*	*Districts*	*Mapping Units*
12.	Central High Land, Pathar and Bundelkhand Upland (Sedimentary Vindhyan)	Chhatarpur, Sagar, Damoh, Panna, Satna, Rewa, Sidhi	424 to 483
13.	Central High Land, Pathar and Bundelkhand Upland (Granite Genessis)	Chhatarpur, Tikamgarh, Panna	484 to 527
14.	Central High Land, Pathar and Bundelkhand (Alluvium)	Bhind, Morena, Gwalior, Shivpuri and Datia	528 to 574
15.	Eastern Baghelkhand Plateau	Jabalpur, Shahdol	575 to 579
16.	Eastern Baghelkhand Plateau (Laterite)	Jabalpur, Shahdol, Satna	580 to 585
17.	Eastern Baghelkhand Plateau (Sedimentary Gondwana)	Sidhi, Shahdol, Ambicapur Bilaspur, Raigarh, Mandala	586 to 649
18.	Eastern Baghelkhand Plateau (Basalt)	Shahdol	650 to 652
19.	Eastern plateau, Mahanadi Basin (Sedimentary)	Balaghat, Rajnandgaon, Mandala, Bilaspur, Durg, Raipur, Rajgarh	653 to 720
20.	Eastern Dandakaranya Plateau (Granite/Gneissic)	Raipur, Bastar, Durg, Rajnandgaon	721 to 777
21.	Eastern Dandakaranya Plateau (Sedimentary)	Bastar, Durg, Raipur	778 to 805
22.	Eastern Dandakaranya Plateau (Laterite)	Bastar, Durg, Rajnandgaon	806 to 819
23.	Eastern Dandakarnya Plateau (Basalt)	Bastar, Durg, Rajnandgaon and Raipur	820 to 825
24.	Eastern Chhota Nagpur Plateau (Granitic-Laterite)	Ambicapur and Raipur	826 to 843
25.	Eastern Chhota Nagpur Plateau (Sedimentary Dhaiwar)	Raipur, Raigarh, Ambicapur, Sidhi	844 to 845
26.	Eastern Chhota Nagpur Plateau (Basatic-Laterite)	Ambicapur	846 to 851

Source: Soil Resource Map of M.P. 1997. National Bureau of Soil Survey and Land Use Planning, (NBSS and LUP), Nagpur.

The mapping units have been assembled after having adopted a 3-tier approach. The upper tier shows diagnostic properties of the dominant soil and its phases. It includes order, sub-order, great group, sub-group, soil depth, particle size of family control section, surface texture, slope, erosion class and other phases of the dominant soil family, such as salinity, flood hazards etc. The middle tier gives landforms attributes and the lower tier gives the mapping unit (polygon) number.

Table 14.2 : Index of Forest Compartments and Soil Map Units (Or Polygon no) Range-Bagnadi Forest Range

Compt No.	*Soil Map Unit*
579, 598, 574, 557	93
565, 563 98	
616, 598, 596	94
690, 591, 574	110

Source : Working Plan, Rajnandgaon, Year 1999.

The soil map shows the soil family association number (polygon numbers), such as 001, 002, 003 ...., 851. These polygon numbers are typical of each mapped unit. The mapped units are organised in a systematic manner in the legend given on all the nine sheets. The legend depicts major characteristics of each polygon number (mapped soils).

The soil map of Madhya Pradesh has been published by NBSS&LUP on 1:5,00,000 scale in nine sheets. The Index map of Madhya Pradesh, a part of soil map provides the location of districts in the states. In order to know the soils of an area, first locate with the help of index map, the respective sheet out of 1 to 9 in which the area of one's interest falls. After locating the area on the relevant sheet, see the soil map unit number(s) appearing in that area. The details of each soil map unit are given in the legend. The soil map unit is an association of soil families. The legend provides brief description of the dominant and associated (sub-dominant) soil families within in a particular map unit, highlighting the soil depth, drainage, texture, slope, erosion, salinity, etc. The taxonomy of the soils, as per USDA System of Soil Classification, and the extent of the area covered by each mapped unit along with its percentage of the total area of the state are given in separate columns.

Optimising the Use of Degraded Forest Land : A Case Study of Range in Rajnandgaon Forest Division (M.P.)

Rajnandgaon forest division has been taken as the study area for explaining the method of optimising the use of degraded forest land. The study area comes under the landform category of 19, 20, 22, and 23. The mapping units are given in the Table 14.1. Each mapping unit is described in terms of attributes such as surface form, parent material, soil depth class, particle sizes class, minerology class, calcareous class, soil temperature regulation class, reaction (pH) class, drainage class, ground water depth, compact layer, (type depth and extent), surface texture class, slope class, erosion class, salinity or sodicity class and extent, surface stoniness, flooding, soil texonomy, organic carbon, calcium carbonate, base saturation, CEC. The study area encompasses 41 number of polygon or map units. The soil characteristics of all map units of the study area are described in the Soil resource map and inventory report prepared by NBSS&LUP, Nagpur. On the basis of this report and maps of the NBSS&LUP Nagpur and forest stock mapping, a soil resource inventory and map for the degraded forest land of the forest division will be prepared. Again, based on the status of soil resource of the degraded forest land, a detailed treatment strategy for each polygon or soil map unit will be developed.

Therefore, the use of degraded forestland can be optimised by the following methods.

- Preparing Soil Resource Map and Inventory of degraded forest land.
- Evolving treatment strategy for every individual soil map units containing degraded forest land.

1. Procedure for Preparing Soil Resource Map and Inventory

The map units or the polygons which fall in the study area are first identified. Then, a soil resource map of the division is prepared on the basis of state soil resource map of 1 : 5,00,000 scale (prepared by NBSS&LUP, Nagpur). This map contains the entire map units or polygons falling with in the forest area of the division. This map is enlarged to 1:250,000-scale map. Now divisional map of 1:250,000 scale indicating compartment number of all the degraded forest is prepared and is super-imposed on the soil resource map to find out the degraded forest blocks/compartments falling in different polygons or mapping units. (Fig. 14.1). Based on this, a table or index containing polygon number or map unit with number of degraded compartments falling in it, is prepared. At anytime, with the help of this table, one can find out the corresponding map unit number or polygon number of the degraded forest blocks or compartments. The soil resource status of the compartment can be determined by referring to the soil characteristics of the respective map unit or polygon number. Then a list containing different mapping units or polygons with their encompassing degraded forest compartments will be prepared for the forest division.

Therefore, in the process of preparing the soil resource map and inventory of the degraded forest land, the following important steps are undertaken.

1. Prepare a indicative map of the division showing the degraded forest block/compartment. The map should be in a scale of 1:2,50,000. Identification of the degraded forest blocks/ compartments should be done on the basis of current forestry stock mapping undertaken in the course of working plan exercise.
2. Prepare a soil resource map of the division or district, initially in 1:5,00,000 scale. Then it is enlarged to 1:2,5,00,000 scale.
3. Superimpose the forest indicative map over the soil resource map of the division. Now prepare a soil resource map of the division showing polygon or soil map unit along with the compartment boundary and number.
4. Prepare a list or index of degraded blocks/compartments with their soil map unit polygon number. Such list can be prepared for the whole division or separate list can be prepared for each forest range.
5. Prepare soil resource inventory for each map unit or polygon encompassing degraded block/compartments, on the basis of Soil Resource Map and Inventory prepared by NBSS&LUP, Nagpur. The inventory of soil resources of the degraded forest block/compartments can be in prepared in tabular form (Table 14.3) or in the form of statement (Table 14.4).
6. Prepare the code of soil parameters for ready reference, on the basis of soil resource report of NBSS&LUP, Nagpur (Table 14.5).

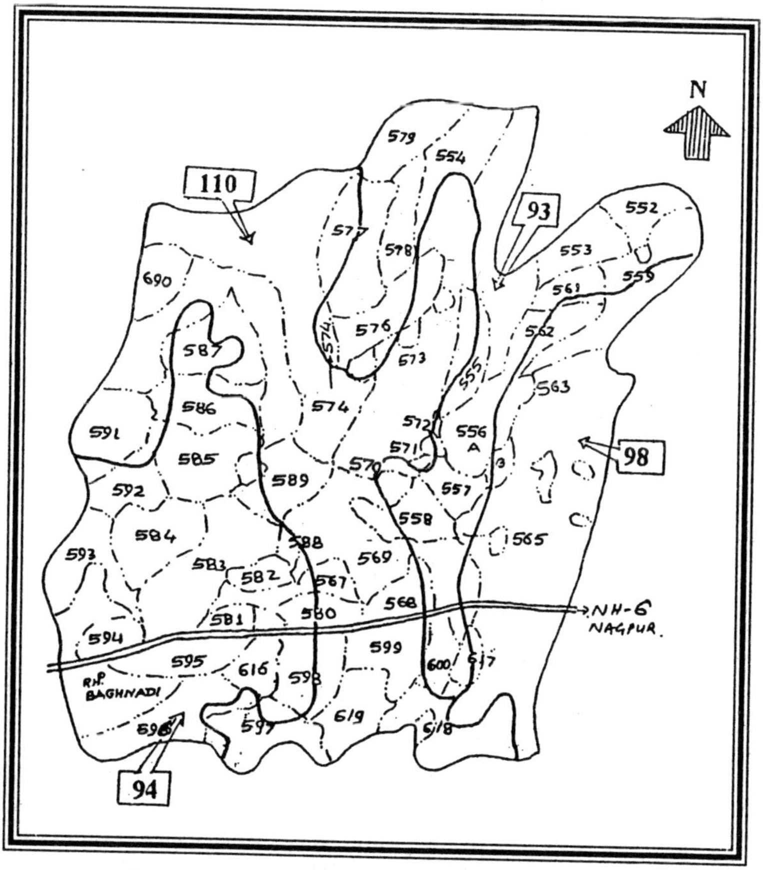

Fig. 14.1 : Map of Baghnadi Range of Rajnandgaon Forest Division (M.P.) Showing : Soil Map Units

Source : Working Plan, Rajnandgaon year 1999.

Index

S.No.	Reference	Symbol
1.	Compartment Boundary	
2.	Soil Map Unit	
3.	Road	
4.	Compartment No.	592

Table 14.3 : Forest Soil Resource Inventory (Short list) Rajnandgaon Forest Division

S. No.	*Map Unit No.*	*Discription*	*Soil Classification*
		(6)	(7)
1	71	Deep, well drained Loamy soil on. Gently sloping plateau with mounds with moderate erosion. Associated with: Deep, Well drained, loamy soils on gently sloping with moderate erosion.	* Fine-loamy, mixed, hyperthermic, typic. Ustochrepts. * Fine, mixed, hyperthermic, Typic Haplustalfs.
2	93	Slightly deep, somewhat excessively drained, loamy soils on moderately steep, sloping hills and hill ranges with severe erosion, associated with: Shallow, well drained, loamy soils on moderately sloping with severe erosion and slightly stony.	* Fine-loamy, kaolinitic, isohy-perthermic, vertic Ustropepts.
3	94	Deep well drained, clayey soils on gently sloping hills and hill ranges with moderate erosion, associated with: Very shallow, somewhat excessively drained, loamy soils on moderately sloping with severe erosion and slightly stony.	* Fine-mixed, isohyperthermic, Vertic Ustropepts. * Loamy, kaolinitic, isohyperthermic, Lithic Ustorthents.
4	98	Deep, well drained, clayey soils on gently sloping hills with pediment with moderate erosion, associated with: Deep, well drained, loamy soils on gently sloping with severe erosion and slightly stony.	* Fine-loamy, kaolinitic, isohyperthermic, Typic Ustropepts. * Fine–loamy, kaolinitic, isohyperthermic, Typic Haplustolls.
5	110	Deep, moderately well drained, clayey soils on gently sloping beds (moderately dissected) with moderate erosion, associated with: deep, moderately well drained, clayey soils on very gently sloping with moderate erosion.	* Fine, Mixed, isohyperthermic, Typic Haplusterts. * Fine, Mixed, isohyperthermic chromic Haplusterts.

Source: Working Plan, Rajnandgaon Forest Division, Year 1999.

Table 14.4 : Soil Characteristics of Polygon/Map Unit 94 of Madhya Pradesh

1.	Range :	Baghnadi of Rajnandgaon Forest Division
2.	Compartments falling in the polygon :	Full : 581, 582, 583, 593, 594, 595, 596 Part : 567, 580, 585, 586, 587, 588, 589, 592, 597, 598
3.	Surface Form :	Hummocky.
4.	Parent Material :	G/N : Granite/Gnesis
5.	Soil depth class :	Dee > 100 c.m.
6.	Particle size class :	Fine
7.	Minerology class :	Mixed
8.	Calcareous class :	Non-calcareous: Calcium Carbonate ($CaCO_3$) % nil
9.	Soil temperature regime class :	Isohyperthermic
10.	Reaction (pH) class :	Slightly acidic: pH: 5.5 to 6.5
11.	Drainage class :	Well drained
12.	Ground water depth (M) :	Deep: < 5 mt.
13.	Compact layer (Type and Extent) :	—
14.	Surface texture class :	Clayey
15.	Slope class :	Gently sloping: 3 to 8% Slope
16.	Erosion class :	Moderate
17.	Salinity and/or sodicity class and extent :	nil
18.	Surface stoniness :	nil
19.	Soil texonomy :	It Uv.

Table 14.5 : Different Parameters and Their Codes

Surface Form

Code	Type
D	Dissected
H	Hummocky
L	Level
G	Gently sloping
U	Undulating
O	Rolling
S	Steep
R	Ridges
V	Valleys
T	Terraces
P	Plateau

Miscellaneous Land

Code	Type
MR	Extremely Rocky land
MS	Extremely Salty land
MZ	Extremely Sandy land

Parental Material

Code	Type
A	Alluvium
C	Colluvium
E	Aeolian
G	Granite
N	Gnesis
S	Schist
D	Sandstone
B	Basalt
L	Limestone
T	Glacial
M	Marine sediments
U	Undifferentiated

Participle Size Class

Code	Class
G	Fragmental
Z	Sandy-skeletal
K	Loamy-skeletal
P	Clayey-skeletal
S	Sandy
L	Loamy
R	Coarse-loamy
M	Fine-loamy
T	Coarse-silty
Y	Fine-silty
C	Clayey
F	Fine
V	Very fine

Minerology

Code	Class
c	Carbonatic
f	Ferratic
g	Gypsic
a	Micaceous
s	Silicaceous
k	Kaolinitic
m	Montmorillonitic
i	Illitic
x	Mixed

Calcareousness

Code	Class	$CaCO_3$(%)
0	Non-calcareous	Nil
1	Slight	<5
2	Moderate	5-15
3	Strong	> 15

(contd.)

Soil Depth

Code	Class	Depth (cm)	Map symbol
0	Ext. shallow	(<10)	I
1	Very shallow	(10-25)	I
2	Shallow	(25-50)	I
3	Sli. deep	(50-75)	t
4	Mod. deep	(75-100)	t
5	Deep	(> 1000)	t

Soil Reaction (pH)

Code	Class	(pH)
1	Strongly acidic	<4.5
2	Mod. acidic	4.5-5.5
3	Slightly acidic	5.5-6.5
4	Neutral	6.5-7.5
5	Slightly alkaline	7.5-8.5
6	Mod. alkaline	8.5-9.5
7	Strongly alkaline	>9.5

Soil Drainage

Code	Class
0	Extremely poor
1	Very poor
2	Poor
3	Imperfect
4	Moderately well
5	Well
6	Somewhat excessive
7	Excessive

Groundwater Depth

Code	Class	Depth (m)
1	Shallow	(<1)
2	Mod. shallow	(1-2)
3	Mod. deep	(2-5)
4	Deep	(<5)

Soil Temperature Regime

Code	Class	MAT(°C)
F	Frigid	< 8
M	Mesic	8-15
T	Thermic	15-22
H	Hyperthermic	22-28
I	Isothermic	<5°C
P	Isohyperthermic	different between MSST and MWST

Slope

Code	Class	Slope (%)
a	Level to nearly level	(0-1%)
b	Very gently sloping	(1-3%)
c	Gently sloping	(3-8%)
d	Moderate sloping	(8-15%)
e	Moderately steep sloping	(15-30%)
f	Steeply sloping	(30-50%)
g	Very steeply sloping	(50%)

Erosion

Code	Class
1	Slight
2	Moderate
3	Severe
4	Very severe

Salinity

Code	Class	Ece (dSm^{-1})
S1	Slight	2-4
S2	Moderate	4-8
S3	Mod. strong	8-15
S4	Strong	15-25
S5	Severe	25-50
S6	Very severe	50

(contd.)

Compact Layer

Code	Class
D	Duric
P	Placic
R	Rock
Y	Clayey
F	Petroferric
G	Petrogypsic
C	Petrocalcic
A	Compacted (man's action)

Surface Texture

Code	Class
s	Sandy
l	Loamy
c	Clayey

Sodicity

Code	Class	
N1	Slight	<5
N2	Moderate	5-15
N3	Strong	>15

Surface Stoniness

Code	Class	Coverage
t1	Slight	<15%
t2	Moderate	15-40%
t3	Strong	>40%

Flooding

Code	Class
1	Slight
2	Moderate
3	Severe
4.	Very severe

Advantages of Soil Resource Map and Inventory for Degraded Forest Land

The following are the advantages of Soil Resource Map and inventory for degraded forest land :

- *For collection of basic soil resource information* : The soil resource status of the degraded forest land can be determined by referring to the soil characteristic table. This information can be stored in the compartment history during the working plan exercise and used while micro-planning.
- *Selection of suitable species for afforestation in degraded forest land* : While undertaking plantation, in degraded forest land, the degree of soil suitability for the species can be analysed.
- *For planning sustainable forest management practices* : The relatively potential area among the list of degraded forest land can be identified on the basis of soil resource characteristic. Forming different categories of areas and treatment series on the basis of agro-climatic zonation can evolve different management regimes.

- *For preparing database in GIS* : Different soil characteristic like, slope, soil depth, drainage, texture etc. can constitute different layers. This, when analysed with other thematic layers, provides a comparative picture of soil resources and growth potentialities with relation to potentiality for plantation.
- *For evolving strategy for soil and water conservation measure* : The soil erosion prone areas in the district or division can be identified with the help of soil characteristics table. Then, effective strategy can be evolved to minimise the soil erosion and maximise water conservation in the degraded forest land.
- *For evaluating plantation in degraded forest land* : Plantation on degraded forest land can be evaluated with the help of soil characteristic table and reason for its success or failure can be determined. On the basis of evaluation corrective steps can be taken to improve the performance.
- *For evolving different forest growth models in different degraded forest land* : Different forest growth models can be evolved after selecting and delineating series of productive areas out of degraded forest area.

2. Evolving Treatment Strategy for Degraded Forest Land

After understanding the status of degraded forest land, it is imperative to evolve an integrated treatment strategy for optimising the use of degraded forest land. Such integrated treatment strategy is required to be developed for each polygon or soil map unit in a different agroclimatic zones. Therefore, a single treatment strategy for the same polygon in the same agroclimatic zone can be prepared. For example, a treatment strategy for polygon no. 94 which falls in agroclimatic zone no. 1 (Chattisgarh plain) should be developed. Such separate strategy for each polygon encompassing degraded forest land should be integrated into a combined strategy for optimising the use of degraded forest land in a forest division.

In evolving treatment strategy for individual soil map units or polygons, it is important to consider for short term and long term measures. Micro treatment of the soil pits before plantation can be taken up as one of the components of short term measures. Species selection for plantation or rehabilitation package, soil erosion control and water conservation measures should be taken as some of the components of long term strategy for optimising soil resources use.

Conclusion

Developing strategies to increase the forest cover, density and productivity in degraded forest land is the key issue, which confronts the country today. In order to address such burning issue, it is imperative to have a systematic soil inventory and soil resource map of degraded forest land on suitable scale for every division/district. Such maps shall provide information on the kind, distribution, and extent of different degraded soil; their nature, characteristics, problems and potentials. Based on soil inventory and soil resource map developed for the degraded forest land of the forest division, a treatment strategy for each soil map unit encompassing degraded forest block/ compartment will be evolved. Such management strategies and practices helps not only to increase forest cover, density and productivity but also to maintain the ecological balance. In nutshell, developing an inventory and a map of soil resources for degraded forest land in a district or division on the basis of NBSS&LUP soil resource map and evolving management strategies and practices will go a long way in sustainable management of degraded forest resources.

15

LEGAL ISSUES WITH RESPECT TO PRIVATE INVESTMENTS IN FORESTRY SECTOR

V.P. SINGH

Over the last few decades, India's forests came under severe pressure to meet the growing requirements of forestry sector of economy. With a high demand of wood, of which 80 per cent is for fuel wood, exploitation of forests far exceeds the limits of sustainability. The result is, degradation of forest resources. A very practical method of facing this challenge is to involve the village communities to protect the forests and to share the benefits therefrom. Joint Forest Management (JFM) is the out come of this philosophy. However, coverage of area under JFM is very small as compared to total degraded forest requiring immediate attention. The process itself is quite slow as it requires time to win the confidence of people and involve them in protection and management of degraded natural resources. Some people, therefore, think that private sector participation can be an answer to arrest degradation of forests and ultimately increase the forest over in the shortest possible time. The private sector participation in forestry will obviously require assigning the forest land to private sector by way of lease or some similar arrangements. This, however, will attract the provisions of Forest (Conservation) Act, 1980, which reads as under Section 2. Notwithstanding anything contained in any other law for the time being in force in a State, no State Government or other authority shall make, except with the prior approval of the Central Government, any order directing—

* * *

(*iii*) That any forest land or any portion thereof may be assigned by way of lease or otherwise to any private person or to any authority, corporation, agency or any other organisation not owned, managed or controlled by the Government.

The Central Government have further clarified that this sub clause shall not be attracted when any forest land or any portion thereof is assigned to any authority, corporation, agency or any other organization wholly owned, managed or controlled by concerned State/Union Territories and or the Central Government. Such government owned or managed or controlled authority/corporation/agency, which have been assigned such forest land shall not re-assign it or any part thereof to other organization or individual. Any scheme or project which involves assigning of any forest land by way of lease or similar arrangements for any purposes whatsoever, including afforestation to any private person or to any authority or agency or organization not wholly owned, managed or controlled by the Government (such as private or joint venture) shall attract the provisions of the Act. It is, thus, clear that no forest land can be assigned by way of lease or any other arrangements to any private person or authority/organization/agency not wholly owned, managed or controlled by Government even for afforestation purposes. Therefore, for private sector participation in forestry, prior approval of the Central Government under the Forest (Conservation) Act will be necessary.

Relevant Provisions of National Forest Policy 1988

Obviously, any proposal for assignment of forest land to private sector has to be examined with respect to National Forest Policy, 1988. The relevant portions of the forest policy on this subject read as under:

4.3.4.2. The holders of customary rights and concessions in forest areas should be motivated to identify themselves with the protection and development of forests from which they derive benefits. The rights and concessions from forests should primarily be for the *bona fide* use of the communities living within and around forest areas, specially the tribals.

4.3.4.3. The life of tribals and other poor living within or near forests revolves around forests. The rights and concessions enjoyed by them should be fully protected. Their domestic requirements of fuel wood, fodder, minor forest produce and construction timber should be the first charge on forest produce. These and substitute materials should be made available through conveniently located depots at reasonable prices.

4.3.4.4. Similar consideration should be given to Scheduled Castes and poor living near forests. However, the area, which such consideration should cover, would be determined by the carrying capacity of the forest.

4.6 Having regard to the symbiotic relationship between the tribal people and forests, a primary task of all agencies responsible for forest management, including the forest development corporations should be to associate the tribal people closely in the protection, regeneration and development of forests as well as to provide gainful employment to people living in and around the forest. This requires special attention to the following :

- One of the major causes for degradation of forest is illegal cutting and removal by contractors and their labour. In order to put an end to this practice, contractors should be replaced by institutions such as tribal cooperatives, labour cooperatives, government corporations, etc. as early as possible;
- Protection, regeneration and optimum collection of minor forest produce along with institutional arrangements for the marketing of such produces;
- Family oriented schemes for improving the status of the tribal beneficiaries; and
- Undertaking integrated area development programmes to meet the needs of the tribal economy in and around the

forest areas, including the provision of alternative sources of domestic energy on a subsidised basis, to reduce pressure.

4.9. The main considerations governing the establishment of forest-based industries and supply of raw materials to them should be as follows :

- As far as possible, a forest-based industry should raise the raw material needed for meeting its own requirements, preferably by establishment of a direct relationship between the factory and the individuals who can grow the raw material by supporting the individuals with inputs including credit, constant technical advice and finally, harvesting and transport services.
- No forest-based enterprise, except that at the village or cottage level, should be permitted in the future unless it has been first cleared after a careful scrutiny with regard to assured availability of raw material. In any case, the fuel, fodder and timber requirements of the local population should not be sacrificed for this purpose.
- Forest based industries must not only provide employment to local people on priority but also involve them fully in raising trees and raw-material.
- Natural forests serve as a gene pool resource and help to maintain ecological balance. Such forests will not, therefore, be made available to industries for undertaking plantation and for any other activities.
- Farmers, particularly small and marginal farmers would be encouraged to grow, on marginal/degraded lands available with them, wood species required for industries. These may also be grown along with fuel and fodder species on community lands not required for pasture purposes, and by Forest department/corporations on degraded forests, not earmarked for natural regeneration.
- The practice of supply of forest produce to industry at concessional prices should cease. Industry should be encouraged to use alternative raw materials. Import of wood and wood products should be liberalised.

- The above considerations will, however, be subject to the current policy relating to land ceiling and land-laws.

Prospects of Private Investment for Re-forestation of Degraded Forests

1. A cursory look at the map of degraded forests will indicate that majority of the degraded forests situated near human habitations. These forests are subjected to over exploitation primarily to meet the *bona fide* requirements of fuel wood, fodder and small timber of the local people and also for sale in nearby markets by the poor people to supplement their income. The villages situated in and around forests are having very poor communication facilities, some of them being located more than 50 kms. away from the nearest metal roads. The productivity of the cultivated lands is generally very low and irrigation facilities are almost non-existent. The 'green' and/or 'white' revolution has not touched them. The fund flow for various forestry and allied activities which used to provide them some employment in lean period is drying up with every passing year due to one reason or the other. There are no industries worth the name in these areas which could provide them employment. The result is that almost all these people are living below poverty line. In order to supplement their income they have to fall back on forests for their *bona fide* requirement of forest produce free of cost or collecting and selling it in the local markets. In such a situation, if degraded forests are assigned to private sector, the bigger question arises as to where from they will meet their *bona fide* needs of forest produce. No doubt, private sector participation can ensure higher wood production but that would not be available to these poor people at a price they can afford.

2. If lands on which people's 'livelihood' are dependent are given to industry, they may have to employ muscle power to keep people at bay, thus escalating social tensions, which are already quite acute in several forest and park areas. The provisions of the Panchayats (Extension to the Scheduled Areas) Act, 1996 have already endowed the Panchayats in the scheduled V areas with the ownership of minor forest produce and the forests are mostly confined to schedule V areas. Local people would obviously, prefer a mixed forest which can provide host of minor forest products so important for their livelihood,

but the industry would prefer high density plantation of a few selected species which will hardly give any minor forest produce. Thus, assigning of degraded forest lands to industry would not only lead to further conflict in the society but will also be against the spirit of this Act.

3. If industry produces its own raw material who would farmers sell to? Where is the market for the farm forestry produce which have caught the imagination of the enlightened farmers and large scale tree plantation have been raised by them on their holdings. Since, the overall demand of the industry is limited and if the same is allowed to raise their own plantation then it could be a set back to the agro-forestry movement which is flourising in certain regions of the country.

4. The prosperity of the Indian farmers is the very basis for the prosperity of the country. The farmers are at a very great disadvantage as their crops are susceptible to great fluctuations in market prices, perishability of commodities and consequent urgency to dispose of the pests and diseases. The only crop that is not susceptible to these adversities is tree farming, which can be taken up either in place of crops or along with them. Leasing of degraded forest to industry would discourage the farmers from taking up tree farming.

5. In almost all the major forestry regions in the country, large chunks of marginal private holdings are available which are not fit for profitable agriculture. In addition, substantial areas of waste lands are available with the Government or the village panchayats which can also be brought under tree cover. The industry should establish links with farmers and Government can act as a facilitator for the purpose of better management of such lands and provide raw material to the industry on sustained basis. The farmers and village communities can produce raw material on private holdings, degraded forests and Government Waste Lands if given a remunerative price, as done in UP by linking of poplar growing farmers with a match factory in north U.P. The experiment shows that, with technological back up, timber size trees suitable for sawing can be raised on farm lands within 8-10 years. In fact, due to farmers enthusiasm for growing poplar its enhanced supply has led to the establishment of several plywood factories in that area, thus providing considerable down stream employment. It is a myth that industry cannot deal with

farmers directly. For sugarcane, the industry has been in touch with farmers for decades and some industries are going for more crops like tomatos in Punjab.

Conclusion

It is, thus, clear that National Forest Policy, 1988 does not envisage assignment of the forest land, degraded or otherwise, to private industries for raising captive plantations. There are large number of Revenue Waste Lands available in all the States, which can be brought under tree cover through intensive management but in view of the existing land ceiling laws, it may be difficult to lease revenue waste lands to the industry. Leasing out large area of such lands to the industry or private individuals may lead to social tensions on account of further increase in the economic disparity among the people. Therefore, in a socialistic society, it may not be advisable to lease out the revenue waste lands to private individuals or the industry by relaxing the provisions of land ceiling laws. It may be of interest to note that as against the collective farming, the group/collective forestry can conveniently be practiced. Therefore, keeping in view the spirit behind the forest policy, land ceiling laws, etc. the following mechanism is suggested for rehabilitation of the degraded forests and the waste lands and at the same time meeting the demand of forest produce on sustainable basis:

1. All the degraded forests and Government/community waste lands within the geographical boundaries of revenue villages may be assigned to village communities and declared as 'village forests' under Section 28 of Indian Forest Act, 1927.
2. The State Governments may make rules for regulation and management of the 'village forests' prescribing the conditions under which the community to which any such assignment is made may be provided with timber and other forest produce and their duties for the protection and improvement of such forest.
3. Notify the catchment of each wood based industry from which it can procure the wood. The movement of wood should be controlled by the Government to ensure that the wood from the catchment area of a industry does not go out of its catchment.

4. Study the market trends, economics of producing wood and finished products and fix up fair price which the industry should be asked to pay for the wood to be procured from the identified catchments.
5. Prepare the plans for management of 'village forests' in consultation with the village community and the industry in whose catchment the village forests are sitauted.
6. The industry should provide the technical and financial (credit) support to the village community involved in management of village forests and also to the farmers involved in the farm forestry programme in its catchment.
7. The required amount of funds for managing the village forests as well as for farm forestry programme should be made available as loans at subsidised interest and the interest rates may be different for village forests and farm forestry programme.
8. In order to remove the bottle necks in movement of timber, the transit rules may have to be amended authorising the village panchayats to issue the transit passes for the forest produce originating from their area and transported to the identified industry. Movement from one industrial catchment to another industrial catchment should be regulated by the Government.

16

STATUS PAPER ON PRIVATE INVESTMENT IN AFFORESTATION OF DEGRADED FOREST AREAS OF MADHYA PRADESH

N. KUMAR

MADHYA PRADESH RAJYA VAN VIKAS NIGAM LTD.

Out of 1,54,506.40 sq. km forest area of Madhya Pradesh about 72,000 sq. km forest area is categorised as degraded forests, which is about 46 per cent of the total forest area of the state. Due to heavy biotic pressure and other distructive factors, about 1 lakh hect. forest area is subjected to degradation process every year. The rehabilitation of these degraded forest areas by afforestation and other necessary measures like soil conservation etc. is necessary for environmental conservation, ecological balance and supply of fuelwood, fodder, timber, minor forest produce and raw material to forest based industries. Rehabilitation of degraded areas of such magnitude is practically impossible through governmental agencies alone due to constraint of financial resources. It, therefore, becomes imminent to involve such private organisations which may be willing to collaborate both in funding and operations. Efforts made in this direction resulted in the formation of two joint sector companies viz. M.P. Plantations Ltd. and M.P. Van Vikas Nigam Ltd. The programme, however, could not materialise for want to sanction from the Government of India who did not approve of leasing out forest areas to joint sector companies.

Keeping in view the Government of India's objections, another alternative of involving private investors on work contract basis was adopted with Madhya Pradesh Rajya Van Vikas Nigam as the implementing agency. The proposal has got the State Government's approval who have also approved the tender notice, terms and conditions of the tender, tender form, agreement of tenderer and work agreement (refer to Appendices IV and V). The key terms and conditions of the work contract are as given below:

(a) The agreement period shall be 30 years.

(b) The Government of Madhya Pradesh shall continue to be the owner of the forest land made available under this scheme. The forest land will be on lease to Madhya Pradesh Rajya Van Vikas Nigam. The contractors shall only be permitted to enter upon the land for execution of the work.

(c) Nistar rights of the villagers shall not be affected.

(d) The contractor shall not be permitted to clear any natural regeneration or tree growth on the land.

(e) Contractor shall provide 5 per cent of the produce obtained in the intermediate harvest to the local villagers for nistar and 2 per cent of the intermediate and final harvest to the 'Nigam' being the supervision charges.

(f) The contractor shall indicate the percentage of produce, from the balance produce, which he shall keep with him and the balance of the produce shall be made available to the Government of Madhya Pradesh through the 'Nigam'.

The tender notice was published in the leading newspapers of the country on 9th May, 1996 with the last date for submission of tenders being 15th July, 1996 in the tender notice specified degraded forest areas in 9 groups under four districts, viz. Betul, Seoni, Chhindwada and Rajnandgaon were proposed. The details of areas proposed are given below :

S. No.	*Name of District*	*Name of Forest Division*	*No. of Comptt.*	*Area Proposed (in Hectare)*
1.	Betul	South Betul	1	220.000
2.	Seoni	North Seoni	12	611.900
3.	Chhindwada	South Chhindwada	6	981.595
		East Chhindwada	13	920.188
		West Chhindwada	1	231.600
4.	Rajnandgaon	Rajnandgaon	8	517.280
		Total	41	3462.563

In response to the tender notice, 11 tenders for seven groups were received. These tenders were examined by a High Power Technical Committee constituted by the State Government under the Chairmanship of the Addl. Principal Chief Conservator of Forests (Development and Planning). On examination of the tenders from technical angle, the committee recommended only four tenders for further examination of their financial bids after eliminating such tenders which did not have requisite technical and financial details along with those which contained elements of violations of the provisions of the Forest Conservation Act, 1980.

The four tenders approved from technical angle were further examined by the committee for their financial bidding. Each proposal was examined for anticipated production of forest produce and market value compared with established standard on the basis of other available technical records for same type of forest products. It was decided to do financial analysis on the basis of standard production of forest produce and market value. After examination, no proposal was found viable. The Committee submitted its report to the Government of MP in which no tender was recommended for sanction.

The following technical and financial lacunae were found in proposals during examination by the committee—

(i) The tenderers have assumed very high availability of produce and value in comparison to the acceptable level. On the basis of such assumptions the share of Government will also work out on the higher side. Again if the assumed quantity of production was not obtained, then the Government would be put to a loss. This might also result in conflict between the two collaborating parties.

(ii) The naturally growing plants in the area shall be the Government property. Removal of these plants at the time of weeding for the proper growth of planted samplings will create disputes.

The committee has given following suggestions in order to make the scheme more practical :

(i) The amount of security deposit equalling to 1 per cent of the total project cost or Rs. 1 lakh whichever is less should be deposited by the tenderer.

(ii) An area should be made available to the tenderer on the basis of phased programme of plantation.

(iii) In the proposed areas, any removal of natural regeneration should be done by the 'Nigam' only at the cost of the tenderer. The area for planting only should be given to tenderer.

(iv) The term weeding should be well defined.

(v) Clear provision should be made regarding quantity of produce to be provided for Nistar and for Van Vikas Nigam.

(vi) Doses of fertilizers and insecticides should be based on soil testing reports.

(vii) The provision of watering should be based on ground water level testing reports.

(viii) The basis of anticipated production of forest produce and sale value should be mentioned.

(ix) In financial bid, the minimum quantity of produce to be given to the Government should be mentioned in terms of CMT/Qntl instead of percentage of production.

(x) A meeting between tenderers and 'Nigam' officials should be organized one week before opening of every tender to settle the problems and confusion.

As per direction of the State Forest Department a meeting for discussion between the committee members and the tenderers was held in the month of December, 1996. In the meeting tenderers expressed following views :

(i) Weeding should be permitted as per silvicultural requirements.

(ii) Except naturally growing plants of more than 15 cm girth at Breast Height or 4 metre height, all plants should be allowed to be removed at the time of area preparation. Likewise removal of all grasses and all coppice shoots in 1 metre wide strip of planted sapplings should be allowed.

(iii) If guarantee of minimum production is insisted, then some percentage of project cost should be provided by the 'Nigam' in the form of grant. As the plantations are totally dependent on nature, the guarantee of minimum production cannot be given.

The Advocate General of M.P. has expressed his approval about the scheme. However, the Chief Conservator of Forest (Central), Government of India has opined that this scheme will attract the provisions of the Forest Conservation Act, 1980.

The Govt. of M.P. Forest Department has, however, rejected all the tenders and directed the 'Nigam' to revise the terms and conditions of the tender to make the scheme more practical. The 'Nigam' has prepared a draft of revised terms and conditions which is under consideration of the Government of M.P.

17

AN ALTERNATIVE STRATEGY FOR REDUCING TRIBAL POVERTY AND GREENING DEGRADED LAND IN MADHYA PRADESH

R. GOPALKRISHNAN

Background

Madhya Pradesh is uniquely placed to make a breakthrough in reduction of tribal poverty. A substantial number of the poor in India are in Madhya Pradesh and a substantial number among the poor in Madhya Pradesh are tribals. Reducing tribal poverty by providing the tribal poor livelihood security is a major challenge before the Government of Madhya Pradesh. The Government of Madhya Pradesh is uniquely endowed with considerable amount of resources, which if imaginatively unlocked can provide such livelihood security to the tribals while at the same time augmenting its forest wealth. Madhya Pradesh has over 60 lakh hectares of degraded forest land which could become an engine of growth to remove the poverty of its tribal population.

Governmental action so far only attempted to make forests provide casual wage employment to the tribal poor and allow some usufruct rights. Government Policy now needs to make a radical departure and use the surplus labour energy of the tribal poor to regenerate the state's depleted forests thereby improving productivity of forests and also getting the tribals out of their poverty trap.

These kind initiatives have not happened elsewhere as Madhya Pradesh has to present the real testing-ground. It has both a considerable area of degraded land and very poor people dependent on natural resources for livelihood.

Tribe-Forest Links: Need for Conceptual Clarity

The widely accepted link between tribes and forest does not require much elaboration at a basic level. There is, however, a need to clarify the conceptual flaw that has crept into much of our forest policies from colonial inheritance. This is mainly on account of a western notion of environmental conservation where man and nature are seen in a dichotomized format. Nature has been given a mechanistic understanding (based on the western philosophic tradition of seeing the world in fragments) and, therefore, as needing "protection" from man. The Indian lived reality as well as our conceptual understanding of this issue has always been to see man and nature as parts of an interacting whole. Parts do not have pre-existing properties but acquire properties from the interaction. It logically follows that what matters is the nature of interaction. If today there is a distorted relationship between man and nature, what has gone wrong is the nature of interaction. What we need to move towards is, therefore, to leave the western, conservationist understanding and seek solutions in an indigenous world-view of inclusion as against enclosures.

Joint Forest Management : Good Step Forward But Does Not Go Far Enough

The Joint Forest Management Programme of the Forest Department has at one level indirectly accepted the above position in that it sees regeneration of forests impossible without community action. It seeks to present tribal people as inherently environment-friendly (a debatable romantic ecology notion if the man-nature interaction has been vitiated) and, therefore, is amenable for being brought under an incentive-framework that can be helpful in the regeneration of forests. It has to be accepted that the Joint Forest Management position of participatory forest management sees participation in an instrumental manner, that it is useful for the forest department to move towards a participatory model for efficiency. It is like saying that, the Forest

Department cannot do regeneration on its own and so let people come and help it out. The one point to carry over from JFM is that indirectly the policy framework accepts the notion that forests cannot be protected without working together with local tribal people.

A Simple Proposal

Against the background and policy ambience discussed above, a simple proposal is being advanced here. The Government of Madhya Pradesh should allocate between 3 to 5 hectares of degraded land to each poor tribal family (nuclear family), after entering into a 30-year Memorandum of Understanding with them for greening the area and for this make available the entire degraded forest land in the State. These families will be called as "Forest Developer Families" (Van Vikas Parivars) and they will be supported by the Government to form Self Help Groups and enable them to benefit from both eligible technical assistance and financial support under SGSY and such funding for natural resource development activities. Each Local District Administration would identify such Forest Developer Families from among landless and marginal peasant tribal families and allow them to choose appropriate land from among the degraded forests land in the district. The Memorandum of Understanding would detail commitments on either side. The technical support group consisting of natural resource management units of the district administration like Forests, Horticulture and Agriculture could suggest the appropriate mix of plants/species that can be grown. While these Forest Developer Families would benefit from technical support as well as financial support from SGSY, no other subsidies to take care of gestation period etc. are envisaged. The output expected is that a large number of such Forest Developer Families (who are getting impoverished by the day) will contribute their resources to turn their earth fertile. By doing so they will create livelihood opportunities for themselves as well as render help to bring the current degraded forest land of Madhya Pradesh. This will be measured on a girdawari kind of system of accounting for Gross Nature Product (GNP) created, beginning from the third year of the MOU. For the Government it will become a unique opportunity to reduce tribal poverty and effect greening.

There is no way a state like MP can make a dent on tribal poverty except by unlocking its forests for anti-poverty action. What is being sought are not rights or *pattas* but opportunities. Opportunities to create forest wealth from degraded earth. Opportunities for creating green cover and opportunities for employment and livelihood that will sustain the forest cover and prevent poverty-driven denudation. Logically there could be no objection to such a formulation. We are merely allowing the coming together of two resources of surplus labour energy and available degraded land to create two "goods" of economic well-being of the poor and increased greening of the state.

In a state like Madhya Pradesh forest is such a powerful engine of growth and it is a commentary on our public policy regime which while advocating both productivity goals and equity goals has so far missed the opportunity for interlinked action. This happened because of our blind acceptance of distant laws often built on colonial notions of locking up "resources" of the "natives" for colonial exploitation, and on a conservationist ethic devised by urban-biased intellectuals insensitive to rural realities. It is time to change the department from such a mindset.

Restating the Logic

- The current state of large-scale degradation of forest does not help anyone and there could be no logic of conservation that can be advanced to preserve degraded forestland as it is—as environmentally degraded space. When we bring in surplus labour energy to make this land productive, the obvious *economic good* in such a proposal need not require special emphasis. Economic returns through increased productivity from such an exercise can be estimated. It may also be mentioned here that government had considered using the capital and technology of the private sector to green such degraded forestland. To permit such an action in a labour surplus economy like ours, where the one-time owners, the tribals languish in penury looking for employment/livelihood opportunities would be to say the least, grossly unfair. We may remember that in the WTO India argued that the labour advantage of developing countries should be non-negotiable. That should apply internally too.

- The proposal obviously expresses a *social good.* All evaluations testify that tribal populations are today among the most marginalised people in Indian society. Even after 50 years of independence, they are struggling for their bare essentials *i.e.* employment, health cover, education, and other basic needs. This is happening in the midst of plenty that has been locked out for them or in the midst of space that can be made plentiful by them if they are given a chance. The increasing marginalisation of tribal society would reduce through provision of better life-chances for them through the employment options suggested in the proposal.
- It is a *political good.* This social marginalisation is acquiring explosive political dimensions. Today, there are threats to the political fabric of the country through separatist movements. The increasingly violent responses that go cloaked in the term of naxalism begs the question as to why the geography of naxalism is coexistent with the locked-out forest resource regions of the country. The naxalite movement is currently feeding on the discontent arising from the denial of basic life-chances to the tribal poor who also comprehend the fundamental injustice meted out to them by a statization of their life-supporting habitat. It may also be pertinent to add here that the tribal populations who have often not had a concept of private property ownership are only asking for opportunities to allow them to live with nature. Denial of even such a basic demand through a position that can only be termed environmental fundamentalism will only result in a backlash of tribal fundamentalism of *"Jal, Jangal, Zameen Hamara"* today to "All Others Keep Out" tomorrow. We already see this in different parts of the country. In our state while we acknowledge the problem as one rooted in such fundamental disparities, solutions so far have been only to equip the coercive arm of the police. Interestingly the police have on several occasions raised these fundamental issues and their inability to deal with the root cause of such disaffection and called for non-police action. In the interest of harmonious policy, it is high time that this challenge is addressed effectively.
- That this measure is an *ethical good* is self-evident from the fact that it give opportunities to the weak whose protection and enhancement is the moral duty of the state.

- That it is an *environment good* is also obvious. Large tracts of presently degraded land will be regenerated. The economic opportunities to the tirbal poor thrown up in this strategy will restrain poverty-driven environmental degeneration and reduce the pressure on environmental, in this case, forest resources.
- That this is a *public policy good* is also self-evident. Tribal populations are at the bottom of the heap in terms of their attainments in social and development indicators. There has to be radical policy alternatives that provide these people with equality of opportunity for employment, education, health care and such other basic needs. Public Policy is equally focused on addressing environmental degradation. Public Policy is also focused on developing participatory models of governance. Now the SGSY and such other anti-poverty schemes are groping for an area-specific strategy to focus anti-poverty investment. The proposal described above would combine all these objectives in a win-win manner.
- It is *legally sound.* The Government of Madhya Pradesh is not proposing any changes in the Forest Conservation Act. It is only devising a better method of forest conservation. It is not asking for reduction of any green cover. It is only advocating and developing a new method for increasing green cover. It is finding a practical option that combines the objective of livelihood security of the tribal poor and the formidable task (increasingly becoming more formidable in the context of the public finance crisis) of greening degraded land and thereby ensuring environmental protection and conservation. No relaxation of the Forest Conservation Act (though it has frailties) is sought here. Instead what is sought is to advance the spirit of that Act to not merely protection but enhancing green cover through direct action by Forest Developer Families.

Suggested Method of Action

- State resolve to develop its degraded land through Forest Developer Families.
- Identification of marginal farmers and landless tribal families by district administration. Issue of Identity Cards to these families.

- Location of degraded land that can be made productive.
- Selection of appropriate land in consultation with Forest Developer Families.
- Photographic documentation of the extent of degradation.
- MOUs to be signed between individual Forest Developer Families and District administration.
- Terms of MOU (to be finalised through discussion).
- Formation of FDF into Self-Help Groups.
- Identification of technical support group of forest and other services to enable these FDFs to choose right plan mix.
- Provision of financial support where required under SGSY.
- Supply of evaluation cards for measurement of *Gross Nature Product (GNP)* created by each Forest Developer Family. This will measure the outcome of the scheme and in most cases we start with a zero-base of degraded land.

FAQ : Frequently Asked/Askable Questions

- **Why don't we just have JFM? Why MOUs with families?**

JFM has limitations. It sees people as a herd. It is a gradual moving out of a regulatory frame and the core is still a group that gets dominated by a paternalistic officialdom. The hearth is different even in tribal villages. They are not operating a common kitchen. The incentive spine is better here because this can trigger individual entreprenuerial energies for greening and income generation. Let people pool energies for collective tasks. The limitations in JFM is already snowballing into a critique of its methodology.

- **But why not a gradual phased transition after some years of JFM, and selected private farmland development for forestry under the new Lok Vaniki etc.?**

JFM has been discussed. The latter does not address the issue of landless who are the poorest. There is now an urgency and tribal poor have waited too long. We cannot jump a chasm in two bounds say the Chinese. Policy-makers have time for incremental planning. Poor have almost no choices.

- **Can people really make the land productive? There have been similar experiments of land allotments in the past?**

They can. After all, the nurseries are created not by forest officials but by tribal forest developer families. If earlier experiments have failed they may need to be interrogated on whether they leveraged the stakeholder principle or whether they were welfare throw-aways. Also the government has the skills and resources to provide technical support as done in Watershed Management Mission programmes.

- **Can we have MOU's with Forest Developer Families?**

Why not? MOU signing is not the prerogative of white men in black suits or brown men in grey suits but also citizens of this country with their democratic government to honour commitments on either side.

- **What about legal issues, most importantly the Forest Conservation Act?**

This scheme works within the provisions of existing law. After all, the spirit of the Act is conservation and conservation for people so that there is no environmental doom from deforestation. If that is the spirit of the Act how can a scheme that promotes greenery and is willing to be held accountable and measurable by increasing the Gross Nature Product be illegal? In fact having a law and not finding resources for greening, thereby accelerating degradation contravenes the spirit of the Law. The proposed arrangement furthers the spirit of the Law, of what brought into existence such a Law in the first place.

- **This is all so obvious. Why did not we do it?**

Yes, it is obvious in Madhya Pradesh. May be other states may not have a situation on such dramatic scale. And why should we look outward for solutions? As we said earlier only MP has this situation of such proportions and so we may have to develop our own initiative. And about it not being on the agenda earlier, may be as somebody said: *"until such time as there is a historian among lions, the history of the lions will always be told from the point of view of the hunter"*.

Follow-up of this Proposal

Discussion of the merits of the proposal in an intersectoral forum.

18

JOINT FOREST MANAGEMENT
EXPERIENCES OF PEOPLE'S PARTICIPATION IN INDIA

MANOJ KUMAR DUBEY

Introduction

Joint Forest Management (JFM) is an approach to optimise the returns, minimise conflicts and link the forestry development works with over all development on sustainable basis. It is a strategy which is going in the right direction for decentralised people-orientated forest management to break the vicious cycle of poverty and resource degradation. It is also a process for empowering the people and strengthening democracy (Kumar, 2000). During the last decade JFM programme emerged as one of the key components of the Indian forest development programme.

This review summaries the past experiences and future strategies for the JFM programme of India.

Philosophy

The National Forest Policy of 1988 places emphasize on creating massive people's movement through the involvement of village communities living close to the forest in protection and development of forests. Pursuant to this policy, the Government of India issued a notification in 1990 requesting the State Governments to involve local communities in the management of forests. It was envisaged that the communities, in lieu of their participation in protection and development of forest areas, will be entitled to sharing of usufruct rights in a manner specified by the concerned State Forest Departments (SFDs). This has led to the development of Joint Forest Management (JFM) programme (Government of India, 1990). The state governments have evolved their own mechanisms of involving local communities in conformity with the proclaimed policy. Refer to Table 18.1 for latest available statistics of JFM in India (FSI, 1999).

Table 18.1 : Facts and figures-People and Forests in India

Forest Cover	63.73 million ha.
Population of Forest tribal/Forest dwellers	About 100 million
National Forest Policy	1988
Introduction of JFM concept	1990
Forest Area under JFM	10.25 million ha.
No. of States involved	22 (out of 25)
No. of Village Protection Comittees	36,130

Source : FSI 1999.

Experiences of JFM

Joint Forest Management in India was started 12 years ago. During this period different authors have reported the following positive as well as negative experiences of the programme:

- JFM has succeeded in eliminating forest degradation in India, but its poverty reduction objective has not fully been evaluated (Kumar, 2002).
- There is a limited amount of awareness about JFM. This limited level of awareness and knowledge constraints the institutionalization of JFM (Rishi, 2002).
- The manner in which information related to JFM is generated, distributed and received within the JFM committees plays a major role in enhancing the gender disparity during JFM decision-making (Kameswari, 2002).
- People's participation has increased due to the many income generating activities and resource development works, and the consequently improved socio-economic conditions (Dasgupta, 2000).

- The committees formed are vulnerable as they have no force of law, and moreover, the constitution and the byelaws of the committees are yet to be framed (Ray, 2000).
- VPC functions well in villages with homogenous social structure in term of caste, religion etc. (Pratima and Jattan, 1999).
- Conflicts between state and local communities as well as conflicts within and between communities have been reduced to a great extent (Kant and Cooke, 1999).
- Mostly women are engaged in collection of fuelwood, fodder and small timber from forest, so they have the direct contact with the forest and for the success of JFM the participation of women is very important (Brahmi *et al.*, 1997).To ensure active women participation Government of India issued a notification in 2000, which stated that 50 per cent members of the JFM General Body shall be women and the presence of at least 50 per cent women members will be a prerequisite for holding the General Body Meeting (Government of India, 2000).

Grey Areas

Lack of clarity of objectives, inadequate benefits to villagers and problems in institutionalization are the main constraints of JFM programme (Partima and Jattan, 2002). Table 18.2 shows the main constraints in implementation of JFM in India.

Table 18.2 : Main Constraints in the Implementation of Joint Forest Management

No.	*Constraints for Forest Department*	*Constraints for People*
1	Staff inadequacy	Negative attitudes of FD
2	Misconceptions of rights and duties	Marketing of non-wood forest products
3	Lack of funds	Lack of awareness of potential gains from JFM
4	Lack of coordination with other development departments	Balance of power between the FPC and FD
5	Poor participation by women	Legal status of the FPC
6	Heterogeneous committee membership	Centralized non-flexible Administration
7	Legal constraints	Political rivalry
8	Lack of good local leadership	Conflict on distribution of employment opportunities and benefits
9	Inadequate extension work, and preferential Species selection by participants	Lack of freedom over species choice

Source : De and Babu, 1998.

Future Strategy

Following strategies have been suggested by different authors to overcome thé shortcomings of JFM programme in India:

- Flexibility, appropriate legal backing and a proper institutional framework can help in accelerating the JFM movement (Jha, 2000).
- Efforts are required at micro level through village workshops to help promote JFM knowledge and awareness by means of study tours, exposure visits, and educational campaigns (Rishi, 2002).
- Village resource development activities may provide an important incentive, aside from leading to improvements in the livelihoods of the people, for sustaining interest in the JFM programme (Sethi and Khan, 2001; Sethi and Singh, 2001).
- JFM initiatives also need to be integrated with other rural development programmes (Dhanagare, 2000).
- Given the stakes involved and complexity of ground realities, monitoring of JFM is critically important for its success (Prasad, 2000).
- Gender policy in JFM needs to be based on a more sophisticated understanding of gender relations and a wider examination of the gendered context of JFM processes and greater involvement of women should be ensured (Locke, 1999).

Conclusion

For achieving the target of sustainable development outlined in Rio Summit, 1992 and followed by Johannesburg Summit, 2002, it would be necessary for local communities to participate in the sustainable use and successful management of natural resources. JFM has been proved to be an effective tool in achieving the target of sustainable development by utilizing people's potential and using the forest department's technical back up.

The performance of JFM varies from village to village and from state to state. So far JFM programme has able to achieve its environmental objectives, but is running behind to attain the social objectives. The need of the hour is to analyze the ground realities and make the programme more people oriented because any concept can survive only if the management is for the people, of the people and by the people.

19

BASOD KALYAN KENDRA
A CHANGE AGENT

SWATI BAKSHI

Introduction

Peoples participation is a central theme of present thinking about forest management. Joint Forest Management (JFM) is strategy designed to bring in improvement upon the growing stock by silviculture operations and regulatory action, so that forest can fulfil its social and environmental obligations. In more precise form it focuses on the rural poor in meeting their daily needs of fuel, fodder, timber and income generation.

As on date, 19 states have issued JFM resolutions and it is estimated that about 35,000 village protection and management groups are protecting and managing over 35 million hectare of state forest land (Khanna and Singh, 1999). In Madhya Pradesh, JFM was started in October 1991. After passing the resolution, work on JFM was started in Raigarh division in the year 1995.

Basod Kalyan Kendra of Village Junvani, East Raigarh forest division M.P., was formed under JFM project for the protection of forest and also for the social and economic upliftment of Basods.

Objectives

The main objectives of this case study is to identify the additional source of employment provided to the Basod families, to identify the status of forest after implementing JFM and to identify the benefits Basods are getting after formation of "the Basod Kalayan Kendra".

Methodology

The tools used for data collection were group discussion and meetings with the Basod families and Forest department officials. This case study was carried out in 1997 in village Junvani.

Description of the Area

Junvani is about 30 km from Raigarh (M.P.) and lies in Raigarh East range. It lies between 21°20′ and 23°15′ North latitude, 82°50′ and 84°24′ east longitude. The temperature in 1994-95 was 25°C in winter and rainfall, 1621 mm. Bamboo is the main crop of the area. Pure crop of Bamboo is however rarely seen. Saja, Salai, Khair, Tun, Mahua etc. are the chief associates in the canopy.

There are about 14 families of Basods (approximately 60 members). They are traditional craftmen and they manufacture number of articles from bamboo.

Before implementation of JFM the Basods's harvested Bamboo illegally from the adjoining forest for making articles. Grazing by animals owned by Basod's was rampant. Due to regular felling and grazing the quality of soil and availability of Bamboo was considerably reduced.

In this degraded forest, plantation of bamboo was done in 1994. In 1995 JFM scheme was taken up in this area and under the shelter of this scheme Forest Protection Committee was formed. Under the supervision of FPC members these Basods families worked in a centre "Basod Kalyan Kendra" which was under the control of the Committee Members. The Committee Members also carried out the protection work of forest. Even after formation of the committee illegal cutting of bamboos continued and now the protection work is again in the hands of Forest Department and some members of the committee were helping them in patrolling the forest. The area is now dense with bamboo vegetation under the guidance and supervision of the forest guard.

Private labours are doing the felling and nistar work. Box shows the details of the Committee Junvani :

Box 19.1 : Details of JFM Committee Junvani

Type of committee	FPC
Year of formation	1995
Protecting area	1037.90 ha.
No. of members	208

Forest Department provides bamboos to the Basods on concessional rate as per the Nistar policy of the Government for making baskets, *supa, tatta* in Basod Kalyan Kendra. VFC purchase these articles from Basod traders after charging supervision fees as "Samiti Seva Shulk" (approximately 10% of the price of all articles). They pay the money to the Basods.

Finding and Analysis

(A) Social Banking

Forest Department is providing one Bamboo at a rate of Rs. 1.70 to each Basod.

In 3 Nos. Bamboo—1 *Tatta* is manufactured

In 20 Nos. Bamboo—7 *Tattas* are manufactured

Forest Department gives each labour Rs. 33.00 as wages for each tatta but the amount of Rs. 3 out of Rs. 33/is deducted into the fund of committee as "Samiti Sewa Shulk". Hence earn only Rs. 30 for each *Tatta.*

These Basods are also preparing *Jhadoo* from dry grass which they collect from forest. They are paying 25 paisa as "Samiti Sewa Shulk" (supervision fee) per *Jhadoo* (broom).

This "Samiti Seva Shulk" is being deposited into the committee's account through which the Basods and other committee members can take a loan for their own requirement.

(B) Benefits after Formation of 'Basod Kalyan Kendra'

The data shown in Table 19.1 indicates the following facts about articles made by the Basods of "Basod Kalyan Kendra", Junvani.

Dry Bamboo's are used as raw material. They used to manufacture 15 items out of Bamboo. The average selling price per item worked out at Rs. 10, the total sale of various Bamboo items worked out to Rs. 45,528 before formation of committee. After formation of committee the Basods are getting the following benefits :

The members are getting an average price of Rs. 11 per product thus showing an incremental income of 30 per cent over the earlier rate. The committee is adding Rs. 3 per item as Service charge which are collected and deposited into the committees account towards saving funds. After adding the Service Charges the average price of per item works out to Rs. 14. Thus generation of income to Rs. 4 over Rs. 10 (An increment of 60% in the price of product) resulted in an average saving of 15 per cent to the committee (see Table 19.1)

Table 19.2 indicates the income generated by families individually in the Junvani village. The income has been earned by sale of various items produced by them. The income earned out of the total bamboo's used in manufacturing various items varies from Rs. 5.52 to Rs. 19.64 per bamboo used in manufacturing process.

The analysis reveals an interesting feature that, there is a need for educating the families to enable them to know the best product mix so that they can get the maximum income out of the limited raw material used by them. This will also save the wastage of bamboos in the manufacturing process.

The per bamboo lowest and highest income also gives a doubt that in those cases where articles returned (made by bamboo) is low or lower than average, these families have not actually used the entire bamboo supplied to them for manufacturing the items and misused it else where. The misuse of forest produce by such families defeats the purpose of JFM. It is, therefore, felt that educating people on this aspect is essential (see Table 19.2).

Table 19.3 indicates the income generated by manufacturing *Tatta* of size 7×5 mt. by the Basods of village Forest Committee Junvani. The income earned by each member including the rate of bamboo before the JFM period was Rs. 20 but after formation of the committee they are getting Rs. 30 for each *Tatta.* Thus this has been increase in the income of the Basods after formation of the committee. They are getting Rs. 10 over the earliest rates (An increment of 80% in the previous rates) (See Table 19.3).

Conclusion

From the above we can conclude that the people take keen interest in protecting forest if motivated. The Basod Kalyan Kendra in Junvani is working very well and smoothly but it might be necessary to create awareness amongst these Basod families about protection of forest which is to be maintained on a sustainable manner along with the day to day need of their livelihood.

Table 19.1 : Table Showing Benefits (articlewise) to the Basod's and committee after formation of Basod Kalyan Kendra—Junvani (1.12.1996 to 16.3.1997)

S.No.	*Name of the Articles*	*Quantity of Articles*	*Purchased by business man (Past)*		*Purchased by to FD through committee present*		*Difference*	*Purchased by the Business man from the committee with seva shulk*		*Income to the committee*
			Rate	*Price*	*Rate*	*Price*		*Rate*	*Price*	
1	Supa Simple	92	10	920	11	1012	92	14	1288	276
2	Supa Jemia	80	11	880	12	960	80	15	1200	240
3	Supa Bijeni	10	15	150	17	170	20	20	300	130
4	Dauri 1. Bitta	210	20	4200	22	4620	420	25	5250	630
5	Dauri 8. Finger	215	15	3225	17	3655	430	20	4300	645
6	Dauri 4. Finger	675	10	6750	12	8100	1350	15	10125	2025
7	Dauri 1. Hand	951	8	7608	10	9510	1902	12	11412	1902
8	Dauri Poriya Big	30	5	150	7	210	60	8	240	30
9	Dauri Poriya Small	35	3	105	5	175	70	6	245	35
10	Basket 30 tami	16	40	640	42	672	32	45	720	48
11	Basket 10 tami	6	12	72	15	19	18	18	108	18
12	Parra Big	87	4	348	5	435	87	6	522	87
13	Chabbeda	128			15	1920		16	2048	128
14	Chandrika	69			15	1035		16	1104	69
15	Tatta	1024	20	20480	30	30720	10240	33	33792	3072
	Total	3628		45528		63284	17756		73635	9335

Source : Range Office Raigarh East.

Table 19.2 : Details of the (benefits) Basod's families of Junavani 1-12-1996 to 16-3-1997

Family No.	*Total Bamboo given*	*Total No. of Articles made*	*Total amount given to the Basod*	*Benefits*
1	210	228	3043	14.4
2	190	138	1978	10.41
3	130	75	718	5.52
4	60	72	854	14.23
5	100	152	1964	19.64
6	230	243	3151	13.7
7	140	234	2730	19.5
8	90	112	1265	14.04
9	220	308	3757	17.07
10	240	168	2851	11.87
11	160	202	2178	13.61
12	200	196	2625	13.12
13	190	275	3105	16.34
14	140	201	2195	15.67
Total	2300	2600	32564	14.158

Source : Range Office Raigarh East.

Proper training must be given to them about using less bamboo in quantity in manufacturing more and better quality articles. This will reduce the over pressure on forest because in one example I have shown that some Basods are not using the full quantity bamboo properly. They are misusing the Bamboo or they are doing illegal use of it. Thus proper monitoring must be done by the Forest Department as well as by the VFC members.

Table 19.3 : Benefits to the Basod's (individual and VFC's after formation of the committee Junvani (1.12.1996 to 16.3.1997)

			Before Formation of the committee		*After formation of the committee*					
S.No.	*Total No. of Bamboo used*	*Total No. of Tatta made*	*Rate (7×5)*	*Net Payment*	*Rate (7×5)*	*Net Payment*	*Cost of Bamboo used*	*After deducting cost of bamboo*	*Samiti Seva Shulk*	*Benefit to the Basod*
1	190	66	20	1320	33	2178	323	1855	198	1657
2	240	88	20	1760	33	2904	408	2496	264	2232
3	140	48	20	960	33	1584	238	1346	144	1202
4	190	66	20	1320	33	2178	323	1855	1968	1657
5	60	21	20	420	33	693	102	591	63	528
6	140	48	20	960	33	1584	238	1346	144	1202
7	210	74	20	1480	33	2442	357	2085	222	1863
8	130	46	20	920	33	1518	221	1297	138	1153
9	220	78	20	1560	33	2574	374	2200	234	1966
10	200	68	20	1360	33	2244	340	1904	204	1700
11	150	53	21	1060	33	1749	255	1494	159	1335
12	120	42	20	840	33	1386	204	1182	126	1056
13	190	66	20	1320	33	2178	323	1855	198	1657
14	160	55	20	1100	33	1815	272	1543	165	1378
15	110	38	20	760	33	1254	187	1067	114	953
16	140	49	20	980	33	1617	238	1379	147	1232
17	140	49	20	980	33	1617	238	1379	147	1232
18	80	28	20	560	33	924	136	788	84	704
19	50	17	20	340	33	561	85	476	51	425
20	70	24	20	480	33	792	119	673	72	601
Total				20480					3072	24531

Source : Range Office Raigarh East.

20

PUBLIC PRIVATE PARTNERSHIP

IN RE-GREENING OF DEGRADED REVENUE/ PRIVATE/FOREST LANDS—COST BENEFIT ANALYSIS

J.B. NUGENT, S.K. DUTTA, M. CHAKRABARTI, D. BHATTACHARYA, A.K. BHATTACHARYA and N. ALI

EXECUTIVE SUMMARY

After the WTO agreement, Indian industry has now gradually to open up to global competition. The paper industry everywhere in the world to a large extent depends on supply of fibres, a considerable amount of which is forest-based. Thus it is beyond any doubt, that Indian paper industry cannot be globally competitive if it fails to procure its requirement of fibres at a competitive rate. Till the eighties, the supply of such raw materials was mostly coming from the forests managed by the 'State'. This supply has been restricted considerably with India signing the Convention on Biodiversity in 1992. From being considered a source of only tangible resources, Indian forests emerged as a vast gene pool. Such a drastic change in attitude led to a complete shift in focus from 'production-centred' management to 'conservation-centred' management of the State forests. Unfortunately, the forest managers of the country failed to strike the necessary balance between supply of tangible resources and intangible resources by the National forest system, leading to scarcity in the availability of tangible resources. Such scarcities have in turn led to considerable degradation of our forests raising doubts about the efficacy of our existing institutional structure in guaranteeing the proper conservation of the forests as well.

The present study was initiated to look into the possibility of striking such a balance between the two objectives through a Public-Private Partnership. Indian paper industry, for quite a long time, has been arguing for a share of the existing degraded forest land on lease, which they may use for producing their requirement of pulp wood at a price that will add to their global competitiveness. Such a measure, it feels, will reduce the present state of uncertainties in supply of fibres as well. The Working Group constituted by the Planning Commission in June 1997 on the 'Prospects of Leasing out Degraded Forest Lands to the Private Entrepreneurs/Forest Corporations' argued against acceding to the demands by the paper industry. An in-depth scrutiny of the arguments of the group in the light of surveys undertaken in the states of Orissa and Andhra Pradesh, reveals a different situation.

The study looked into the implications of leasing out degraded forest land to private sector entrepreneurs in general and the paper industry in particular, for production of pulp-wood at four levels:

- The communities,
- The forest department,
- The paper industry, and
- The country as a whole.

It also attacks the premise that community is dependent and would be affected.

While considering the communities, it is observed that they

- are mostly lying below the poverty line;
- do not get enough alternative employment opportunities in their localities;
- do not depend much on the degraded forest land for their livelihood due to a highly reduced supply of resources from such lands;
- cannot optimally utilize the agricultural land owned by them due to lack of access to necessary complementary inputs like

credit, good quality seeds, on the one hand and to remunerative marketing options for their products, on the other; and

- do not have access to other productive assets like skills, technological knowledge etc.

Under such circumstances it is imperative that they are provided with opportunities to higher income, higher scopes for employment along with access to proper health care and educational facilities, so that they can improve their stocks of physical as well as human capitals.

Leasing out of degraded forest land to private entrepreneurs would considerably increase the scope of employment generation—it is estimated that one hectare of plantation of pulp wood species creates an employment potential of around 450 mandays over its entire rotation period. In addition, the villagers stand to get around 1.2 MT of fuelwood every year per hectare, free of cost.

Let's now look from the perspective of the Forest Department. In the effort to strike a balance between the objectives of production and conservation, a portion of the existing degraded forestland may be leased out to the private enterprises having real interests in putting them to productive use. The rest of the forestland may be put to use exclusively for conservation of biodiversity and generation of complementary intangible benefits. Towards translating it, we may conceptualise a new format of JFM that incorporates a three-way cooperation among communities interested in secured livelihood, the Forest department, keen on conservation to protect the national interests and the private enterprises interested in increased and assured supply of forestry resources as their raw materials. The *Panchayats* may also be roped into such a collaborative mechanism. Such plantations on degraded forestland will also act as green boundaries around the standing well-stocked forests and help facilitate their protection from timber mafias.

To consider the perspective of the Paper Industry, it is observed that the present gap between demand for and supply of forest based fibres is around 5-6 million tonnes. In anticipation that the ongoing research activities will help achieve a higher yield rate, leasing out around 0.8 to 1 million hectares of degraded forest land to the private entrepreneurs, may help the tide over the crisis. Given the cost advantages that the Indian units already enjoy in terms of their labour costs, leasing out such lands will ensure added advantage in terms of cost of fibre over their global competitors. Such an advantage may well be converted into a larger share in the global export market. If such opportunities beckons, the paper industry may require another 1 million hectares of degraded forestland to produce the necessary surplus output for exports. Under such a mechanism we may jointly produce some tangible benefits through increased productivity and some intangible benefits like carbon sequestration, water conservation and soil conservation services. The prospect of harvesting the benefits of Kyoto mechanisms and entering the carbon trading market in the coming days looks very bright. Activities on such leased land will be in addition to their present efforts *vis-à-vis* promoting and providing extension services in farm forestry to the communities.

However, to enhance their credibility, it is imperative that the Industry requires some 'code of conduct' which will ensure that they will not compromise with the faith the nation has put on them by leasing out a good amount of land, still constitutionally remaining under the ownership and management of the country. These could be facilitated and coordinated by national level industry organizations like CII towards ensuring that they continuously keep the nation informed about the gains that are accruing to the communities, the inter-linked sectors, the forest department and the country as a whole in the context of the lands being leased out to them. An 'in-house certification system' by IPMA along with the development of a mechanism of 'social audit' of the units in the industry by independent organizations will help achieve such an objective.

Some changes in the legal framework are also necessary to be introduced by the Government. We have proposed some changes in the Forest Conservation Act, 1980, which may be helpful in ensuring a smooth transition from the present regime of 'conservation only' to one that strikes a balance between 'production and conservation' and ensures participation of the communities in the real sense of the term.

Thus to conclude we argue that

- a portion of the degraded forest land may be leased out to the paper industry to take care of the apparent contradictions arising out of India simultaneously signing the Convention on Biodiversity and the World Trade Organization Protocols.

- The captains of the paper industry have to play a proactive role in changing their perceived role from that of 'opportunism' to one that makes themselves accountable towards ensuring the well-being of the nation.
- The State has to shed its inhibitions of looking at forests only from the perspective of conservation and consider the importance of increasing the productivity of land under forests as well.

We may summarise the benefits that may accrue to the different stakeholders if around 1.2 million hectares of degraded land are leased out to the paper industry.

* Area of plantation	: 1.2 million hectares
* Area of plantation in one year (12/7)	: 1.72 lakh hectares
* Average mandays over one rotation	: 450 mandays
Therefore, annual mandays required (450×1.72)	: 774 lakh mandays
Total annual wage (@ Rs.50/manday)	: Rs.387 crores
* If one person works for 330 days/year	
Total annual employment generation (774/330)	: 2.35 lakh
* Total fuelwood, @ 1.2 MT/Ha, (1.72×1.2)	: 2.1 lakh MT/Yr
Value @ Rs.500/MT	: 10.5 crores/Yr
* If MAI of natural forest is 0.7 to 1.0 cu mt.	: 7 MT
Wood on 7 years	
Harvesting pressure reduction on natural forest	: 30,000 MT to 2 lakh Ha/Yr
* Pulp wood yield (1.71 lakh Ha × 80 MT/Ha)	: 137 lakh MT/Yr
If 4 MT wood = 1 MT of paper	
Total paper production (137/4)	: 34 lakh MT
* Foreign exchange saved by reducing imports 34 lakh MT paper @ USD600/MT	: USD 2040 million or : USD 2040 million or ➢ USD 2 billion/Yr or ➢ Rs. 979200 lakh/Yr or ➢ Rs.10000 crores/Yr
* Kyoto carbon sink @ 30 USD/hectare/year	: USD 36 million or ➢ Rs.173 crores/Yr
* Soil and water conservation	: Difficult to measure

In all the total benefit works out to be well above Rs.10500 crores per year. This gains are in addition to the prospective gains in the form of water and soil conservation, protection to the existing forests and the consequent protection of our gene pool therein.

The increased production by the paper industry will have its spillover effects on the sectors enjoying backward and forward linkages with it. All will be bubbling with opportunities of increased production and consequent value addition to the economic health of the country. The employment potential of the economy will increase manifold.

Such a 'trinity'—strong bondage of partnership across Community, State and Private sector—will go a long way in opening up the space for a fruitful private-public partnership in re-greening the country and motivate such collaborations in other sectors of the economy to bring about the required change in the country and make the dream of becoming a 'developed' country by 2020 an absolute reality.

INTRODUCTION

The single criterion that distinguishes mankind from other living beings is its capacity to produce—transform resources directly available from nature into something else which are not available naturally. Thus while the other living beings sustain on consuming water, fruits, grass or fleshes, we are happy consuming Coca Cola, orange marmalade or sausages. Unlike other living animals, most of us do not live in caves or tree branches. We cannot do without clothings either. Such demand profiles result from our ability to 'produce'. It is apparent that in her pursuit to produce, mankind has used the natural resources and simultaneously, dumped the unwanted bye-products in nature. Nature has acquired the role of both a source and a sink.

A typical text book in Economics identifies four factors of production. They are:

- Land,
- Labour,
- Capital, and
- Organization.

When land is identified as those resources directly available from nature, labour refers to the human beings' capacity to work. Whereas Capital is defined as the sum total of every means of production and organization that makes the production decision. To elaborate, any natural resource—land cover on the mother earth, water, air, natural forests come under the category "Land". The capacity of a human being to work is obviously dependent on his physical strength and skill. Such strength and skills are contingent upon either direct or indirect consumption of natural resources. Capital emerges from a production processes. In the ultimate analysis, it is also derived directly or indirectly through the consumption of natural resources. Organization as the decision-making unit derives its reward by allocating natural resources across different uses. Thus we may safely argue that production refers to a process that ends up utilizing natural resources, either directly or indirectly.

In the present day parlance, a higher level of development is associated with higher level of production. It is also observed that a higher level of production is often associated with a lower level of direct consumption of natural resources. In other words, a more capital-intensive technology apparently uses very little amount of natural resources. However, as one goes backwards in tracing the origin of the inputs used, the journey always ends up in hitting a natural source.

Every production process involves use of energy. Technological improvements have always been associated with enlargements in the consumption of energy, even though some may argue that the consumption of energy per unit of output produced has declined. The truth is absolute consumption of energy has been increasing rapidly, ever since we realized the virtue of substituting labour with capital. There is no denial of the fact that labour also requires energy for its sustenance. The absolute consumption of energy under a labour intensive technology regime is, acceptedly, much lower. Appreciation of this concern leads one to realize that human beings cannot produce energy. Sun is the principal source of energy in our solar system and only living beings endowed with chlorophyll have the capacity to trap them. The food we eat to energize ourselves are derived from the plants, either in the form of the plants themselves or in the form of fleshes of animals that sustain on plants. The traditional sources of energy—coal, petroleum and fuelwood—are either directly or indirectly derived from sunlight. Non-traditional sources of energy generated using nuclear, wind and hydro-power are the only exceptions to this general rule, even though there is no denial of the fact that these are also part of our natural resource base.

We may, therefore, argue that the existence of human beings in the future is contingent upon the availability of natural resources. Nothing can be 'produced' without them. Some of these resources—the living ones—grow, while some others—the inanimate ones—do not. When the human population on the mother earth was too low to influence the growth of other living creatures and the inanimate resources were extracted at a rate too small compared to their global reserves, we were not much concerned about protecting and consequently, conserving the natural resources. Some people even today are hopeful of developing alternatives to some of these inanimate resources—petroleum, for example, using, of course, some other natural resources that either grow naturally or are available in unlimited quantities. However, some others are not convinced with such prospects. And the number of persons joining the second group has been rising consistently since the mid-1980s. They are talking of 'sustainable development'.

Production of paper and paperboards is an important component of the debate on 'sustainable development'. The debate is apparently very straightforward. Production and subsequent consumption of paper and paperboards is considered an indicator of development. To put it simply, higher level of development is associated with higher per capita consumption of paper and paper products. On the other hand, such products are made of natural fibres and hence involve use of higher amount of natural resources. How then to balance between these two apparently opposite impacts? The answer is not simple to provide. The present report is an attempt to identify some possible solutions that may help balance this apparent contradiction.

Of all the natural resources endowed upon mankind, soil cover on the mother earth has been one of the most important basic resources which plays a strategic role in determining the living standards of human beings. Soil cover, in fact, is the base for all sorts of productive activities. Any type of transformation requires soil cover in varying degrees. Historically, soil cover has been put to increasing number of varied uses. In the context of the present study we shall refer only the soil cover as land. Given the fixed amount of land available on the mother earth and the simultaneous increase in population as well as both quantitative and qualitative changes in the goods and services demanded over time, the pressure on land has been increasing tremendously. In earlier days when economic activities were mainly concentrated around agricultural activities and the population of the world was not that high like what we have today, land was not considered to be scarce. Forests were felled to make land available for agriculture as well as for human settlement. In the course of time man graduated into industrial production which necessitated further diversion of land from agriculture and/or forests to meet the requirement of industries as well as that for urbanization. Simultaneously, the agricultural sector grabbed further forestland to make up for the loss caused by such diversion. Thus we have gradually shifted from slash and burn agriculture to multiple cropping on the one hand and on the other, land meant for agricultural production have been getting diverted to non-agricultural use at a very alarming rate. Under such circumstances it is necessary that a proper planning is undertaken in allocating the available land in an optimal manner among different possible uses to balance the requirements of all types of economic activities on the one hand and the natural resource base on the other.

Conceptually, perspective land use planning involves altering the land area allocations over alternative uses through suitable technological and institutional devices such that

(i) the supplies of the various commodities and service which follow from the stipulated land use pattern broadly conforms to the projected demand for such items,

(ii) the uses are sustainable in the sense that the current uses by one group does not jeopardize the uses of another group or those of the future generation, and

(iii) the bio-mass production, i.e. the streams of output which follow from the stipulated land use pattern (and even income and employment following from the streams of output) are maximized.

From an earth covered only with forests and rivers with patches of deserts here and there, even a few thousand years ago, we have arrived at a level of 'development' where land is to be used for the following purposes:

(a) maintenance of forest cover upto a certain specific level for ensuring ecological stability;
(b) production of cereals and other food crops for ensuring food security;
(c) production of fodder for the live stock that ensures supply of other food stuffs necessary for human consumption;
(d) maintenance of water bodies for inland fisheries, irrigation and supply of water for drinking and even for protection of bio-diversity;
(e) cultivation of commercial crops for use as inputs by manufacturing enterprises;
(f) provision of shelter to the world's population both in rural and urban areas;
(g) extraction of minerals; and
(h) establishing industrial enterprises.

Given these multitude of possible uses of land and the fact that very often there are possibilities of emergence of a new use outcompeting the existing uses, and consequent conflicts, land use planning *per se* is to be considered an important exercise facing the policy makers and planners of an economy.

To understand the implications of such conflicts *vis-à-vis* land use planning we have developed a conceptual framework presented as Appendix I to this report. The framework distinguishes between land put under two uses, namely, to produce forest and non-forest products and concludes that several policy measures may be taken up by an economy to resolve the apparent conflict *vis-à-vis* the allocation of land between uses for production of forest and non-forest products. They may be considered singularly or in a combination of some or all of them. They are:

P1. Importing land for a particular use from another country which has land under such use in excess of the requirement of their own people. However, such decisions are possible if the land use pattern for the world as a whole is sustainable at a given point of time.

P2. Increasing the productivity of land under a particular use and thereby reducing the requirement of quantity of land under a particular use. Such a measure requires technological innovations.

P3. Considering putting land under joint use—producing both forest and non-forest products simultaneously from the same plot of land. Such a measure requires both institutional and technological innovations.

The following sections will be evolved in the light of the three policy alternatives developed here. We shall have opportunities to identify an optimal combination of these measures *vis-à-vis* the problem of regreening the degraded land of the country.

In order to lead this study to a meaningful conclusion, we have divided it into 7 sections. The following section poses the debate in the perspective of the conflicting uses of land, with particular reference to the long-standing request to the government by the captains of paper industry towards allocation of degraded forest lands to them on a long term lease and the persistent refusal by the powers by citing the enormous possible social costs involved in such a transfer on lease. The third section is devoted for the arguments raised by the Working Group constituted by the Planning Commission in June 1997 on the 'Prospects of Leasing out Degraded Forest Lands to the Private Entrepreneurs/Forest Corporations' against acceding to such demands by the paper industry. Section IV makes an in-depth scrutiny of the tenability of such arguments in the light of surveys undertaken in the states of Orissa and Andhra Pradesh. Section V specifically looks into the ecological implications of the leasing out of degraded forestlands. The subsequent section constructs a framework for analyzing the fundamental problem faced by the paper industry today and the implications of a changed policy for their sustenance. The final section concludes the whole discussions with some recommended measures that may be taken up by both the Government and the leaders of paper industry.

THE DEBATE IN PERSPECTIVE

In view of the decision of the Planning Commission to bring 25 per cent of the land mass of the country under tree cover by the end of the Tenth Five Year Plan, further to be extended to 33 per cent at the end of the Eleventh Five Year Plan. Bringing the wastelands of the country under tree cover has assumed significant importance in the portfolio of the tasks ahead for the country. The target—to achieve this objective by 2012 has been estimated to afforest around 43 million hectares of land, of which 15 million hectares will be degraded forest land, at an investment of Rs. 48562.50 crores (a little over 10 billion US$) over ten years. As the government alone will not be able to muster the required financial resources, relevant private-public partnership is considered to be one way out. The corporate sector—paper industry in particular—is interested to bear a considerable portion of the costs in return for a long term lease to develop captive plantation to grow pulpwood species. They argue that such actions will generate enough employment for the people living in and around these land and hence increase their livelihood security. Moreover, such an effort will allow them to procure the main raw materials—pulp wood—at a cheaper price and with considerable degree of certainty to help increase their competitiveness in the global market. Such an arrangement will also increase the employment potential in the regions. Pro-poor activists, on the other hand, fear that such a move may further alienate the poorer section of the society from the meagre landholdings under their possession and jeopardise their livelihood. Unfortunately, none of these arguments have been backed by any objective study that looked into them from a comparative perspective. Against the backdrop of this debate and the apparent void in the existing literature, the present study will look into the *Cost-benefit analysis of alternate uses of degraded land.* It is obvious that the existing stock of degraded land in the country is used in many ways. Some portions are lying unused as well. If brought under forest cover, such lands will cease to have their existing use, thus inflicting a cost to the existing users—most of whom are anticipated to belong to the marginalized group. The act of conversion will have the potential to provide some benefits to the existing users as well. It is thus intended to look into three alternate scenarios:

- degraded land already converted into captive plantations by paper industry,
- degraded land converted into community forestry,
- degraded land lying as it is,

and find out the comparative net benefits—both tangible and

intangible—acruing to the people, paper industry and the nation as a whole.

From the light of thc findings we would chalk out future policy prescriptions on two issues: firstly, identifying efficient 'property rights regime mix' and the relevant institutional structures—at operational, collective and constitutional choice levels—necessary to establish such regimes and secondly, the road map for Indian paper industry towards attaining a competitive position in the global paper market.

Fig. 20.1 below portrays the backward and forward linkages

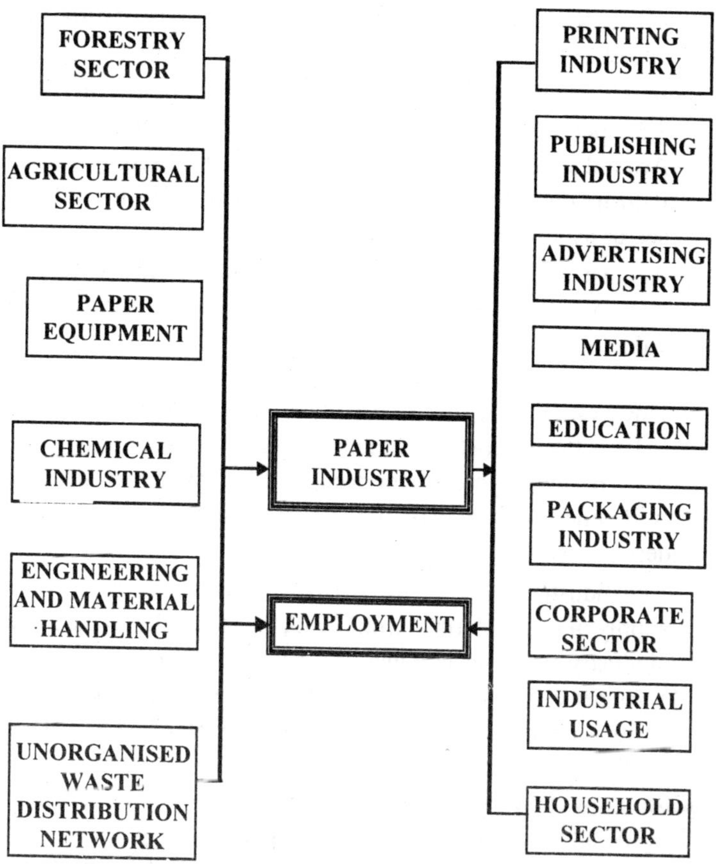

Fig 20.1 : Forward and Backward Linkages in Paper Industry

across the sectors *vis-à-vis* the paper industry. It identifies the relevant stakeholders whose interests are involved with the growth of the paper industry in the country. Whereas the entries arranged vertically indicate direct linkages: we have inserted a box labelled *employment.*

Such a box signifies the extent of employment potential that are generated by the paper industry directly and also indirectly through the linkages across the sectors.

Some salient features of Indian paper industry are as follows:

- As per the Annual Survey of Industries, 1999, paper industry accounted for around 2.31 per cent of the output and gross value added of the total industrial sector of the country.
- In the process it accounted for 2.81 per cent of the capital and 3.52 per cent of labour employed by the industrial sector. Around 3 lakh 52 thousand persons were employed by the paper industry out of around a crore persons employed by the industrial sector as a whole.
- Against a national average of gross value added at Rs. 1.98 lakhs per employee, the paper industry recorded a corresponding value of Rs. 1.3 lakhs.

However, the figures above relate to the direct employment potential and the productivity of labour therein *vis-à-vis* Indian paper industry. The employment generated by the related industries and sectors, if added, will increase the overall employment potential to a considerable extent.

The industry, however, has been facing a serious crisis in terms of its supply of raw materials. Until 1980, the raw material base in the paper industry was concentrated around bamboo (65%) and wood (25%). As per estimates available for the current period, 48 per cent of the units are dependent on wood/bamboo based raw materials whereas the rest of the units use non-wood-based raw materials, like bagasse, straw and waste paper. Raw material choice is also dependent on the product mix of a unit. Newsprint and industrial grades of paper can be made from waste paper and agri-residues, whereas most of the writing and printing grade papers require forest based materials for their production.

It is worthy to note that the raw material cost is very high in India. It constitutes around 40-50 per cent of the cost of sales. The high share of raw material costs is mainly due to two reasons.

1. High cost of fibres, mainly hard wood, and
2. Low fibre usage efficiency.

Consequently, in spite of having a comparative advantage in terms of low costs of labour and consumables, a representative Indian unit gets outcompeted by a unit based in USA due to high costs of raw materials, capital (high interest rates) and energy (The Indian Paper Industry: Enhancing Competitiveness, Report prepared by ICRA Advisory Services, September, 2002).

Such high cost of fibres is attributed to

- Reduction in supply from the traditional source—'State' forests since 1980s with increased restrictions on extraction of forest products.
- Too much emphasis on conservation of forests to increase the productivity of intangible values, in accompaniment with utter neglect of increasing the productivity of tangible values from out of the 'State' forests. Such neglect is clearly demonstrated by the fact that no recent estimates are even available on the annual rate of increment of Indian forests. The latest estimate puts the value at 0.7 m^3 per hectare per year, which is abysmally low, compared to global standards. A low productivity of forests accompanied by continuously increasing demand for forest products puts pressure on the existing forests to supply resources beyond their sustainable yields—a potential cause for degradation of existing forests. The possibilities of producing higher tangible values per unit land, if put to alternative uses, is also adding fuel to the debate as to whether more than 20 per cent of the land—a scarce resource by all standard in the country—can really be reserved simply for the production of intangibles? The country is probably yet to reach a level of

development that induces it to value 'intangibles' more than the 'tangibles'.

We present some statistical results obtained by analyzing data from the Human Development Report 1998 of the state of Madhya Pradesh and the State of Forest Report 1999 prepared by the Forest Survey of India. Both the documents present data for the 45 districts of the then state of Madhya Pradesh. Now some of these districts are part of the newly created state of Chhattisgarh.

The Spearman rank correlation coefficient is found to be significantly positive between

Poverty ratio across the districts and the percentage of geographic area of the district under dense forests (0.503, significant at 1% level) as well as under open forests (0.465, significant at 1% level).

Such values of the correlation coefficient indicate that districts having larger percentage area under dense and/or open forests have a tendency to record a higher poverty ratio. However, it should be mentioned that such result does not say anything about the causal relationship between the two.

On the other hand, the same coefficient has been observed to be significantly negative between

- Life expectancy index across the districts and the percentage of geographic area of the district under dense forests (-0.312, significant at 5% level).
- Income HDI across the districts and the percentage of geographic area of the district under dense forests (-0.428, significant at 1% level) as well as under open forests (-0.459, significant at 1% level).
- Poverty ratio across the districts and the change in forest cover between 1997 and 1999 (-0.305, significant at 5% level).
- Education HDI across the districts and the change in forest cover between 1997 and 1999 (-0.307, significant at 5% level).

One may infer that a larger proportional area under dense forest is associated with lower life expectancy and lower income HDI. On the other hand, a higher poverty ratio may lead to lower level of afforestation or a higher level of deforestation. Interestingly, a higher level of education HDI also appears to have similar impact on the extent of change in forest cover.

Thus we may infer that the existing policy mix *vis-à-vis* the utilization of forestry resources shows a tendency to discriminate against the people living in districts having a higher proportion of their geographical area under forests. The reasons are not difficult to seek.

- Increasingly stringent policies and legal framework are restricting access of the people residing in and around forests to the resources available therein.
- A forest policy that puts too much emphasis on conservation and apparently neglects the need to increase the productive capacity of the forests is reducing the scope of increased value addition for the people of these districts. Had such lands been put to alternative uses, the economic value addition would have been much higher, resulting in obvious reduction in poverty ratio.
- Not much infrastructural development has taken place in these districts, apparently, in the name of not disturbing the environment.

How is the country as a whole appreciating the role of the people residing in these districts? Is it sacrificing enough towards protection of the forests? The 42nd Amendment to the Constitution has brought forests under concurrent list. The Forest Conservation Act, 1980 does not allow any state to convert forest land to other uses without the permission of the Union Government. The Supreme Court judgment in Godavarman case prohibits felling of forests without following the prescriptions of the Working Plan. The Working Plans, incidentally, have to be approved by the Union Government after the respective State governments prepare them. Such rights should always be complemented by responsibilities. In consideration to its rights on this vast extent of landmass under forests, the 'Nation' is also expected to shoulder the relevant responsibilities. The extent of responsibilities shouldered by the 'Nation' may be measured by the plan allocations made to the protection of forests and wildlife over the years. Table 20.1 below provides an insight into the plan allocation pattern along different sectors.

Table 20.1: Share under different development heads in total plan outlay (%)

Development Heads	*95-96*	*96-97*	*97-98*	*98-99*	*99-00*	*00-01*	*01-02*
Agriculture and allied activities of which	3.63	3.38	3.40	3.14	3.13	2.75	2.60
Animal husbandry	0.13	0.13	0.12	0.08	0.11	0.07	0.11
Fisheries	0.14	0.12	0.13	0.13	0.12	0.11	0.09
Forestry and wild life	0.29	0.26	0.26	0.28	0.32	0.26	0.26
Plantations	0.14	0.11	0.16	0.14	0.14	0.16	0.10
Rural Development	8.17	6.35	6.56	6.29	5.37	3.85	3.42
Irrigation and flood control	0.33	1.05	0.33	0.39	0.36	0.37	0.37
Energy	27.17	25.29	26.07	27.10	27.19	24.03	25.95
Industry and minerals	14.50	13.09	13.29	8.99	6.68	6.52	6.11
Transport	16.04	18.56	16.02	16.27	17.81	19.20	17.34
Communications	13.11	13.00	13.74	15.22	15.47	18.71	15.58
Science and Technology	2.06	2.37	2.42	2.7	2.97	2.94	2.77
Social services	13.98	15.20	16.97	18.55	19.66	20.23	20.65

It is really painful to observe such a low allocation of resources for the protection of forests and wild life. It is even further painful to note the declining share of plan resources allocated to forestry and wild life over the years. One should not forget, even at the cost of repetition, that land under forests account for around a fourth of the landmass of the country and around 100 million people (roughly,10% of the population) are dependent on forests for their livelihood. It is well known that people residing in and around forests in general and those belonging to the tribal communities in particular do depend on activities related to

- Animal husbandry,
- Fisheries, and
- Plantations.

other than being engaged in cultivation (mostly as landless labourers) to complement their earnings from collection and sell of forest products. One finds the share of allocation to all these sectors declining over the 9th Plan period. On the other hand, the share of allocation is increasing rapidly for the transport and communications sectors. Such emphasis on these sectors should be considered in association with the fact that the government has virtually opened up these sectors to unrestrained investments from private investors—both domestic and foreign.

Such plan designs point exclusively to the responsibilities borne by the 'Nation' as a whole to protect the forestry resources of the country. The Mid-term Appraisal of Ninth Five-Year Plan does not hesitate to mention that

> "Forests are managed by states primarily with state funds supported by external donors. For the forestry sector, during the Eighth Plan (1992-97) the state plan outlay was Rs. 3500 crores, whereas the funds received from the Ministry of Environment and Forests as centrally sponsored schemes were only about Rs. 500 crores, or less than 20 per cent of the total. This would be even less than 15 per cent if one took into account the transfer of funds from District Rural Development Agencies (DRDA) to district field agencies for forestry related works. For the Ninth Plan the state plan outlay is Rs. 6300 crores, whereas the contribution of the Centrally sponsored schemes would not be more than Rs. 1500 crores. Most externally aided projects in the forestry sector are negotiated with the states and external aid is reflected in the state budgets.
>
> " Investment in forestry in India is likely to decline sharply in the next two years, however, because of a ban imposed by countries like Japan and Denmark on assistance to India. Rough estimates indicate that annual external funding in the forestry sector will decline from the present level of Rs. 844 crores to just Rs. 300 crores by 2002-03 unless special efforts are taken to start new projects. There is likely to be a decline in assistance from the World Bank too as no new projects have been signed after UP (1997) and Kerala (1998), nor are there any in the pipeline.
>
> "External assistance is not without strings. DFID spends too much on documentation. According to its own report…., DFID produced 160 consultancy reports for a small Himachal Pradesh

project of Rs. 40 crores in two districts. Many of these reports have never been read, let alone acted upon by Forest Department (FD) staff. It caused tremendous burden to the local staff. The gainers were the British Universities.........." (p. 332).

On the other hand we observe efforts through legal and administrative measures to curtail the traditional rights of the forest users. The Supreme Court Judgement in 2000 prohibiting harvesting of forest produce from wildlife sanctuaries is a case in point. The proposed Bio-diversity Bill

(a) only discusses access and benefit sharing in terms of 'transfer of biological resources to agencies outside India',
(b) [refuses to recognize the] ...rights and knowledge of local communities,
(c) [does not]...ensure that their consent is taken for the wider use of this knowledge, and further,
(d) [does not] ensure that they receive equitable benefits from such use (all provided for under Article 8(j) of the Biodiversity Convention)',
(e) does not address ownership of resources, leaving benefit sharing issues unresolved,
(f) does not devolve any authority to local communities to create a local stake in conservation and sustainable use.

Obviously, the incentive-disincentive structure resulting out of the existing 'State property-rights regime' is unfavourable towards those residing in and around forests. Such an imbalanced reward-sacrifice matrix cannot ensure 'sustainable' use of the forestry resources. The 'nation' as a whole needs to ensure a higher sacrifice towards the well-being of the people who are expected to protect and preserve the forests.

It appears that probably, the policies followed since independence in respect of the use of the forest resources under the ownership of the 'State' gave rise to a reward-sacrifice structure that settled with gradually shrinking reward for the people residing in and around forests (expected to take up larger responsibility in protecting the forests), whereas the sacrifice by the nation towards protection of the forest—they are considered to be the property of the nation—shrank simultaneously in relative terms, if not absolutely. The results from Madhya Pradesh and Chhattisgarh mentioned above indicate such a tendency. Even Joint Forest Management were introduced in most of the states with financial support (loan/grants/aid) from bilateral/ multilateral agencies. Now that such funds are shrinking, the prospects of sustainability of JFM have become uncertain in many of the states, like West Bengal. There are several possible alternative options to change the reward-sacrifice matrix. Some of them are:

1. Identification of special programme/schemes by the Planning Commission to ensure 'development' of the people living in and around forests and consequent increased allocation to protection of forests and environment.
2. A change in the formula for distribution of the divisible pool of central taxes across the states proposed under the Finance Commission award. States having percentage of its geographic area under forests in excess of the national average should be compensated adequately, whereas those having percentage forest cover that is below the national average should sacrifice. The argument considers the fact that the opportunity costs of maintaining the land under forests—gains forgone by not converting forests into some other economic uses like agriculture, industrialization, urbanization etc.—are considerably high if considered from the perspective of a particular state.
3. The present thrust in the forest policy is only on 'protection and conservation'. The issue of increasing the 'productivity' of the forests has almost been swept under the carpet. A judicious balance between these two concerns may raise the economic benefits of the states without compromising the requirement for 'protection and conservation'. Suitable policy changes may be developed to encourage participation of the private sector in such an endeavour.
4. A "forest tax" may be conceptualized at the union level with the tax rates increasing as the distance from forest increases. The

proceeds from the realization of such taxes may be redistributed to areas having forests for development of the people residing therein.

The present report will deal exclusively with the policy alternative enumerated under item 3 above—introducing suitable policy changes to encourage participation of the private sector towards increasing the productivity of forests. In the next section we look into a document that is considered to be the present stand of the Union Government towards such a proposal.

Prescriptions from the 'Saxena Report'

Planning Commission set up a Working Group in 1997, under the chairmanship of Dr. N.C. Saxena, the then Secretary, Department of Wastelands Development, Government of India to examine the prospects of leasing out of degraded forests to private entrepreneurs/ Forest Corporations. Hereinafter, the report by the group will be referred to as SR (Saxena Report). SR strongly argued not to lease out degraded forests to private entrepreneurs. Simultaneously, SR also maintained that non-forest barren/degraded lands may be utilized for plantation of pulp wood by private entrepreneurs. The conclusions in SR were centred around the following arguments:

- *Involvement of industry should be encouraged in reclamation of non-forest barren/degraded land as:*

A1. There is a considerable availability of such lands in the country, estimated at 20 million hectares;

A2. Governments, specially at the states, are interested in leasing out such lands to the private sector enterprises;

A3. Such land is available in contiguous patches and hence is amenable to economies of scale;

A4. Degraded land of this nature does not support the livelihood needs of the poor and hence their conversion into other use will not exacerbate social tensions.

- *Involvement of industry should not be allowed in reclamation of barren/degraded forest land as:*

B1. Such an act will jeopardize the livelihood requirements of an estimated 100 million people in the country and Joint Forest Management (JFM) has been successful enough to regenerate a considerable amount of such degraded forest land through involvement of the local people during the last decade;

B2. Any act of leasing out degraded forest land to the paper industry will attract attention from other land-intensive industries like coffee, rubber etc. Even smaller paper units will also demand their share in the cake. Such large scale leasing will have ecological implications;

B3. The requirement of the paper industry may be met from out of the existing inventory of uncultivated wasteland. Expansion of farm forestry has ensured a steady supply of pulp wood, as is evidenced by the sharp fall in the price of eucalyptus prices. However, there is a locational problem. The pockets of expanding farm forestry are located far away from the region where most of the paper units were established. The processing units may be split, with the pulp-making units set-up around the farm forestry areas. The pulp produced may be transported to the paper-making units. Given the fact that pulps are imported from abroad, such dispensation will surely work out to be cost-effective;

B4. Lakhs of farmers will be deprived of a market for agroforestry;

B5. 'Millions of voiceless forest dwellers' will be denied access to the NTFPs and the other biomass that they gather;

B6. Their hardship will be more confounded in view of the phenomenon of 'market failures' existing in most of the farm products;

B7. Expansion of farm forestry is confined within the realm of non-cultivable agricultural waste land, it will never compromise the food security of the country as is often alleged;

B8. Paper industry will be unduly subsidized if they receive land almost free of capital cost;

B9. Leasing out of forest land to private enterprises is unconstitutional and illegal as the process involved contradicts with the *Panchayat*

Act, Forest Conservation Act and the spirit of the present Forest Policy. Most of the forests being located in Schedule V areas, the constitutional safeguards guaranteed to the tribals will also be violated;

B10. The paper industry is looking for lands that have a minimum of 1 metre soil depth. Such land will regenerate on their own without much costs. Hence, regeneration will be cheaper economic option than plantations;

B11. The industry has no credible plans to resolve the demands that the local communities have over such lands. Any leasing of forestlands is likely to result in hardships and oppression for the local communities who have historically depended on such lands for meeting their basic needs.

B12. However, the report did not fail to mention that the unit cost of reclamation is considerably high.

Observations on the 'Saxena Report'

The present section examines the tenability of the arguments put forward in SR and notes that most of them are not backed by hard facts. We take them up for scrutiny one after another in the paragraphs that follow.

A1. 20 million hectares of degraded land are available outside the domain of the 'forestry' sector.

Even though there is no debate that a considerable amount of the Indian land mass has gone degraded—land that cannot be or are not used for productive purposes, the quantum of such degraded land is not known for certain. One estimate suggests that one in every fifth hectare of land—63.85 million hectares in all—falls under the category of wastelands. Estimates by the Food and Agricultural Organization of the United Nations (FAO) put 43 per cent of the country's total land area under severe degradation and another 16 per cent has been characterized as very severely degraded. V.B. Eswaran lists several estimates of degraded land made so far. The figures vary from a low of 38.4 million hectares to a high of 187 million hectares. The variation across the estimates are mainly influenced by variations in terms of :

➢ definition of degraded land/wastelands used for the estimates, and

➢ relevant time period of estimates.

It will be prudent to categorise the degraded land under two heads—

1. Land which can be reclaimed and used for productive purposes, and
2. Land which cannot be so used for any productive puopose.

The first type of land is often defined as wastelands. In the context of the present report we accept the definition used by the March 2000 *Wastelands Atlas of India* prepared by the National Remote Sensing Agency for the Department of Land Resources which defines wasteland as: 'Degraded land which can be brought under vegetative cover with reasonable effort, and which is currently under-utilized and land which is deteriorating for lack of appropriate water and soil management or on account of natural causes. Wastelands can result from inherent/imposed disabilities such as by location, environment, chemical and physical properties of the soil or financial or management constraints.' However, confusions remain even then. One of the major problems in the determination of degraded land is whether or not that land is not really being used by local people. In the past, some of the "wastelands" were hardly "waste" and were an essential part of local communities strategies for survival. Thus, separating "degraded but heavily used land" from "degraded and no long used land" will be a real challenge for future work for the policy makers. The NRSA estimates could not take care of this challenge.

U.R. Rao (Former Chairman, Indian Space Research Organisation and Secretary, Department of Space, India) in a lecture given on the 27th November, 1995, put the estimated quantity of degraded land at 53.3 million hectares out of which around 58 per cent are non-reclaimable, implying the possibility of 42 per cent i.e. around 22.5 million hectares of them being reclaimable. The NRSA estimate puts the share of non-reclaimable land at around 30.5 per cent of the total degraded land—at 19.46 million hectares (inclusive of land under categories 3,9,11,12,13—our definition—of the table 4.1 below), implying roughly a shade over 44 million hectares of degraded lands are recoverable. The inventory of recoverable degraded land consists of land under the following uses:

- Gullied and/or Ravinous land
- Land with or without scrub
- Land affected by salinity/alkalinity—coastal/inland
- Shifting Cultivation Area
- Underutilized/degraded notified forest land
- Degraded land under plantation crop
- Sands—Inland/Coastal
- Mining/Industrial wastelands.

Table 20.2 : Category-wise Wastelands of India

(Area in sq.kms.)

S. No.	*Category*	*Total Wastelands*	*% to total Geo-graphical Area covered*
1.	Gullied and/or Ravinous land	20553.35	0.65
2.	Land with or without scrub	194014.29	6.13
3.	Waterlogged and marshy land	16568.45	0.52
4.	Land affected by salinity/alkalinity—coastal/inland	20477.38	0.65
5.	Shifting cultivation area	35142.20	1.11
6.	Underutilized/degraded notified forest land	140652.31	4.44
7.	Degraded pastures/grazing land	25978.91	0.82
8.	Degraded land under plantation crop	5828.09	0.18
9.	Sands—Inland/Coastal	50021.65	1.58
10.	Mining/Industrial wastelands	1252.13	0.04
11.	Barren rocky/stony waste/sheet rock area	64584.77	2.04
12.	Steep sloping area	7656.29	0.24
13.	Snow covered and/or glacial area	55788.49	1.76
	Total Wasteland Area	638518.31	20.17

Source: 1:50,000 scale wasteland maps prepared from Landsat Thematic Mapper/ IRS LISS II/III Data

Note: 1,20,849.00 sq.kms. in Jammu & Kashmir is not mapped and hence not considered for calculating the percentage.

Source: http://dolr.nic.in/category.htm

According to a Planning Commission document the current estimates of culturable wastelands are 13.9 million hectares[5]. The FAO estimate mentioned above also suggests that around 82% of the potential arable land in the country is actually in use. The spread of degraded land across the states is depicted in Table 20.3 below.

Table 20.3 : State-wise Westelands of India (1986-99) [sq. kms]

S. No.	*State*	*No. of dists. covered*	*Total Geog. area of dist. covered*	*Total W.L. area in dists. covered*	*% to total geog. area*
1	Andhra Pradesh	23	275068.00	51750.19	18.81
2	Arunachal Pradesh	13	83743.00	18326.25	21.88
3	Assam	23	78438.00	20019.17	25.52
4	Bihar	55	173877.00	20997.55	12.08
5	Goa	2	3702.00	613.27	16.57
6	Gujarat	25	196024.00	43021.28	21.95
7	Haryana	19	44212.00	3733.98	8.45
8	Himachal Pradesh	12	55673.00	31659.00	56.87
9	Jammu&Kashmir*	14	101387.00	65444.24	64.55
10	Karnataka	27	191791.00	20839.28	10.87
11	Kerala	14	38863.00	1448.18	3.73
12	Madhya Pradesh	62	443446.00	69713.75	15.72
13	Maharashtra	32	307690.00	53489.08	17.38
14	Manipur	9	22327.00	12948.62	58.00
15	Meghalaya	7	22429.00	9904.38	44.16
16	Mizoram	3	21081.00	4071.68	19.31
17	Nagaland	7	16579.00	8404.10	50.69
18	Orissa	30	155707.00	21341.71	13.71
19	Punjab	17	50362.00	2228.40	4.42
20	Rajasthan	32	342239.00	105639.11	30.87
21	Sikkim	4	7096.00	3569.58	50.30
22	Tripura	4	10486.00	1276.03	12.17
23	Tamil Nadu	29	130058.00	23013.90	17.70
24	Uttar Pradesh	83	294411.00	38772.80	13.17
25	West Bengal	18	88752.00	5718.48	6.44
26	Union Territory	20	10973.00	574.30	5.23
	TOTAL:	584	3166414.00	638518.31	20.17
	*Un-Surveyed Area (J&K)		120849.00		
	Total Geo. Area		3287263.00		

Although, category-wise information are not available, the table suggests that more than one-fourth of the degraded land of the country is concentrated in Himachal Pradesh, Jammu & Kashmir, and the eight north-eastern states. Thus we may argue:

- There is a considerable confusion existing in the available datasets about the exact quantity of reclaimable degraded land in the country; and
- Degraded land is highly concentrated in the states situated across the Himalayas. Availability of separate quantification of degraded land from Uttaranchal would have inflated the share of the Himalayan states further.

The SR itself is aware of this shortcoming while it writes its verdicts. It notes, "Information needed for assessing their potential, like ownership, extent and type of degradation, and present and possible future uses has never been collected on a systematic basis for the entire country. This makes it difficult to estimate how much of the 55 million hectares of the non-forest and non-cultivated land is degraded". SR adds further, "It may be useful to undertake actual state-wise ground surveys so as to establish the extent of degraded lands in the country". It then argues, "...we make the arbitrary assumption that one-fifth of this land (55 million hectares estimated by the National Commission on Agriculture) is not degraded, giving a total of 44 million hectares of uncultivated degraded land. Even this may not be entirely available for tree plantation, as part may be required for schools, *panchayat* buildings, cattle resting place, wells and dwellings. In dry areas, the proportions available would be higher. We estimate the area of degraded non-forest land available for tree growing as 75 per cent of 44 million hectares, i.e., 33 million hectares. Adding 27 million hectares of degraded forests.... the total comes to 60 million hectares." Development of a policy document on utilization of degraded land that bases its argument on a hypothetically fixed amount of degraded land available for reclamation arguably reduces the credibility of such report. The estimated availability of 20 million hectares of degraded non-forest land to be leased out to the private sector is highly hypothetical and may not be reflecting the ground realities. Moreover, there is no information available on the ownership status of the degraded land and the status of their degradation. Unless one is certain that all these 33 million hectares of degraded lands are owned by the state, how can one recommend them to be leased out to the private sector? If a considerable amount of these lands are under private ownership, they cannot necessarily be leased out to other private players simply at the dictat of the state! *It will not be out of place to mention that the areas we surveyed revealed that only 8.5 per cent of the degraded non-forest land lies under the ownership of the state, the rest are all privately owned.*

A2. Governments, specially at the states, are interested in leasing out such lands to the private sector enterprises

SR dwells at length about the keenness of the state governments in leasing out degraded non-forest lands to private parties attended with the subsequent lack of interest shown by the captains of the private sector in lapping up such offers. However, it did not care much to go into the reasons behind such apathy shown by industrial houses. The reason, apparently, is very straight forward. Such lands are not amenable to economic uses—their use involves returns not commensurate with the investments necessary to reclaim them. A blanket policy prescription not supported by any attempt at categorizing such land in terms of their ownership, locations in respect of agro-climatic zones and capability of providing economic returns may not be effective in attaining the desired objectives. A detailed audit of the degraded non-forest land is necessary to classify them across their ownership, degree of degradation, suitability for any particular use, present use status and agro-climatic zones, before any policy prescriptions are suggested regarding their ultimate use.

A3. Such land is available in contiguous patches and hence is amenable to economies of scale

One wonders if such assertion is backed by any scientifically collected data set, even though one may get such impressions while visiting certain locations in Rajasthan and Gujarat.

A4. Degraded land of this nature does not support the livelihood needs of the poor and hence their conversion into other use will not exacerbate social tensions

Such a statement also demands a thorough scrutiny, given the dependence of some nomadic communities on such land for grazing their cattle.

We shall now take up the arguments put forward by SR in

advocating against leasing out of degraded forest land to the private sector.

B1. SR argues that, involvement of industry in conversion of degraded forest land will jeopardize the livelihood of 100 million people in the country and Joint Forest Management (JFM) has been successful enough to regenerate a considerable amount of such degraded forest land through involvement of the local people during the last decade.

The report seems to assume:

- Around 100 million people are depending on degraded forestland for their livelihood.
- The livelihood of these peoples will be jeopardized if degraded forestlands are leased out to private industries.
- JFM could manage the degraded forest land successfully.

However, SR has not clarified following points:

What is implied by 'dependence' on degraded forestland? Is it associated with primary stakeholder (people living in the domain of forest) or both primary and secondary stakeholders (people living far away from forests). The latter also have some dependence (tangible/intangible) on forest. On the other hand, the former is not a homogenous group—different sub-groups within it have different degrees of dependence on forestland. SR remained silent in pinpointing the groups and identifying degrees of their dependence on forests.

Degraded forestlands generates two types of benefits for the primary users—tangible benefit, viz. collection of timber, fuelwood and NTFPs and intangible benefits; viz. recreational and religious uses. We have defined *'dependence'* (of primary users) as percentage of total annual income generated from appropriation of tangible benefits from degraded forestland. From our study of three in Orissa and one village in Andhra Pradesh (AP) we found that on an average only around 2.48 per cent of their livelihood is dependent on tangible benefits from degraded forestland. Such generalization as has been made by SR is sometimes deceptive if the heterogeneity of the households in terms of their socio-economic status are not given due recognition. As is obvious, households are not homogenous. The situation becomes clear if we consider the variation across them in terms of their land holdings. From the following table (Table: 21.4) it can be seen that

Table: 21.4: Dependence on Degraded Forestland Across Economic Groups

Land Holding (in Hectare)	*% of Families*	*Dependence (= % of Total Income from Degraded Forestland)*		
		Maximum	*Minimum*	*Average*
Land less	41.25	35.56	0.00	4.31
0.01 to 2.00	46.25	4.37	0.00	1.28
2.01to 4.00	6.25	2.30	0.40	1.40
4.01 to 6.00	2.50	0.79	0.49	0.64
6.01 to 8.00	2.50	0.56	0.00	0.28
8.01 to 10.00	0.00	0.00	0.00	0.00
10.01 to 12.00	1.25	0.00	0.00	0.00
Total	100.00	35.56	0.00	2.48

1. Landless people depend most on degraded forestland.
2. Dependence on degraded forestland decreases with increase in landholding.
3. Among the landless, while some of the households (1.25%) are highly dependent (35.56 %) on degraded forestland some are not getting tangible benefit from degraded forestland at all.

However, such low level of dependence on degraded forest land need not necessarily be attributed to a low level of demand for resources from the forests. As we shall find later in detail, the *low dependence* phenomenon is influenced by supply bottlenecks. We found that:

I. Supply of tangible benefits from degraded forestland is very low compared to demand for tangible benefit by local community.
II. Marketing facilities for the products collected by them are conspicuous by absence and the consequent income they derive by selling them is very low.

So, they have two options under the present situation : either to tap other sources of supply of tangible forest products, thereby degrading some of the existing well-stocked forests or to increase their dependence on non-forest products. An alternative may be had in terms of increasing the supply—both in quantitative (if used for subsistence) and value (if sold in the market) terms—of tangible forest products from the existing degraded forests.

Choice between these options will depend on resource endowment status of the people. Our study found that most of these people are suffering from abject poverty. People who are living in these villages are suffering from lack of basic facilities like health, drinking water, education, and proper transportation facilities even after independence. They have some land but have no access to capital input as well as irrigation facilities necessary to cultivate their lands. Moreover, suitable channels for marketing their agricultural products are absent. However, villagers were given some vested land under Land Reform programme and in Orissa we found that most of these lands are too awfully degraded [some distributed plots are even situated inside canals and ponds!] to be put to any agricultural use. So it is clear that existing situation does not necessarily secure the livelihood of the people. Most of the privately owned degraded lands are either lying fallow (in Orissa) or are being converted into pulp wood plantations (in Andhra Pradesh). One reason behind the lands lying fallow may be due to the fact that net benefit (revenue-cost) from cultivating degraded land with food crops is negative in Orissa. Conversion of degraded and even some productive agricultural land in Andhra Pradesh, on the other hand, may be explained by the fact that pulp-wood plantation on own land yields the highest net benefit (Table 20.5). It may be noted that out of 40 sample households we

Table 20.5 : Net Benefits Per Hectare Per Household (Rupees per year) Orissa and Andhra Pradesh

State		*Orissa*	*Andhra Pradesh*	*Total*
Total land (Ha.)		138.57	197.72	336.29
Cultivation		-1027.22	2120.31(-558.78)	171.84 (-580.60)
Wage		59.28	0.00	59.28
Plantation	Own	4272.49	3139.37	3301.24
	Leased Out	1250.00	0.00	1250.00
	VSS	0.00	N.E	N.E
Kendu leaves		254.6	13.64	49.43
Fuelwood		2.01	0.00	2.01
Medicinal plants		6.57	0.00	6.57
Religious and cultural use		0.00	0.00	0.00
Cashew		1126.25	0.00	1126.25
Grazing		N.E	N.E	6.07

* N.E= Not Estimated

visited in Andhra Pradesh, 38 recorded negative net benefits from agricultural activities. As the estimated net benefit from two big cultivators are lumped with those from the rest 38 households, the resultant estimate turns positive. The figures in the parentheses (in respect of cultivation) are estimates with those two households being dropped. As we estimate the net benefit across the use and ownership of degraded land, the villagers have been found to reap the highest amount of net benefit when they planted pulp-wood on their own land followed by leased out land to paper mills for plantation of pulp wood.

The major source of earning for families in Orissa is through leasing out 84.92 ha. (51.09%) of their family degraded land for captive plantation as they have no capability to invest on their own in agriculture lands. The benefit from the leasing out land for captive plantation is higher (people are getting Rs. 1250 per hectare in a year if leases out for captive plantations) than the net benefit derived out of cultivation of other crops (due to low productivity and market failure). Moreover this activity (captive plantation) creates some employment opportunity to the villagers—6 per cent households have got employment in this activity to look after plantation for 12 years even though 12 per cent households in Orissa got employment during plantation period. The pattern of employment generation from degraded land across uses is given in Table 20.6. People appears to be more

Table 20.6 : Annual Average Mandays Generated from Degraded Land Across Uses and Ownership Categories (Orissa and AP)

Use/Ownership	*Annual Mandays Generated/Hectare*			
	Private	*State*	*Forest Department*	
			Degraded	*Wellstocked*
Cultivation	48	—	—	—
Plantation	25*	—	387	—
NTFP	12	18	6	—
Religious and Cultural Use	—	—	—	—
Cashew	13	—	—	—
Grazing	—	—	—	—

*The estimated employment generation from plantation refers to that generated in term of plantation activities during the first year. Generally, a plantation matures in a period of five-seven years, depending on location and species planted. Employment generated in subsequent years for maintenance and protection of the plantations and ultimate harvesting has not been taken into account. Employment estimates for plantation in degraded land under the management of FD refers to that generated by VSS.

interested in leasing out their degraded land to pulpwood growers as they are getting secured and stable return along with employment opportunities. In AP, people also are ready to convert their cultivable land to pulpwood plantation due to market failure in respect to channelizing their agricultural products. However, this option is open to only those who have some land.

The study found that above-mentioned factors are pushing most of the people towards the first option. As we have seen in Orissa dependence on well stocked forest is gradually increasing. People are using well-stocked forest for various purposes like collection of NTFPs, harvesting timber to construct their house, make ploughs and *Dhenki* and for religious purpose. So human pressure on well stocked forest is increasing. This may lead to increasing pressure on well stocked forest. To reduce the dependence of livelihood of the people on tangible goods it is necessary to increase their dependence on non-forest products through well-designed development programmes. However, so far the development activities have failed to reduce this dependence of the people on forest products.

The report also assumes to solve these dual problems of regreening and livelihood security of people through introduction of JFM. From our study of 19 JFM/Non-JFM (16 sites in northern part of West Bengal and 2 sites in Orissa and 1 site in AP) sites we can argue that, the performane of JFM is a site-specific phenomena. A large number of Forest Protection Committee (FPC) or Eco Development Committee (EDC) have stopped their activities (Poobong, Khuntimari, Jaldumur, Shalugara and Rava FPC, in West Bengal), whereas some FPCs and EDCs (Jayanti and Bhutia Basti EDC) are not functioning well (CREATE: 2002). The employment generating activities like plantation, thinning, felling have been stopped due to lack of financial support by the government. According to revised guidelines (2000) for JFM issued by Ministry of Environment and Forest working plans are getting priority to the JFM microplans (Sarin, 2003). In Orissa, people are not getting 50 per cent benefit from major/final harvest and 100 per cent of the intermediate produce which were committed in 1993 JFM resolution and even Government of Orissa is claiming 50 per cent benefit from community forest where Forest Department did contribute nothing towards generating and protecting the forests (Sarin, 2003). Even monopoly rights of collection and marketing of 29 Non-Timber Forest Products (NTFP) were leased to a private trader. Moreover, Acts involving more and more restriction on use of forest resources by the people are being introduced since 1970s, like Wild Life Protection Act (WPA), 1972, Forest Conservation Act, 1980 etc. Recent Supreme Court judgement of 1998 and 2000 practically banning all felling and collection of NTFPs in forest indicates the concern of the powers that be in protecting forest resources without taking the issue of the plight of the people who have been traditionally using them for livelihood into account. In India the ownership of forest largely vests on government. In the various government orders, we find that none of this order/ resolution is in favour of giving ownership to community over forest. Moreover, sustainability of JFM hinges upon an agreement among the powers that be in acceding to the long standing demand of the forest dependent people to be armed with legal backing to harvest and sell forest products and fulfilment of such demand will ensure a higher level of collective action from the communities (Chakrabarti *et al.*, 2003). So, success of JFM crucially depends on managing above-mentioned problems. And government has so far been not much successful in doing so. Introduction of JFM should not be panacea for all situations.

B2. Any act of leasing out degraded forest land to the paper industry will attract attention from other land-intensive industries like coffee, rubber etc. Even smaller paper units will also demand their share in the cake. Such large scale leasing will have ecological implications.

Such a statement clearly reveals the adhocism prevailing in crafting 'policy prescriptions'—a phenomenon that is being sustained since independence. As we have clearly mentioned in Section II, land use planning involves balancing the use of land across often competitive uses so as to ensure the maximum benefit to the entire society. A policy decision on the use of forest land cannot simply wish away the demand from other uses simply because their consideration will render the issue more complex. The SR was expected to take into account the demands from different sections into account and develop a comprehensive land use strategy for the country as a whole.

In the next section we shall dwell at length the apparent threat of 'ecological implications' and the truth that emerges out of ground realities.

From our study of three villages of Orissa and one village of Andhra Pradesh we found that 162 hectare of land was lying degraded under forest department as compared to around 50 hectare of degraded non-forest land under the ownership of state and private individuals. As per Table 20.1 above underutilized/degraded notified forest land constitute 4.44 per cent of the total geographical area of the country—a little more than 14 million hectares to be precise. The State of Forest Report, 1999 also reveals that area under degraded forests are increasing over the last decades, whereas that under close forest are declining.

So, what could be done with this huge amount of degraded forest land? SR seems to have found the answer in using JFM alone to regreen them. As we have argued earlier (B1) introduction of JFM can only ensure success under some specific conditions. One of these conditions is provision of finance for JFM. It has been learnt that funds available for forest improvement activities in Hill circle in West Bengal has come down to Rs. 6 crores in 2003-04 from Rs. 23 crores in 2002-03. As a result, area of plantation will come down to 100 hectare in 2003-04 from 550 hectare usually done annually. Given the sorry state of financial affairs prevailing in most of the states it is doubtful if their priorities will ever be directed towards regreening of the degraded forest land in stead of funding other activities.

So, is there any alternative way ? Will there be no ecological impact if the degraded forestland remains as it is? The decision situation before forest department can be depicted through following decision tree (Fig. 20.2). The decision tree shows that in existing situation best possible alternatives before forest department (FD) is introduction of JFM and/or inviting industry as FD seems to have failed to maintain the health of the forest. From the diagram it can be seen that involvement of industry in regreening of degraded forestland under some specific terms and conditions may not be differ much from JFM as far as type of plantation is concerned, though they may be different in distributional terms.

B3. The requirement of the paper industry may be met from out of the existing inventory of uncultivated wasteland. Expansion of farm forestry has ensured a steady supply of pulp wood, as is evidenced by the sharp fall in the price of eucalyptus prices. However, there is a locational problem. The pockets of expanding farm forestry are located far away from the region where most of the paper units were

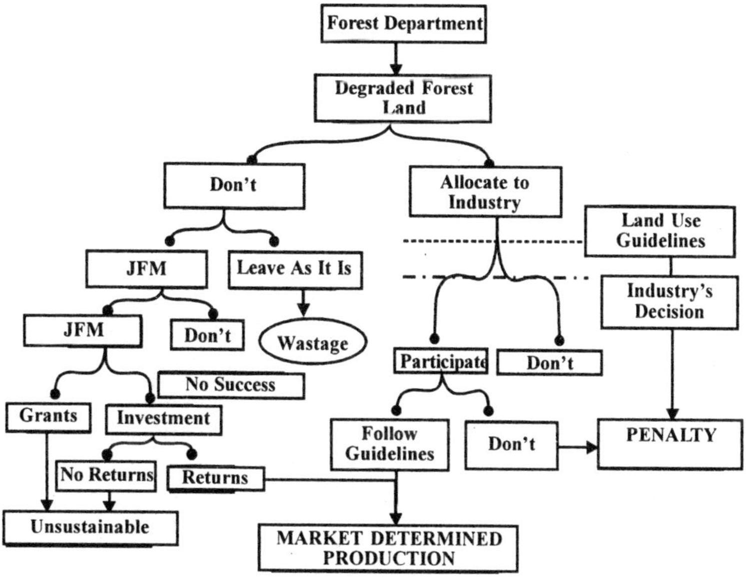

Fig. 20.2 : Degraded Forestland : The Decision Game

established. The processing units may be split, with the pulp-making units set-up around the farm forestry areas. The pulp produced may be transported to the paper-making units. Given the fact that pulps are imported from abroad, such dispensation will surely work out to be cost-effective.

The above arguments are apparently based on following assumptions:

1. The demand for pulpwood in domestic paper industry will follow the present trend in the long run.
2. With the growth of farm forestry, there is already a supply glut in the pulpwood market resulting a downward pressure on the prices pulpwood, viz., Eucalyptus. And entry of the big producers will further worsen the situation for lakhs of small farmers.
3. There is a locational problem. The places supplying pulpwood are mostly located far away from paper mills. This implies that transportation costs will be higher and thus the possibility of vertical disintegration of paper mill units across locations may be considered.

4. However, as paper industry depends on imported paper pulp such a disintegration will still be cost effective and hence viable.
5. Indian paper industry will produce at the scale it is operating now, implying the present scale of operation as optimum.
6. India never aspires to join the export market in pulp, paper and paper boards.

But how far are these assumptions are tenable? Is the paper industry following what is known in growth literature as 'golden path'? Available data on pulp production on India as well as World (from 1961-2001, *Source:* FAOSTAT) do not seem to support these. The pulp production in India has gone through significant structural changes during 1980s and 1990s. The report has not cared to develop any methodology of estimating domestic as well as global demand for pulp in any future date, keeping such structural changes in mind. Fig. 20.3 below locates the structural changes observed in the production of pulp in India.

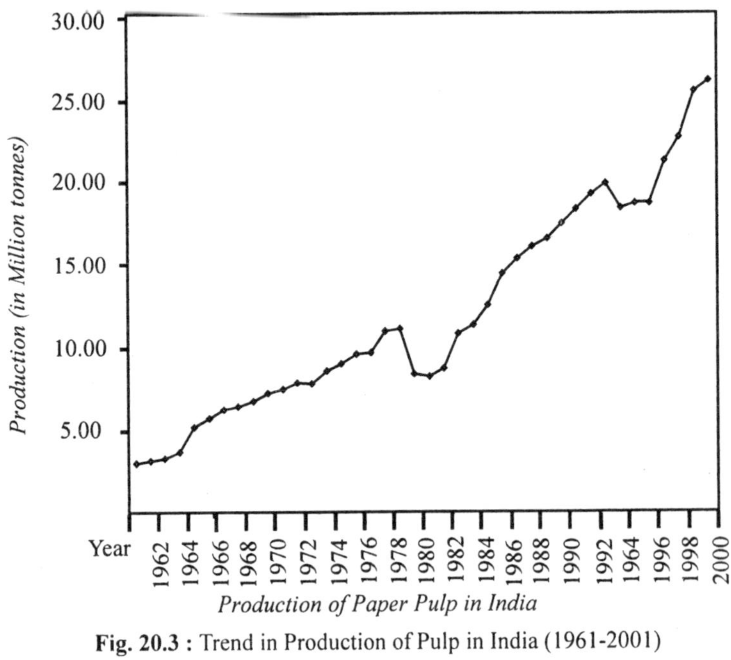

Fig. 20.3 : Trend in Production of Pulp in India (1961-2001)

Secondly, only around 5 per cent of domestic demand for paper pulp is met through import at present (2001). So comparison of cost effectiveness *vis-à-vis* imported pulp and thus justifying vertical disintegration of paper mills does not hold much ground.

Thirdly, the world export trade of pulp is now worth 1.64 billion dollar (in 2001) of which Indian share is just 0.07 per cent. India is not even a regular exporter of pulp rather the export pattern of pulp has periodic feature implying that production *plus* import of paper pulp is just sufficient to meet the existing scale of operation of paper mills. So a permanent supply glut in the market of pulpwood is hard to apprehend, if India is keen to participate vigorously in world export trade of paper pulp and paper and paper products in the coming future.

Fourthly, about the falling price of Eucalyptus, data collected from planters and paper mills do not lead to such definite conclusion though it appears that the supply price of Eucalyptus supplied by the Forest Department has been really falling in respect of a particular unit (see Box 20.1). The price at mill gate of eucalyptus supplied by the contractors appears to have not declined during the last three years. However, it is difficult to say anything about the trend in prices received by the farmers. Enquiries in the field revealed that generally, the farmers receive prices net of the transport costs, whereas the paper mills pay a fixed price, irrespective of the distance from which the products have been brought. Thus farmers located at a longer distance from the mills receive a lower price for their products. However, is the presence of intermediaries between paper mills and the farmers—a potential source of 'market failure' observed in the markets of agro-products—may be a cause of concern in the days to come.

BOX 20.1 : Is Price of Eucalyptus Really Falling ?

Year	*Price of Eucalyptus in Rs./Ton across paper mills*							
	Gogubaka VSS	*T*		*B*	*S*	*J*	*I*	
	Anticipated Net Revenue	*Govt.*	*Contractor*				*AST*	*BDT*
1999-00				1509				
2000-01		1825	1100					
2001-02		1825	1200		2444	1804	1525	2542
2002-03	2000	1500	1200				1539	2565
2007	2500							
2012	3000							

Identity of paper mill units have been suppressed for obvious reasons.

B4. Lakhs of farmers will be deprived of a market for agroforestry.

As we argued in the context of our discussion on B3, like all other agri-products, market for agro-forestry products is also characterized by the presence of 'traders' control'. Under the present disposition, it is very difficult to ensure genuine access of farmers in the market for agro-forestry products, even if paper mills are not allowed entry into production of pulp wood in degraded forest lands. Rather a strong coordination between the farmers (sellers) and the mills (buyers) may ensure mutual benefits to both the parties at exchange provided certain conditions are fulfilled. We shall elaborate those conditions later in the report.

B5. 'Millions of voiceless forest dwellers' will be denied access to the NTFPs and the other biomass that they gather'.

It appears from the statement above that 'millions of voiceless forest dwellers' have been provided access to the NTFPs and the other biomass they gather. It is time we realize that the access, if any, enjoyed by these millions to forest products are simply by default, as the forest department officials are not in a position to enforce the rights conferred on them by the Indian Forest Act, 1927 and subsequently strengthened with the enactment of Wild Life Protection Act and Forest Conservation Act. The recent Supreme Court judgements banning felling in reserve forests and collection of any materials from inside protected areas have already ensured that access to NTFPs and other biomass by the 'millions of voiceless forest dwellers' shrinks considerably.

B6. Their hardship will be more confounded in view of the phenomenon of 'market failures' existing in most of the farm products.

We have already mentioned in the context of our discussion on B4 that the phenomenon of 'market failure' is also visible in the context of agro-forestry products.

B7. Expansion of farm forestry is confined within the realm of non-cultivable agricultural waste land, it will never compromise the food security of the country as is often alleged.

The high degree of 'market failure' observed in most of the farm products (as has been observed by SR as well) has played a vital role in conversion of land devoted to farm products to agro-forestry. This phenomenon has been observed quite evidently in Andhra Pradesh, where farmers have shifted from production of green grams and maize to plantation of eucalyptus in anticipation of a larger return in the coming years.

B8. Paper industry will be unduly subsidized if they receive land almost free of capital cost.

SR should have worked out the cost that the paper industry should pay towards leasing in such land, instead of simply bypassing the issue. Such payments on the part of the paper industry could have been used for the development of the forest dwellers and reduce their dependence on forest products. Such a process would have improved the standard of living of the people for whom the report has shown a considerable amount of 'concern'.

B9. Leasing out of forest land to private enterprises is unconstitutional and illegal as the process involved contradicts with the Panchayat Act, Forest Conservation Act and the spirit of the present Forest Policy. Most of the forests being located in Schedule V areas, the constitutional safeguards guaranteed to the tribals will also be violated.

This is a source of real concern. Constitutional safeguard given to the tribals in the Schedule V areas is often considered as a means to protect them from potential threats of alienation from land rights. However, insights from our discussions with the villagers often tend to provide an opposite feeling. We have come across tribal households in Orissa, who are earning incomes low enough to be considered as lying below poverty line, in spite of owning 30 to 40 acres of land. Examples from Tamil Nadu suggest that many tribal families simply migrated to urban areas in search for non-farm employment abandoning their land in the villages. They could not sell their land before migrating in view of the existence of the 'constitutional safeguards'. Given the present day thrust on shift from subsistence farming to production for the market, agricultural activities have no longer remained intensive of land endowment alone. Endowments in other inputs, namely, water, chemical inputs, credit, market intelligence etc. play vital roles in determining the return to an agriculturist. Most among the tribals do not have any access to these complimentary endowments. A proper re-look into the so called 'constitutional safeguard' is necessary to find out if they have become counter-productive in securing the livelihoods of the marginalized tribal communities.

The other relevant acts that appear to be contradicting with such proposals of leasing out forest land also deserve a reconsideration in the light of the conceptual framework mentioned in Section II. We shall discuss the legal aspects in a separate section.

B10. The paper industry is looking for lands that have a minimum of 1metre soil depth. Such land will regenerate on their own without much costs. Hence, regeneration will be cheaper economic option than plantations.

This argument has attracted the attention of many academics, policy-makers as well as practicing foresters. Success of the Arabari model in naturally regenerating *Sal* plantations through coppicing without involving much costs is often cited in support of this statement. However, certain issues remain unanswered.

- Costless natural regeneration assumes total absence of human interference. Given the concern in the SR about the dependence of 100 million people on the forests in the country, how can one ensure complete prevention of human interference in such degraded forests? Even if such programme is designed that ensures total absence of human interference on such degraded forest land, what will be the extent of opportunity costs involved in terms of loss of livelihood that will accompany such a measure?
- Had it been possible to introduce costless natural regeneration in degraded forest land, why had such a vast quantity of forest land remained degraded today? Why do we have to seek loans from multilateral agencies and grants from donor countries to implement 'forestry projects' across the states?

Assessments grounded in reality reveal that 'Arabari model' of natural regeneration involving little expenses is too difficult to be replicated successfully all over the country.

B11. The industry has no credible plans to resolve the demands that the local communities have over such lands. Any leasing of forestlands is likely to result in hardships and oppression for the local communities who have historically depended on such lands for meeting their basic needs.

Indian industry has undoubtedly developed a negative image about itself over the last four-five decades among the citizens of the country. Common people consider them as 'exploiters' as they buy inputs from the society and 'deceivers' while they sell their products back to the people. Such an image has often been used to argue in favour of ensuring 'command and control' by the State under the implicit assumption that a democratic 'State' can neither 'exploit' nor 'deceive' her citizens. Experiences since independence, however, reveal that activities of the State have often been influenced by the interest of a selective group of people who in many situations even took up policies that jeopardized the interest of the common people. Painting such a demonic image of the functioning of private enterprises, these votaries of State dominance could hide their inefficiencies to a considerable extent.

However, the arguments offered above should not be considered as claiming the private sector to be innocent. It also colluded from time to time with the decision makers to further their interests. Private sector enterprises have developed 'industry associations' across sectors and/or products to press for their demands with the State. IPMA is one such example at the product level. Examples of CII or FICCI come to mind when we think of such associations at the national level. However, such associations have failed, if not miserably, to develop any internal 'code of conduct' that regulates the activities of their individual members. No effort has been made by them to ensure supply of 'standard' products at 'standard' prices under a 'standardized' input procurement policy. Thus even though a particular unit may be following some such norms, they were either not visible at the national level or such units were often forced to give up such policies in the face of undue advantages accruing to their competitors who did not care to stick to any standard. Standards, if any, were always expected to be coercively thrust on them by the State. Hence, the problem of negative image—a problem that industrial enterprises find hard to shrug off their shoulders.

What have been the experiences of degraded forest land being leased out? As per the existing norms such land can be leased out only to Forest Development Corporations or public sector enterprises producing paper and paper products. We cite a success story experienced by the Mysore Paper Mill, Bhadravati—a public sector enterprise that was given 30000 hectares of degraded forest land on lease by the Government of Karnataka (Appendix II).

It is revealed that

- Beginning with a situation having a yield of less than 5 tons per hectare in 1980, MPM has achieved yields which are considerably higher—Eucalyptus (Dry Zone): 25-40 MT/Ha, Acacia (Wet Zone) : 80-100 MT/Ha, Pines (Wet Zone): 90-100 MT/Ha
- On an average 45.5 MT of pulp wood were extracted per hectare and 97777 MT of raw materials have been supplied every year.
- The plantation and subsequent harvesting activities generated employment to the tune of 3.24 million mandays using 4000 hectares of land. On an activity wise break up, it is estimated that plantation activities generate 570 mandays per hectare and another 240 mandays per hectare are generated during extraction of harvest.
- Villagers receive firewood free of costs in the form of the lops, tops and barks on harvest.
- This endeavour has reduced the pressure on the existing natural forests.
- Encouraged by the results, the organization has extended their technology to farm forestry activities taken up by individual farmers and now contemplates to introduce a 'produce-sharing' contract with the farmers.

However, it is also true that some private sector enterprises have also developed the necessary technologies. We reproduce two reports one from the ITC Bhadrachalam unit (Appendix III) and the other from the experiences of BILT SEWA unit in Orissa (Appendix IV) that throw some light of the positive impacts of plantation projects taken up by private enterprises on the community. However, there are doubts if all the units under IPMA have developed such capabilities.

The process described above indicates:

- ➢ A typical private sector enterprise is also capable of taking care of the interests of the community if it has to interact with them directly.
- ➢ It has the capability to increase the productivity of degraded private land.

- Its programme generates considerable amount of employment for the members of the communities.
- It has the will to work hard to design such projects.
- The project is also simultaneously economically viable.

One should note the effort put in by the enterprise in increasing the productivity of eucalyptus plantations, which is well above the national average. Thus we may argue that it is not proper to accept the argument put forward by SR against the possible impact of leasing out forest lands to the private enterprises without considering the experiments carried out by the enterprises. More careful study of the experiments carried out all over the country is necessary before arriving at such 'blanket' conclusions. There should, however, be a note of caution. The success story, however, does not necessarily indicate that all the paper units have the necessary willingness or capability to take up such challenge. IPMA may have to take the necessary lead in sharing such experiences and capabilities among the other members and motivate them to innovate some such collaborative projects that also look into the interests of the communities who have been using such lands for their livelihood.

So far we have not considered the possible 'ecological impact' of conversion of degraded forest land—an issue raised a number of times by SR. Forests provide several services to humanity. Some of them are tangible and others are intangible. For example, we may consider the supply of timber and firewood by the forests. They are tangible benefits. Forests are also considered to be providing such services as stores of gene pool, soil conservation, water conservation, carbon sink etc. They are clubbed as intangible benefits as it is difficult to quantify such benefits in terms of their values. Such benefits also have 'public good' characteristics in the sense that their use can neither be divisible nor excludable. An example may clear the points. Forests help carbon sequestration. It absorbs carbon di-oxide from the atmosphere and releases oxygen. Such a service by the forest purifies the air around us. Thus forests are necessary for our survival even if we do not consume the other material products obtained from it. Such a service provided by the forest cannot be distributed across only those people who are ready to pay for it. Those who do not pay for such service also enjoy them. Thus it is very difficult to fix a price for such service and charge them from the users. On the other hand, timber collected from forest goes to one who pays a price for it and is consumed by her alone. Under such conditions the service provided by forest as a store of gene pool is also intangible. The same applies in the context of soil and water conservation services provided by forests. Thus it is very difficult to quantify the values of such services and hence renders the planning process very difficult. SR also suffers from such a problem. It cautions us about the possible danger of being deprived of such services as are obtained from the degraded forests right now, in case they are converted into plantations of pulpwood—so called 'ecological implications'. Section V takes up this concern for an in-depth scrutiny.

THE ECOLOGICAL IMPLICATIONS

Ecology means the 'natural relationship (or the study of the relationship) between plants, animals and people, and the places in which they live'. All the living population of a particular area cannot be functionally separated from the environment; rather both the community and the environment function together as a system what is known as ecological system or ecosystem (Whittaker, 1960; Odum, 1966, 1971). According to modern ecologists, an ecosystem 'is a functional unit where the biotic components of the environment interplay' (Mishra *et al.*, 1970).

Generally, from a structural standpoint, four essential components are found in an ecosystem. They are:

(i) abiotic components,
(ii) producer components,
(iii) consumer components, and
(iv) decomposers (Dutta, 1985, 445).

Any instability/change/breakdown in this structure will cause unstability to the total ecological set of up of a particular zone or region. Among the various important ecosystems, forest is considered a very important and complex one, where, well-ordered and lengthy ecological succession takes place (Tucker, 1999, 202). A variety of plants represent a natural forest. Forest positively controls the following attributes of an ecosystem (Dutta, 1985, 476; Verma, 1999, 192; Dury *et al.*, 2002, 3):

- Biodiversity of an area.
- Soil quality.
- Soil erosion.
- Water cycle.
- Climatic temperature.
- Carbon cycle.

The present section is an attempt towards understanding the 'ecological implications' of a possible conversion of degraded forest land into plantation of pulpwood. During the course of the study we collected detailed information about the floral inventory and soil characteristics across degraded land put under different uses and maintained under different ownership structures. Thus we came across degraded land owned by the State, Forest Department and individuals. On the other hand, while some of them are lying fallow, others are put under different kinds of vegetative growth.

As has been mentioned earlier, the present data covers those collected from two sites each in Orissa and Andhra Pradesh. Table 20.7 below provides insights into the issue of biodiversity in these

Table 20.7 : Plant Species Richness and Plant Species Diversity Index

Land Use Type	*State*	*Simpson's Index*			*Shannon Index*		
		Shrub	*Sapling*	*Tree*	*Shrub*	*Sapling*	*Tree*
Well stocked Forest	Orissa	00.00	15.00	04.02	00.00	2.70	01.98
	A.P.	Not studied			Not studied		
Degraded Forest	Orissa	00.00	08.00	00.00	00.00	01.82	00.00
	A.P.	00.00	09.00	00.00	00.00	02.15	01.38
Community Forest	Orissa	00.00	03.52	01.00	00.00	01.78	00.00
	A.P.	Not Applicable			Not Applicable		
Captive Plantation	Orissa	00.00	01.95	01.06	00.00	00.72	00.01
	A.P.	Not Applicable			Not Applicable		
VSS Plantation	Orissa	Not Applicable			Not Applicable		
	A.P.	00.00	02.05	01.01	00.00	00.72	00.01
Individual Plantation	Orissa	00.00	01.21	00.00	00.00	00.19	00.00
	A.P.	00.00	01.00	00.00	00.00	00.00	00.00
Individual Degraded Land	Orissa	00.00	00.00	00.00	00.00	00.00	00.00
	A.P.	00.00	03.96	00.00	00.69	01.37	00.00

regions. We estimated two well-known indices often used to quantify the extent of biodiversity in a particular location. They are: Simpson's Specie Richness Index and Shannon's Specie Diversity Index. Such indices were estimated separately for trees, shrubs and saplings. The answer to the question as to whether particular specie is facing a threat of extinction or not can be captured through estimation of number of singletons and doubletons. A particular specie is considered a singleton if it is found only once in the sample plots. A doubleton is observed two times. Obviously, high numbers of singletons and doubletons would imply a higher number of species being threatened to extinction. Tables 20.8 and 20.9 provide the relevant information on these parameters.

Table 20.8 suggests

- The degraded forests have significantly high biodiversity in respect of saplings. However, the biodiversity indices are insignificant in respect of trees and shrubs.
- Well stocked forests in Orissa recorded the highest biodiversity, both in terms of trees and saplings, out of the different categories of land surveyed—an observation which is quite obvious. It should be mentioned that we did not cover any well stocked forest from Andhra Pradesh in our study and hence any comparison across the states are not possible.
- The community forest studied in Orissa, being a patch of coppiced *Sal* recorded unit value of Simpson's index. It also suggests existence of moderate biodiversity in respect of saplings.
- Interestingly, individual degraded land in Andhra Pradesh also recorded moderate biodiversity in saplings.

To summarize we may argue that tree covers are present only in well stocked forests, community forests, and plantations, both, captive (on land leased out to paper mills) and individual plots. The land under other uses—degraded forests in particular—do not have any tree cover.

In terms of singleton and doubleton species, well-stocked forest provides shelter to highest number of species falling under these categories in Orissa (singleton: H-13, Sap-6, T-8 and doubleton: H-4, Sap-4, T-2) followed by community forest (singleton: H-11, Sap-6, T-1 and doubleton: H-4, Sap-2, T-0). Captive plantations (singleton: H-6, Sap-1, T-0 and doubleton: H-2, Sap-0, T-1) and individual degraded land (singleton: H-5, Sap-0, T-0 and doubleton: H-4, Sap-0, T-0) is trailing behind on this count in Orissa. However, no singleton or doubleton species was recorded from degraded forest land.

In Andhra Pradesh, degraded forestland is providing shelter to maximum number of singleton and doubleton species (singleton: H-9, Sap-4, T-4 and doubleton: H-5, Sap-1, T-0) followed by the individual degraded land (singleton: H-11, Sh-2, Sap-2, T-0 and doubleton: H-6, Sh-0, Sap-0, T-0) and individual plantation (singleton: H-12, Sap-0, T-0 and doubleton: H-3, Sap-0, T-0). VSS plantation has the lowest number of singleton and doubleton (singleton: H-7, Sap-0, T-1 and doubleton: H-2, Sap-0, T-0). However, the condition is site specific. In Andhra Pradesh the process of degradation of the forest is still on and consequently we get some more variation than the other types of lands. We may conclude that any type of conversion of degraded forestland has some potential ecological implications in terms of loss of plant diversity.

Master species list of flora recorded across degraded land put under different uses have been provided in Appendix V.

We shall now look into the estimates of the bio-mass of the standing flora across different uses of land. Table 20.5 throws light on the situation prevailing in Orissa while Table 20.6 explains the situation prevailing in Andhra Pradesh. This estimates provide us with some idea about the extent of tangible services provided. It observed that in Orissa

- Around 108 cubic metres of biomass may be derived from the standing crops per hectare of the forest managed by the communities.
- A little over 14 cubic metres can be had per hectare of degraded forest are now under JFM.
- Well-stocked forests recorded around 466 cubic metres of biomass per hectare. These forest are also managed under JFM.
- Captive plantations on land leased out to paper mill unit have 453 cubic metres of standing biomass per hectare.
- Plantation of pulpwood species by individuals on their private holdings recorded around 82 cubic metres of standing biomass per hectare.

Turning on to Andhra Pradesh, we observe that

- The natural forests under the VSS recorded a standing biomass of 73 cubic metres per hectare.
- Degraded forests handed over to the VSS and converted to

Table 20.8 : Plant Diversity and Biomass in Orissa

Orissa	*Plants Type*	*Species richness*	*Density/ha*	*Mean DBH (metre)*	*Mean Height (metre)*	*Biomass/ha. (m^3)*	*Singletons s*	*Doubletons*
Hundivuta (Community)	Herb	27	529000	0	0	0.00	11	4
	Sapling	15	3800	0.044	2.43	56.13	6	2
	Shrub	1	40	0.029	1.20	0.13	1	0
	Tree	1	40	0.1899	11.47	51.95	0	0
JFM (Degraded)	Herb	37	48080			0.00	0	0
	Sapling	5	2400	0.0330	1.75	14.36	0	0
	Shrub	0	0	0	0	0.00	0	0
	Tree	0	0	0	0	0.00	0	0
JFM (Well stocked)	Herb	37	337920	0	0	0.00	13	4
	Sapling	12	2000	0.0478	2.48	35.59	6	4
	Shrub	0	0	0	0	0.00	0	0
	Tree	23	510	0.1566	10.95	430.03	8	2
Individual Plantation	Herb	33	122400	0	0	0.00	5	1
	Sapling	2	1828	0.0609	3.83	81.53	0	0
	Shrub	0	0	0	0	0.00	0	0
	Tree	0	0	0	0	0.00	0	0
Captive Plantation	Herb	30	857000	0	0	0.00	6	2
	Sapling	2	1480	0.1067	7.50	396.81	1	0
	Shrub	0	0	0	0	0.00	0	0
	Tree	2	82	0.1254	13.89	56.24	0	1
Individual Degraded Land	Herb	28	764000	0	0	0.00	5	4
	Sapling	0	0	0	0	0.00	0	0
	Shrub	0	0	0	0	0.00	0	0
	Tree	0	0	0	0	0.00	0	0

Table 20.9 : Plant Diversity and Biomass in Orissa

Andhra Pradesh	*Plants Type*	*Species richness*	*Density/ha*	*Mean DBH (metre)*	*Mean Height (metre)*	*Biomass/ha. (m^3)*	*Singletons*	*Doubletons*
VSS	Herb	39	690000	0	0	0.00	9	5
	Sapling	11	1873	0.0520	2.78	44.21	4	1
	Shrub	0	0	0	0	0.00	0	0
	Tree	4	16	0.2445	9.50	28.53	4	0
VSS Plantation	Herb	22	235000	0	0	0.00	7	2
	Sapling	2	1560	0.0696	7.67	182.00	0	0
	Shrub	0	0	0	0	0.00	0	0
	Tree	2	426	0.1263	17.28	368.71	1	0
Individual Plantation	Herb	30	495000	0	0	0.00	12	3
	Sapling	1	2000	0.0753	8.55	304.45	0	0
	Shrub	0	0	0	0	0.00	0	0
	Tree	0	0	0	0	0.00	0	0
Individual Degraded Land	Herb	34	689000	0	0	0.00	11	6
	Sapling	5	2000	0.0805	2.08	84.65	2	0
	Shrub	2	200	0.0305	1.25	0.730	2	0
	Tree	0	0	0	0	0.00	0	0

plantation of pulpwood species have an estimated standing biomass of more than 551 cubic metres per hectare.

- Individual degraded land records a corresponding figure of 85 cubic metres.
- Individual plantations of pulpwood species recorded around 305 cubic metres of standing biomass per hectare.

We may summarise the findings in arguing that plantation of pulp wood species has the potential to generate a higher amount of biomass and consequently a higher amount of tangible benefits compared to other uses. It should, however, be noted that the estimates have not been normalized to take care of the differences in the age of different plantations.

From the study of soil, it is observed that well stocked forest now under JFM in Orissa and VSS plantation in Andhra Pradesh is apparently good for growing different crops like paddy, cotton, vegetables, etc according to their pH characteristics (see crop pH Table 20.10). But other types of land can grow lesser number of crop

Table 20.10 : Approximate Soil pH Ranges Suitable for Various Plant Species

pH 4.8-5.2	*pH 5.8-6.2*		*pH 6.3-6.7*[1]
Azaleas	Apples	Peaches	Alfalfa
Blackberries	Beans, lima	Peanuts	Asparagus
Blueberries	Beans, snap	Peppers, pimiento	Cabbage
Grass, carpet	Beans, velvet	Pine, yellow	Carrots
Grass, centipede	Cantaloupes	Radishes	Ladino-grass
Hydrangia, blue	Corn	Rice	Lettuce
Irish potatoes	Cotton	Small grain	Onions
Juniper, Irish	Cowpeas	Sorghum	Peas
Pine, longleaf	Crimson clover	Soya beans	Red clover
Watermelon (<5.8)	Cucumber	Squash	Spinach
	Grass, most kinds	Strawberries	Sweet clover
	Greens, mustard, etc.	Sudan grass	Timothy
	Iris, blueflag	Sweet potatoes	White clover
	Kale	Tobacco	
	Lespedeza	Tomatoes	
	Parsnips	Trefoil, birdsfoot	
	Red top	Vetch	

1: On highly organic soils, it is not advisable to lime to a high pH. Good plant growth on these soils may be obtained where the pH is around 5.5. Approximately pH ranges refer to those suitable for mineral soils in humid regions where lime is often added. Many of these species, for example alfalfa, apples, iris, cotton, rice, sorghum, sweet clover and timothy, grow well on calcareous soils with pH values up to 7.5 to 8.0.

Source: Jones, Ulysses. S., 1987; *Fertilizers and Soil Fertility* (2nd Edition); pp. 229.

varieties. The humus layer is better in well-stocked JFM forest followed by community forest and captive plantation in Orissa. It is the lowest in degraded lands owned by individuals in Andhra Pradesh (see Table 20.11 and Chart 20.1). The top soil cover is better in Individual (AP) plantation followed by captive plantation (Orissa) and well stocked forest (Orissa).

Table 20.11 : Depth of Humus Cover and Top Soil Across Land Types

Land Type	*Humus (cm.)*	*Top Soil (cm.)*
Well Stocked Forest (Orissa)	0.98	6.55
Degraded Forest (Orissa)	0.40	1.8
Community Forest (Orissa)	0.95	5.7
Captive Plantation (Orissa)	0.75	8.18
Individual Plantation (Orissa)	0.45	5.9
Individual Degraded Land (Orissa)	0.05	2.6
Individual Degraded Land (AP)	0.55	8.5
VSS Degraded Forest (AP)	0.45	1.8
Individual Plantation (AP)	0.4	11.65
VSS Plantation (AP)	0.67	7.6

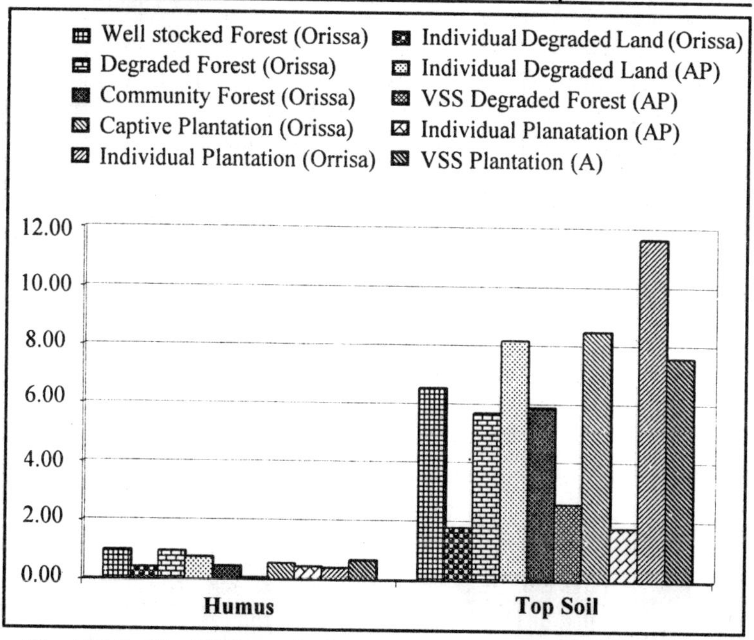

Chart 20.1 : The Difference in the Presence of Humus Layer and Top Soil across the study sites

As one looks into the water conservation capacity it is found that in Orissa it is more in the top soil of captive plantation followed by well stocked forest and community forest (see Table 20.12 and Table 20.13). However, B layer is better in terms of retention of water in degraded forest followed by well stocked and community forest. In Andhra Pradesh top soil of individual plantation and B layer of VSS have greater water retention capacities than others.

Charts 20.2 through 20.13 represent the variation in the amount of organic carbon trapped in the ground across different uses. From these analysis it is clear that the well-stocked forest in Orissa is capturing more carbon in top and sub-soil, followed by captive plantation and community forest. The capacity to trap carbon in soil is the worst in degraded forestland. In captive and individual plantation it is different across the species and age of plantations. We have calculated the carbon content of the trees and interestingly it has been detected that 7 years old *Acacia auriculoformis* is containing more carbon (0.75%) than *Eucalyptus* sp. (0.40%) of the same age.

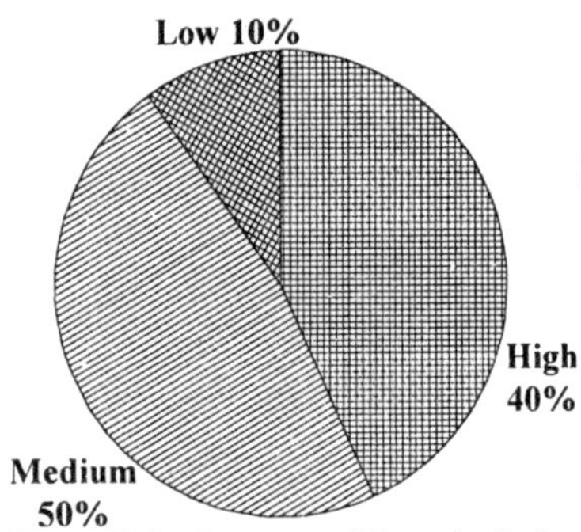

Chart 20.2 : Presence of Organic Carbon in Top Soil: Well Stocked Forest

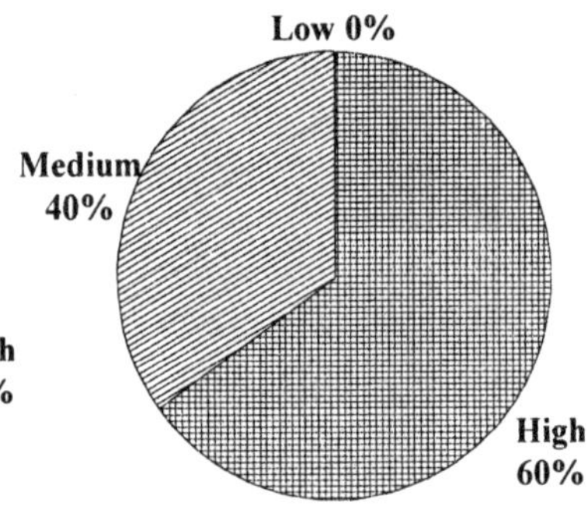

Chart 20.3 : Presence of Organic Carbon in Sub Soil : Well Stocked Forest

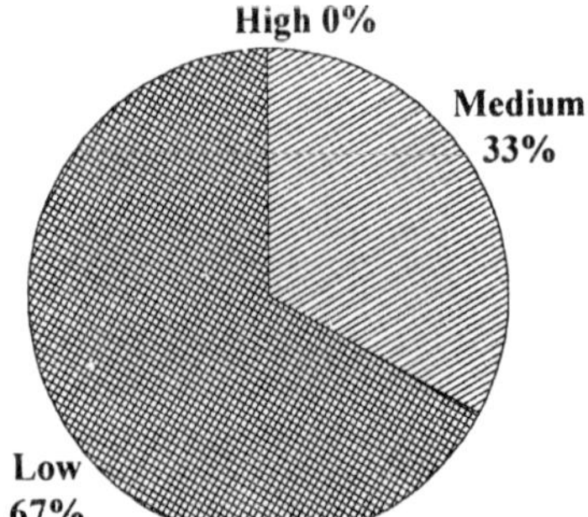

Chart 20.4 : Presence of Organic Carbon in Top Soil: Degraded Forest

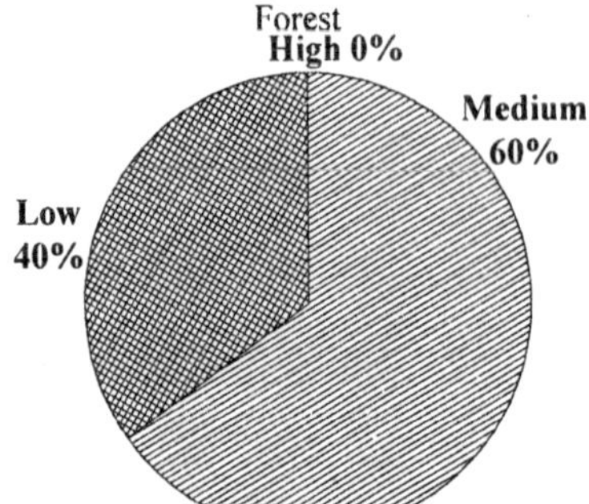

Chart 20.5 : Presence of Organic Carbon in Sub Soil : Degraded Forest

Table 20.12 : Chemical Properties of Soil at Orissa Site

Soil content	*JFM Well Stocked Forest*		*JFM Degraded Forest*		*Community Forest*		*Individual Plantation*		*Individual Degraded Land*		*Captive Plantation*	
	A horizon	*B horizon*	*A horizon*	*B horizon*	*A horizon*	*B horizon*	*A horizon*	*B horizon*	*A horizon*	*B horizon*	*A horizon*	*B horizon*
Average soil moisture (gm)	0.47	0.87	0.38	0.70	0.44	0.87	0.48	0.73	0.38	0.68	0.52	0.95
Soil pH	5.4-7.9	7.0-7.9	5.2-5.5	5.4-5.7	5.9-6.4	5.9-6.4	5.3-6.0	5.2-6.0	5.3-6.1	5.3-6.1	4.0-6.0	4.4-6.3
ECE	0.10-1.10	0.07-0.16	0.07-0.09	0.08-0.09	0.07-0.09	0.06-0.07	0.11-1.12	0.07-0.12	0.12-0.26	0.12-0.16	0.06-0.13	0.08-1.09
Zn	0.68-1.63	0.57-1.71	0.71-1.18	0.27-1.35	0.23-0.69	0.05-0.51	0.68-1.12	0.50-1.02	0.78-1.58	0.68 0.93	0.27-0.77	0.23-1.09
Fe	4.90-36.80	8.12-23.30	3.31-13.69	14.14-21.52	4.71-19.26	4.50-9.15	2.69-13.59	3.13-10.88	5.74-13.39	4.67-14.20	3.87-10.13	3.30-9.91
Mn	15.35-39.48	16.27 72.18	20.10-36.34	37.7-51.84	16.59-52.66	11.27-32.44	15.93-34.64	12.84-26.16	16.16-34.64	14.40-36.50	14.58-38.38	9.76-33.58
Cu	0.90-3.15	2.06-6.83	1.44-2.39	0.00-3.03	0.00	0.00	0.75-2.17	0.79-2.60	1.38-2.64	1.11-2.52	0.88-1.35	1.07-4.46
OC	Low-High	Medium High	Low-Medium	Low-Medium	Low-Medium	Low-Medium	Low-Medium	Low-Medium	Low-High	Low-Medium	Low-Medium	Low-High
P	2.00-12.00	2.00-12.00	2.00 5.00	2.00-3.00	3.00-9.00	2.00-8.00	5.00-12.00	6.00-20.00	8.00-12.00	5.00-9.00	3.00-12.00	3.00-15.00
K	405.00->500.00	85.00->500.00	195.00-305.00	145.00-235.00	185.00-495.00	45.00-405.00	45.00-290.00	60.00-290.00	140.00->500.00	155.00-380.00	30.00-280.00	45.00-185.00

Table 20.13 : Chemical Properties of Soil at Andhra Pradesh Site

	VSS		*Individual Plantation*		*Individual Degraded Land*		*VSS Plantation*	
	A horizon	*B horizon*	*A horizon*	*B horizon*	*A horizon*	*B horizon*	*A horizon*	*B horizon*
Average Soil Moisture (gm)	1.16	4.26	2.83	2.61	1.44	2.75	1.43	2.69
Soil pH	4.7-6.0	5.6-7.1	5.5-6.5	5.6-6.5	4.8-6.4	4.8-6.5	5.5-6.9	5.4-6.9
ECE	0.08-0.37	0.08-0.41	0.76-0.18	0.11-0.19	0.06-0.22	0.02-0.35	0.04-0.36	0.03-0.08
Zn	0.24-3.81	0.44-1.83	0.44-5.89	2.52-3.56	0.29-2.57	1.43-2.42	1.78-3.16	1.48-2.62
Fe	54.12-237.80	59.92-164.38	41.12-769.40	62.74-85.92	28.26-228.60	18.94-55.82	32.50-168.76	24.44-313.00
Mn	17.82-197.92	42.56-141.20	6.92-168.54	90.96-147.28	18.72-150.64	5.55-30.78	11.16-34.40	3.66-14.39
Cu	0.21-4.83	0.61-5.13	0.40-5.17	4.44-8.21	0.27-5.10	3.09-3.48	2.50-5.93	2.29-4.16
OC	Not Analysed in this site							
P	27.10-67.78	27.10-101.68	61.00-176.26	122.02-169.48	115.24-230.40	230.40-264.40	264.40-305.00	271.00-359.20
K	8.18-84.54	9.67-35.86	1.57-53.14	40.80-70.56	21.20-96.72	41.04-62.92	40.90-90.80	36.20-59.30

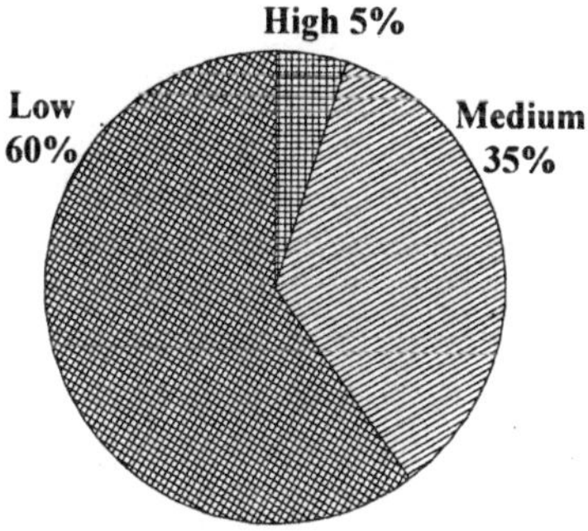

Chart 20.6 : Presence of Organic Carbon in Sub Soil: Captive Plantation

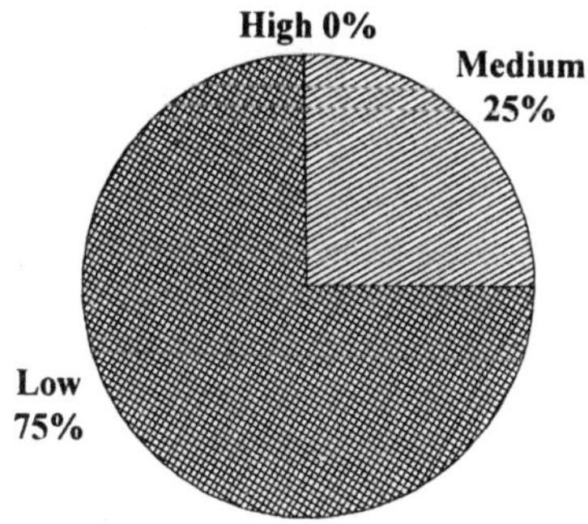

Chart 20.7 : Presence of Organic Carbon in Top Soil: Captive Plantation

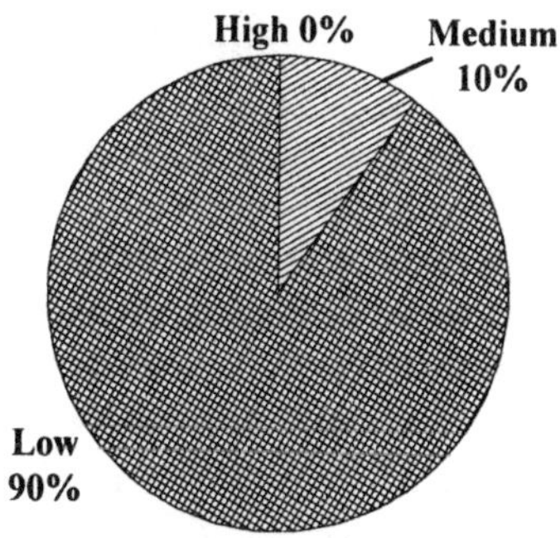

Chart 20.8 : Presence of Organic Carbon in Sub Soil: Community Forest

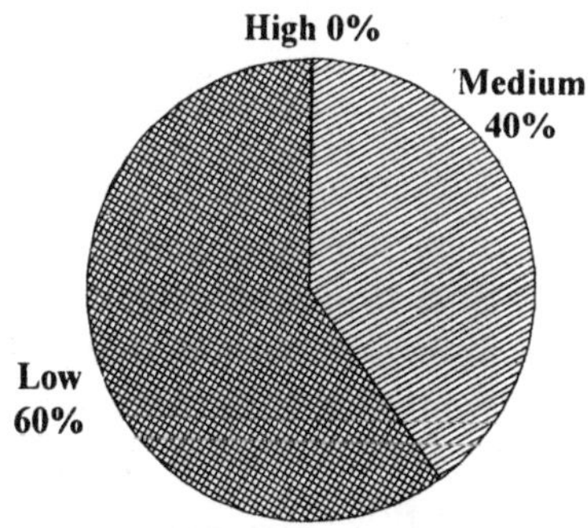

Chart 20.9 : Presence of Organic Carbon in Top Soil: Community Forest

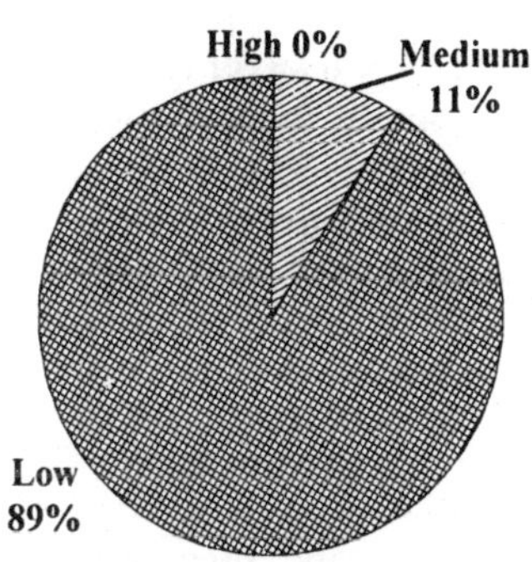

Chart 20.10 : Presence of Organic Carbon in Sub Soil: Individual Plantation

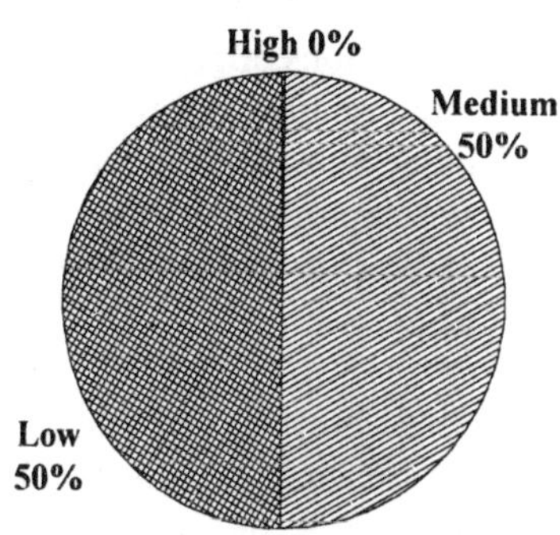

Chart 20.11 : Presence of Organic Carbon in Top Soil: Individual Plantation

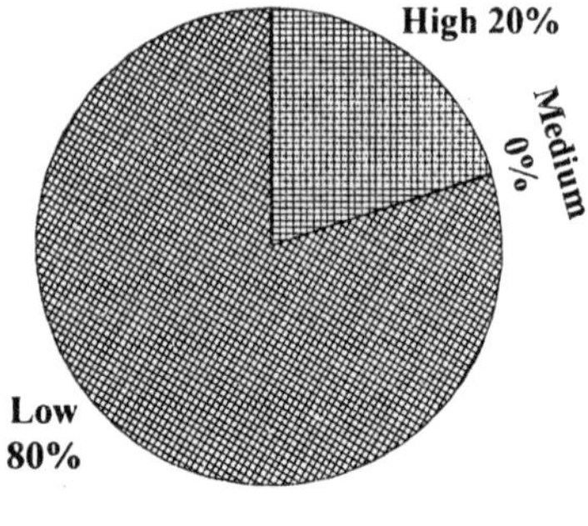

Chart 20.12 : Presence of Organic Carbon in Sub Soil: Individual Degraded Land

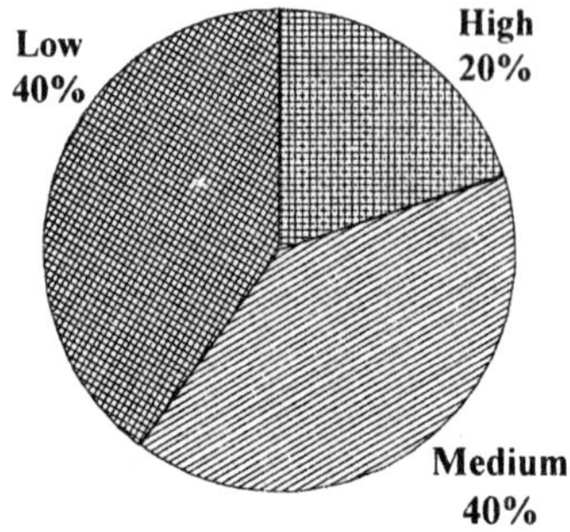

Chart 20.13 : Presence of Organic Carbon in Top Soil: Individual Degraded Land

To summarise the findings from the quantitative data we may argue that

- The conversion of degraded forests into plantation of pulp wood species will definitely reduce biodiversity in the plots.
- Both humus and top soil layers have been found to be on the lower side in the degraded forests, compared to the corresponding figures for plantations. However, well stocked state forests and community forests (both in Orissa and AP) recorded a better situation.
- Water retention capacity is found to be more or less identical across well-stocked forests, community forests and captive plantations.
- Plantations of pulp wood are found to have a larger capacity to trap organic carbon to the ground, compared to the natural and older forests. This finding is supported by existing literature. We reproduce a diagram (Fig. 20.4) which shows that the carbon sequestration property of a plantation is dependent on its age.

It has been argued that the rate of carbon uptake in the above ground biomass is initially strong when the trees are young, i.e. under about 20 years of age, reaches a maximum and then weakens (National Emissions Trading-Crediting the Carbon-discussion paper-3, 1999, p. 15). The newly growing plants have more efficacies to work as a good carbon sink.

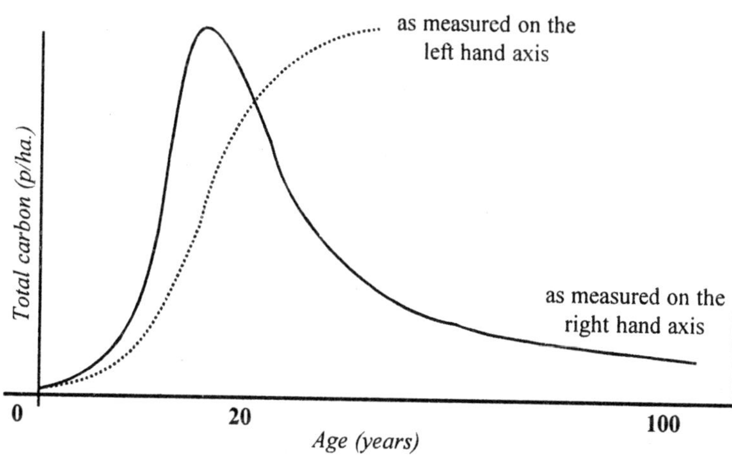

Fig. 20.4 : Age—Carbon Sequestration Relationship

Source: Greenhouse Challenges Sinks Workbook (1998).

So, protecting existing forest and growing new ones will help limit greenhouse emissions and at the same time will help to conserve the biodiversity. According to the prevailing Australian policy existing forest can be used as carbon 'offsets', as well as new carbon sinks should be generated through commercial and farm forestry (Dury *et al.*, 2002). Beneficial action can be taken to include:

- Protecting existing native trees and shrubs from loss or damage by fire, land clearing or animals.
- Encouraging regeneration of native trees and shrubs, for example by fencing stock out of some areas.
- Establishing permanent plantings of native trees, for example as farm windbreaks, shelterbelts.
- Establishing commercial forests and farm forests for further harvest.

Experiences from paper manufacturing units elaborated in Appendices II through IV also highlight the positive impact of pulpwood plantations on faunal biodiversity (Mysore Paper Mills) carbon sequestration (ITC and BILT) and water conservation.

Table 20.14 summarizes the possible impact of conversion on several ecological parameters we have to be concerned with in the future.

Table 20.14: Possible Ecological Impact of Conversion of Degraded Forestland

Types of Benefits		*Possible Sign and Strength of Impact*	
		Degraded Forest	*Plantation*
Tangible Benefits			
Timber	Soft	+	+++
	Hard	+	+++
Bamboo		-	+
Fuel wood		+	+
Fodder		+++	+
Other NTFPs		++	-
Intangible Benefits			
Carbon Sequestration		+	+++
Soil Conservation		+	+
Water Conservation		+	+
Bio-Diversity		++	+

The table clearly reveals that there is a trade-off across the impacts. The decision to convert degraded forests into pulp wood plantations will be contingent upon the values we impute to the different 'ecological' services rendered by forests. If we value biodiversity more than the requirement of increased biomass or more efficient carbon sequestration capabilities, degraded forests should be left untouched to facilitate natural regeneration, or planted through JFM with multiple species. Else we may give a nod to their conversion through plantation of pulp wood species. It should be noted that degraded forest land even under JFM are being converted into mono-culture eucalyptus plantations in Andhra Pradesh.

In the following sections we take up the different parameters that should be considered while making such value judgments.

THE FUTURE AHEAD

India signed three important international documents during the last decade. They are:

1. The Convention on Biodiversity of 1992,
2. The WTO agreement of 1994, and
3. The Kyoto Protocol of 1997.

Ratification of these documents implies certain obligatory functions on the part of the country. Without going into the details of each of the agreements we may note that the country is now obliged to

- Reduce its prospecting of resources from the forests in order to guarantee a larger importance to biodiversity conservation;
- Reduce protection to industries in terms of either heavy tariff or quota on importables or subsidies on exportables; and
- Reduce the level of green house gas emission within a specified period of time.

All these requirements have a significant impact on the future of the paper industry of the country. The reasons are quite obvious. Steady supply of raw materials, namely, bamboo and other soft-wood from the state-owned forests, very often at highly subsidized rates, is ruled out in the days to come. Thus the players in the industry have to procure raw materials from sources beyond the forests. Unless a proper strategy is in place, this may increase the cost of raw materials and the subsequent cost of production of pulp and paper. An increased cost of production and the subsequent increase in prices at which the products are sold may turn the Indian units unviable on the face of liberalised world trade scenario. Further, the existing units also find it difficult to compete with potential imports of paper and paper products from other countries if they do not increase the scale of their operations. As we shall elaborate later, the units in the paper industry in countries that are potential exporters in the world market are operating at scales which are more than ten times the scale of operation prevalent among the largest units in the country. The demand for raw materials will see a quantum jump if the scale of operation is to be increased accordingly. What's the possible way out in view of the weaknesses in terms of availability of raw materials facing the paper industry in the country and the threats in terms of meeting the obligations of WTO agreements? The opportunities probably lie in harnessing the provisions of the Kyoto protocol, provided some changes are bargained in favour of countries like India in exchange of the constraints imposed through WTO. The Kyoto Protocol represents the culmination of negotiations begun more than a decade ago to address the threat posed by the build-up of green house gases in the earth's atmosphere linked to human activity. Obligations under the protocol will become binding after it has been ratified by at least 55 countries, including industrialised countries that account for at least 55 per cent of the total carbon dioxide emissions. As of today (24th February, 2003), 105 countries have ratified the treaty, making up 43.9 per cent of the Annex I[6] emissions. India has deposited the instrument of accession on 26th August, 2002 although it belongs to neither of the Annexes. The protocol allows parties included in Annex B to participate in emissions trading for the purposes of fulfilling their commitments under Article 3 of the Protocol. Article 6 states that "for the purpose of meeting its commitments under Article 3, any party included in Annex I may transfer to, or acquire from, any other such Party emission reduction units resulting from projects aimed at reducing anthropogenic emissions by sources or enhancing anthropogenic removals by sinks of greenhouse gases in any sector……". "The net changes in greenhouse gas emissions from sources and removal by sinks resulting from direct human-induced land use change and forestry activities, limited to afforestation, reforestation and deforestation since 1990, measured as verifiable changes in stocks in each commitment period shall be used to meet the commitments…." (Article 3 paragraph 3). Even though

(i) a detailed mechanism for functioning of the emissions trade market is yet to emerge and

(ii) the role of the non-Annex I countries in such a market is till to be identified,

---

6. The Kyoto Protocol commits Annex I parties to individual, legally binding targets to limit or reduce their green house gas emissions, adding up to a total cut of at least 5% from 1990 levels in the commitment period 2008-2012. The Annex covers emissions of six main greenhouse gases, namely, CO2, CH4, N2O, Hydroflurocarbons (CFCs), Perflurocarbons (PFCs) and SF6. The individuals targets for Annex I parties are listed in Annex B of the Protocol.

such potential opportunities may be kept in mind while we plan to tackle the possible threats to our national interest through appropriate changes in domestic policies. It may be noted that the European Union Common Agricultural Policy framed in 1992 took care of almost all contingencies that were to be faced later on signing the Agreement on Agriculture under WTO regime in 1994. Chart 20.14 below provides a schematic structure of a SWOT (Strength-Weakness-Opportunities and Threat) Analysis of the paperindustry in the light of the three international protocols India already signed.

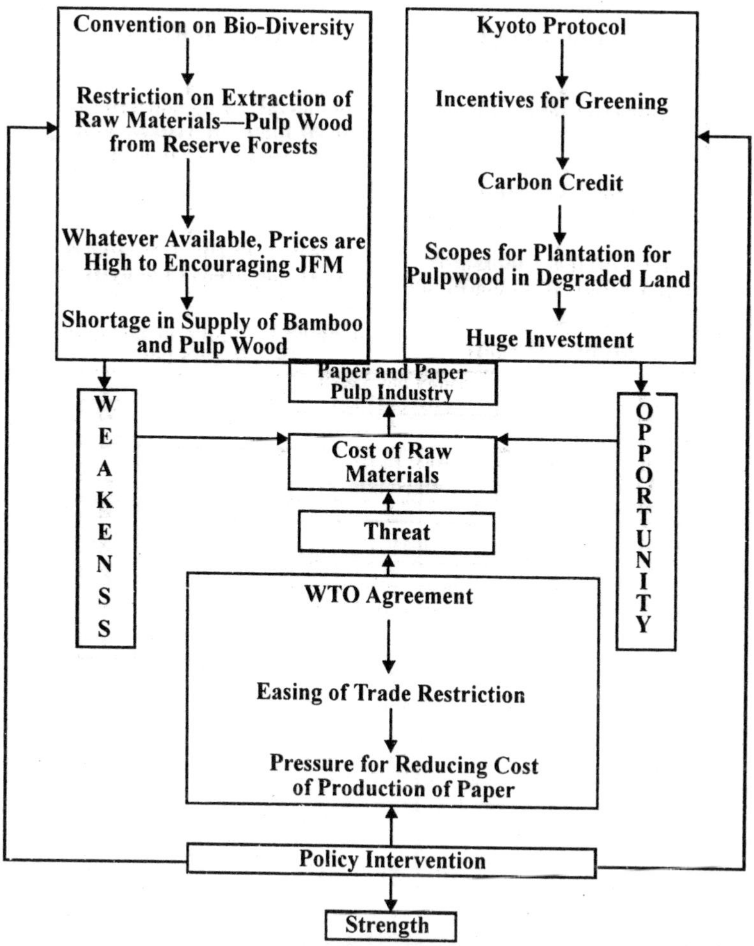

Chart 20.14: Flow Diagram of the Present State of Paper Industry in India

The paper industry in India has a long history. The first mill was commissioned in 1832. In 1950-51 there were 17 paper mills with an installed capacity of 0.137 million tons. The number of mills increased to 406 with the installed capacity increasing to 6.2 million tons. It appears that even though the number of mills increased manifold during the post-independent period, the average installed capacity per mill has not yet doubled during this period. A state-wise breakup of the paper mills is given below (Table 20.15). The states are arranged in descending order in terms of average capacity of the existing paper mills. The stylised feature that comes out of the table is that most of the units operating at higher scale are located in the four states of Orissa, Karnataka, Tamil Nadu and Andhra Pradesh, even though Maharashtra, Gujarat and Uttar Pradesh record bigger values in respect of both number of units and total installed capacity.

Table 20.15 : Average Capacity of Indian Paper Industry : A State-wise Breakup

State	*Number of mills*	*Installed Capacity (TPA)*	*Average Capacity (TPA)*
Assam	2	220500	110250.0
Kerala	5	215600	43120.00
Nagaland	1	33000	33000.00
Orissa	9	270850	30094.44
Karnataka	14	345000	24642.86
Tamil Nadu	31	639250	20620.97
Andhra Pradesh	22	414550	18843.18
Maharashtra	71	1034050	14564.08
Madhya Pradesh	21	290650	13840.48
Gujarat	68	935800	13761.76
Uttar Pradesh	73	870780	11928.49
Punjab	37	375162	10139.51
Haryana	15	149140	9942.67
Pondicherry	1	9000	9000.00
Himachal Pradesh	6	53200	8866.67
Bihar	4	25000	6250.00
Jammu & Kashmir	1	5000	5000.00
Rajasthan	7	12195	1742.14
West Bengal	18	22600	1255.56
Total	406	5921327	14584.55

That the bigger units are located in a contiguous region comprising of the states of Orissa, Karnataka, Tamil Nadu and Andhra Pradesh is not a simple accident. A look at the distribution of state owned forests across the states reveals that these four states together account for more than 22 per cent of the forest cover of the country. 24.23 per cent of the geographical area of these four states are recorded as forest land against a national average of 23.28 per cent, implying a slightly larger proportion of their land mass being not available for any other possible uses, compared to the average situation prevailing in the country. On the other hand, whereas around 16.72 per cent of the recorded forest area in our country is devoid of any tree cover, the corresponding figure for these four states together comes out to be a whopping 22.77 per cent. The implications are easier to comprehend. This region accounts for a large chunk of degraded forest land in the country. The species available in these forests were also suitable as inputs for these units. And who does not know that till the declaration of the New Forest Policy in 1988, government forests were the cheapest source of raw materials to the paper industry? The 'Saxena Report' also agrees on this count. Now that the priorities have changed suddenly with India signing the Convention on Biodiversity, the paper industry should be properly rehabilitated.

The main problems faced by the Indian paper industry are as follow :

1. Existing low average scale of operation across the units, and
2. Uncertainties in respect of supply of raw materials in view of the changed policy of the State to forests.

It is obvious from the issues raised above that whereas the first problem assumes greater significance in the face of our commitment to WTO, the second problem results from our commitment to CBD. Let's elaborate.

For Indian paper industry to be globally competitive, even on an assumption that the industry won't export but would like to produce at a cost that is competitive enough to offset exports from the rest of the world, it is necessary that they take advantage of the scale factor.

A higher production capacity necessarily means a lower per unit cost of production. Average paper mill sizes in different regions of the world are given in Table 20.16.

Table 20.16 : Average Paper Mill Size

Region	*Average capacity (000 tonnes per annum)*
India	14.58
Asia	17.59
Latin America	50.67
Europe	82.24
Africa	86.06
Australasia	173.81
North America	183.57

Increasing the scale of operation in paper industry requires considerable amount of financial investments. It also simultaneously depends a lot on the availability of cellulosic raw materials at a competitive price. As per an estimate the projected shortfall between demand for and supply of paper in the domestic market is estimated to be 2.54 million tons in 2005-2006, leading to an enlarged requirement of wood supply to the tune of 13.2 million tons of debarked standing trees.

Wherefrom can these requirement be supplied? The necessary conditions to be attached to such supply are:

1. There should be no uncertainties in supply—uncertainties in supply will not render increasing production capacity to be efficient, return on investment will be low.
2. The supply will have to be at a competitive price—the cost of production will have to be low, necessitating increase in productivity of land.

It is obvious that forests meant for conservation of bio-diversity cannot supply these requirements. Supply from such land will not be cost-effective in view of low productivity of these forests, being more geared towards protection of bio-diversity and hence multiple specie plantations. Such supply has to come from lands put specifically to the production of pulp wood species and managed properly to ensure high yield. We have already noted in Section II that there are three possible choices. They are :

1. Importing land—paper industry should lease in lands for production of cellulosic raw materials in other countries and import them to feed their paper production units within the country.
2. Increasing productivity in domestic sector—the industry may lease in land from private individuals and produce their requirements.
3. Putting land under joint use—in view of our observation in the earlier section that a particular type of forest cannot generate all the services expected from it—there is a trade-off across types—the entire recorded forest area in the country may be divided into two parts, one taking care of the concern for bio-diversity and hence exclusively protected and the other being devoted to production of other services like tangible benefits and carbon sequestration.

The first choice is contingent upon the economics and the willingness of the exporting countries involved. The second choice is constrained by the availability of private degraded land. An improvement in the marketing channels for other agricultural products and consequent increase in return from them may attract the farmers to look for alternative crops. The third choice looks apparently unacceptable in view of the existing policy mix developed nationally. However, we have documented quite clearly in the earlier sections that the so-called negative impacts of leasing out degraded forest lands to private enterprises may be exaggerated. Field level enquiry did not support most of these concerns. Rather, a changed policy regime will ensure sustenance of the paper industry and help reap the potential benefits of participation in the carbon trade market without compromising our commitment to CBD.

We discuss a possible road map in the concluding section.

A POSSIBLE ROAD MAP

In the introductory section (elaborated in Appendix I) we mentioned three possible policy options available to our economy in view of the scarcity of land on the one hand and the increased demand for products generated by both forestry and non-forestry activities. To repeat, they are

P4. Importing land for a particular use from another country which has land under such use in excess of the requirement of their own people. However, such decisions are possible if the land use pattern for the world as a whole is sustainable at a given point of time.

P5. Increasing the productivity of land under a particular use and thereby reducing the requirement of quantity of land under a particular use. Such a measure requires technological innovations.

P6. Considering putting land under joint use—producing both forest and non-forest products simultaneously from the same plot of land. Such a measure requires both institutional and technological innovations.

In the following paragraphs, we shall try to look into the implications of these policy alternatives from the perspective of leasing out degraded forest land to private sector entrepreneurs in general and the paper industry in particular, for production of pulp-wood. The implications may be studied at four levels:

1. The communities;
2. The forest department;
3. The paper industry; and
4. The country as a whole.

To consider the *communities*, it is observed that they

- are mostly lying below poverty line;
- do not get enough alternative employment opportunities in their localities;
- do not depend much on the degraded forest land for their livelihood in view of a highly reduced supply of resources from such lands;
- cannot optimally utilize their agricultural land due to lack of

access to necessary inputs on the one hand and to remunerative marketing options for their products on the other; and

- do not have access to other productive assets.

Under such circumstances it is imperative that they are provided with opportunities to higher income, higher scopes for employment along with access to proper health care and educational facilities, so that they may improve their stocks of physical as well as human capital.

Leasing out of degraded forest land to private entrepreneurs may considerably increase the scope of employment generation—it is estimated that one hectare of plantation of pulp wood species creates an employment potential of around 450 mandays over its entire rotation period. Assuming an average rotation of 6 years, annual average mandays generated per hectare of plantations is estimated as 75 days. In addition, the villagers will receive around 1.2 MT of fuelwood every year per hectare, free of cost. At an average price of Rs. 500 per MT, the benefits accruing to the community works out to be Rs. 600 per hectare per year. If fuelwood requirements of the communities are met from such plantations, their dependence on standing forests for such purposes will also be minimized and they will have the incentives to protect them as sources of other necessary NTFPs. Such plantations on degraded forest land will also act as green boundaries around the standing well-stocked forests and help facilitate their protection from timber mafias.

Thus from the perspective of the communities, policy alternative P1 is ruled out, as importing land for plantation purposes will involve exporting the benefits mentioned above to the communities residing around the land so imported.

Let's now look from the perspective of the *forest department*. The National Forest Policy of 1988 enumerates both conservation of forests and increasing their productivity as objectives to be achieved simultaneously. Unfortunately, the policy document failed to address to the apparent contradiction between these two objectives and did not specify any guideline to balance between them. Eventually, the 'conservation' aspect of the policy—increasing the intangible services of forests—has been emerging as the sole issue of concern, with the issue of increasing productivity—the supply of tangible benefits—receiving the least attention from the forest managers. It may be noted that these two services of forests are never substitutes of one another, rather they are highly complementary to each other. Examples from several countries may be cited where the forest managers have been able to strike a balance between the two. If we think about a possible division of labour between the Forest Department and the private entrepreneurs—the establishment of Forest Development Corporations across the states were motivated by such thinking—with the former taking care of the protection aspects and the latter looking after the ways to increase productivity, the necessary balance may be struck. Consequently, a portion of the existing degraded forest land may be leased out to the private enterprises having real interests in putting them to productive use. Such a decision will also help rejuvenate the sagging fortune of Joint Forest Management in the country today. Most of the FPCs/EDCs in the country today are suffering from severe funds crunch in view of the shrinkage in flow of funds from the donor agencies. We may conceptualise a new format of JFM that incorporates a three-way cooperation among communities, the Forest department and the private enterprises interested in increased and assured supply of forestry resources as their raw materials. The *Panchayats* may also be roped into such a collaborative mechanism. Involvement of the relevant stakeholders in the management of the land under forest will also reduce the cost of protecting the forest resources from possible plunder by timber mafia.

To consider the perspective of the *paper industry,* it is observed that the present gap between demand for and supply of forest based fibres is around 5-6 million tones. Assuming an annual growth of demand by 10 per cent, the increased requirement will leap up to 13-16 million tones. An average yield of 100 MT per hectare, in anticipation that the ongoing research activities will help achieve such a yield rate from an average yield reported at around 50 tons per hectare, will involve leasing out around 0.8 to 1 million hectares of degraded forest land to the private entrepreneurs. Given the cost advantages that the Indian units enjoy in terms of their labour costs, leasing out such lands will ensure added advantage in terms of cost of fibre over their global competitors. Such an advantage may well be converted into a large share in the global export market. It may be noted that as of now, India imports around 2 per cent of her requirement of paper and paper boards in volume terms. However, such imports are a little over 6 per cent in value terms. If such opportunities beckon, the paper industry may require another one million hectares of degraded forest land to produce the necessary surplus output for exports.

The increased production by the paper industry will have its spill over effects on the sectors enjoying backward and forward linkages with it. All will be bubbling with opportunities of increased production and consequent value addition to the economic health of the country. The employment potential of the economy will increase manifold.

All these benefits accruing at the levels of community, forest department, paper industry and the sectors aligned with it, will necessarily improve the health of the economy as a whole.

Thus it is imperative that we have to choose a combination of P2 and P3 mentioned above. Choice of P2 implies increasing the productivity of Indian forests. A choice of P3 signifies the fact that we may jointly produce some tangible benefits though increased productivity and some intangible benefits like carbon sequestration, water conservation and soil conservation services. The prospect of harvesting the benefits of Kyoto mechanisms and entering the carbon trading market in the coming days looks very bright.

However, as has been pointed out earlier, the captains of the paper industry also have to pull up their socks. To enhance their credibility, it is imperative that they come out with some 'code of conduct' which will ensure that they will not compromise with the faith the nation has put on them by leasing out a good amount of land, still constitutionally remaining under the management of the country. They should ensure that

- they develop and subsequently maintain a proper contractual arrangement *vis-à-vis* sharing of the produce from such land with the communities;
- such land leased out to them are not used for purposes other than production of pulp wood;
- they make continuous endeavour through research and development activities to increase the yield of pulpwood;
- they increase their fibre usage efficiencies over the years to ensure optimal use of the land occupied under lease agreements;
- their ownership should not change in such a manner that the national interests are jeopardized;

Proposed Changes in the Forest Conservation Act, 1980 for Sustainable Development of the Forests

S. No.	*Issues*	*Present Provision in FC Act*	*Proposal*
1.	Even though Forest Conservation Act does not define forest, the definition of forest has been extended to include all area recorded as forest in Government records according to guidelines of FC Act, 1980. Honourable Supreme Court has further extended the definition of forest by incorporating the dictionary meaning of forest	FC Act does not define forest. As per guidelines Section 1.1, on Forest Conservation Act, 1980, definition of forest land is as follows:— 1.1 Definition: (i) The term 'Forest land' mentioned in Section 2 of the Act refers to reserved forest, protected forests or *any area recorded as forest in the government records.* All proposals for diversions of such areas to any non-forest purpose, *even if the area is privately owned,* would require the prior approval of the Central Government. In addition Hon. Supreme Court has widened the definition of forestland by incorporating the dictionary meaning of forestland.	Forest Conservation Act, 1980 Section 2. *Explanation:* A new Sub-Clause C will be inserted as below *For the purpose of this Act "forest land" will mean land which has been notified as reserve forest and protected forest under "IFA 1927".*
2.	Infrastructure facilities like approach road, electrification, drinking water facility, small tanks, school building etc. in the villages located in and around forest attract the provision of FC Act thereby hindering their development. Due to application of this Act in these villages, the otherwise scarce developmental inputs tend to be diverted to other villages away from forest.	As per Forest Conservation Act Section 2 Explanation (b) "Non Forest Purpose" does not include any work relating or ancillary to conservation, development and management of forest and wildlife, namely, the establishment of check-posts, fire lines, wireless communications and construction of fencing, bridges and culverts, dams, waterholes, trench marks, boundary marks, pipelines or other like purposes.	The Forest Conservation Act, 1980 Section 2 Explanation (b) should be amended as below: Non-forest purpose "does not include any work relating or ancillary to conservation, development and management of forests and wildlife, namely, the establishment of

S. No.	*Issues*	*Present Provision in FC Act*	*Proposal*
		Guidelines section 3.2 (vii) (b) also provides that compensatory afforestation will not be insisted on for forest land diversion cases up to 1 ha.	check-posts, fire lines, wireless communications and construction of fencing, bridges and culverts, dams, waterholes, trench marks, boundary marks, pipelines and *community development works like Nistar tank, Stop dam, School building, primary health centre, approach road, rural electric line involving diversion of forest land up to 2 ha. For such works involving forest area more than 2 ha. and less than 5 ha. and for laying of electric lines, telephone lines, pipelines etc. involving diversion of forest land in linear strips, full power of sanction is delegated to the State Government.*
3.	Forest Villages were established prior to 1980 by State Government with the objective of helping forestry activities. Conversion of these forest villages to revenue villages have not been permitted under Forest Conservation Act thereby depriving them of development inputs. Similarly encroachment on forest land prior to 25.10.1980 should not	FC Act Section 2 Sub-Section (i) provides that without prior permission of Central Government no authority shall make any order directing that any reserved forest (within the meaning of the expression "reserved forest" in any law for the time being in force in that State) or any portion thereof, shall cease to be reserved. FC Act Section 2 (iii) provides that without prior permission of Central	Forest Conservation Act, 1980, under Section 5, a new Sub-Section 3 should be added as below: *Provision of this Act shall not apply on forest villages, and encroachment on forest land prior to enactment of this Act.*

S. No.	*Issues*	*Present Provision in FC Act*	*Proposal*
	require clearance under FC Act, on the ground that de-notification of such land has not taken place prior to 25.10.1980.	Government no authority shall make any order directing that any forest land or any portion thereof may be assigned by the way of lease or otherwise to any private person or to any authority, corporation, agency or any other organisation not owned, managed or controlled by Government.	
4.	Due to resource crunch, large extent of degraded forests are awaiting to be rehabilitated. Some mechanism to involve the tribal population who have lived in symbiotic relationship with forest since ages should be developed to sustain their interest in forest regeneration.	FC Act Section 2 (iii) provides that without prior permission of Central Government no authority shall make any order directing that any forest land or any portion thereof may be assigned by the way of lease or otherwise to any private person or to any authority, corporation, agency or any other organisation not owned, managed or controlled by Government.	Forest Conservation Act, 1980 Section 2 Sub-section (iii) amended to add *"the community management of an identified piece of forest land as per sanctioned micro plan and assignment of usufruct right on forest produce to its forest dwelling members by the community, will not be considered as a lease".*
5.	Boulder, sand, Murram, soil required for domestic consumption of people living in and around forest provision of Forest Conservation Act is attracted.	The FC Act, 1980 Section 2 explanation mentions Non-forest purpose means the breaking up or clearing of any forest land or a portion thereof.	FC Act, 1980, Section 2 Explanation on *Non-forest purpose will exclude, use of boulder, sand, Murram, soil, depending upon availability through a prescribed procedure, from adjoining forest for domestic purpose, by people living in and around forest area, provided such use does not involve felling or damage to trees.*

- they continuously keep the nation informed about the gains that are accruing to the communities, the linked sectors, the forest department and the country as a whole in view of the lands being leased out to them. A 'certification system' developed by the industry along with the development of a mechanism of 'social audit' of the industry by independent organizations will help achieve such an objective; and
- the industry association is simultaneously caring for the requirements of the member units in need of help and vigilant enough to continuously monitor the activities of a member unit and strictly enforce discipline on an erring member if need arise.

Some changes in the legal framework are also necessary to be introduced by the government. We propose below some changes in the Forest Conservation Act, 1980 which may be helpful in ensuring a smooth transition from the present regime of 'conservation only' to one that strikes a balance between 'production and conservation' and ensures participation in the real sense of the term of the communities.

Thus to conclude we argue that

- A portion of the degraded forest land should be leased out to the paper industry to take care of the apparent contradictions arising out of India simultaneously signing the Convention on Biodiversity and the World Trade Organization Protocols;
- The captains of the paper industry have to play a proactive role in changing their perceived role from that of 'opportunism' to one that makes them accountable towards ensuring the well-being of the nation; and
- The State has to shed its inhibitions of looking at forests only from the perspective of conservation and consider the importance of increasing the productivity of land under forests as well.

Such a 'trinity' will go a long way in opening up the space for a fruitful private-public partnership in regreening the country and motivate such collaborations in other sectors of the economy to bring about the required change in the country and make the dream of becoming a 'developed' country by 2020 an absolute reality.

21

RECOMMENDATIONS OF THE WORKSHOP

The various issues pertaining to the involvement of private sector in the sustainable development of degraded forests and wastelands were discussed thoroughly by clubbing the participants into groups with the following assigned themes:

Group 1: Ecological and Silvicultural issues pertaining to the development of forests and forest produce including NWFP.

Group 2: Legal issues, Government rules and relevant Acts with respect to private investment in forestry sector.

Group 3: Socio-economic issues and JFM perspectives of private sector investment.

Group 4: Financial issues including pattern of sharing and investment.

Group 5: Management and administrative issues with respect to institutional framework.

The observations and recommendations of each group are summarized below:

GROUP 1: ECOLOGICAL AND SILVICULTURAL ISSUES

Observations

- Priority between ecological and productivity objectives.

- Such a process involves considerable technological and social ingenuity.
- Initiation of the process of PPPP by forest department to achieve the goal of 33 per cent forest cover by 2012.
- Field level experiences have forced FD officials to go on the defensive approach–more emphasis has been placed on protection of the resources over the concern for productivity.
- Confusions prevail on what to balance among different objectives? No benchmark data set in the country, which can give complete and reliable information on these issues.

Recommendations

- It is necessary to strike a balance between productive, silvicultural and ecological objectives.
- Lists of inventories and impact assessment on ecology are to be made involving the communities, foresters and other stakeholders.
- Joint Forest Management programme may be considered the beginning of this process and can be proved a good platform for this concept.
- Silviculture of different commercial and non-commercial species of all strata should be studied and well documented for the future reference.
- Methodology to value forest products (including NTFP's) at all strata should be developed.
- Very difficult to suggest a road map instantly for the afforestation of degraded forests/wasteland through PPPP.

GROUP 2: LEGAL ISSUES, GOVERNMENT RULES AND RELEVANT ACTS

Observations

- What are the present Government rules and relevant Acts for private investment in forestry sector?

- Do we require amendments/interventions in the existing policy, acts, rules and regulations? If yes what and who is the competent authority for it and at what levels?
- What is the present legal regime for land, usufruct, and resource sharing for the purpose and finance sharing?
- What are the main problems faced by plantation industries and available solutions within the legal framework?
- Proposed financial mechanism to promote investment in forestry sector for the rehabilitation of degraded forests/wastelands via PPPP.

Recommendations

- Changes in the current National Forest Policy is not necessary to facilitate PPPP in re-greening of degraded forestland as provisions enshrined in clauses 2.2 and 4.9 of the above policy are adequate and amendments are not required.
- Private investment in Protected Areas should not be allowed because of the prohibitory nature of Wildlife Protection Act, 1972.
- Even though Indian Forest Act does not have any provision for allowing private participation, provisions in the Forest Conservation Act, 1980 are adequate to take care of this requirement.
- MoEF should frame guidelines for issuing approvals for private investment in forestry sector based on following suggested parameters:
    - Forest areas situated in V and VI schedule or otherwise,
    - Level of participation of local communities,
    - Density of forest crop, site quality and range of growing stock,
    - Parameter for usufruct (nistar),
    - Period for lease permitted for this purpose, and
    - Identification of sensitive areas.
- Government may levy a portion of final felling in the form of lease rent.

- State Land Ceiling Act, Panchayat Act and Tribal Act should be amended to encourage private investment for the rehabilitation of degraded forests/wastelands.
- Crop of plantation origin should not require permission of collector under land revenue code.
- Forest produce (including NTFP's) from the rehabilitated forest land/wasteland should be exempted from transit regulations.
- It is proposed to constitute a separate forest financial corporation to facilitate private investment for the rehabilitation of degraded forests/wastelands.
- Forest Development Corporations may be allowed to raise the cheap capital for the afforestation of degraded forest/wasteland.
- Long gestation period, low Internal Rate of Return (IRR), uncertain market condition, lack of consultants, and trained manpower and scarcity of reliable data are the main constraints for the rehabilitation of degraded forests/wastelands. Separate Cells may be created in different institutes like ICFRE and SFRI's for this purpose.

GROUP 3: SOCIO-ECONOMIC ISSUES AND JFM PERSPECTIVES OBSERVATIONS

- Threats and opportunities to the JFM fund as a result of private investment.
- Fund for community forestry by corporate sector under Corporate Social Responsibility (CSR).
- Scaling of the success of PPPP from social point of view.
- Resource and employment generation through private investment.
- Status of Schedule 5th Area—right to access agreement is not possible.

Recommendations

- In Schedule 5th Area Corporate involvement can be encouraged with the association of Local People, Government and NGO.

- Transparent market mechanism for the forest products including NTFPs should be developed to encourage farmers to grow and protect forest crops and to eliminate middle men.
- Need to re-structure corporate intervention strategy of raw material procurement within and outside the organisation.
- Involve Panchayats and activate Van Sub-Committees for the protection of afforested area and usufruct sharing.
- For the ultimate success make and implement this PPPP into sustainable rather project type mode.

GROUP 4: FINANCIAL ISSUES, PATTERN OF SHARING AND INVESTMENT

Observations

- Pattern of investment sharing *vs.* usufruct sharing.
- Main constraints in Institutional financing.
- Fluctuating rate of Interest.
- Direct lending by NABARD to the corporate/FD for afforestation of DF/W.
- Long term fiscal policy for the private investment in forestry sector.
- Priority for lending in forestry sector.
- Tax concessions for promoting tree cultivation in DF/W.

Recommendations

- Creating a separate corpus for forestry sector by raising the public money from green thinkers and environment friendly corporates.
- Proposal for variable interest rate suiting to economic scenario and rate of interest to be fixed as per the degree of land degradation.
- Development of demonstration plots to ensure future strategies for the rehabilitation of DF/W.

- Generating awareness and interest for this scheme, among financers and Banks through audio-visual means.
- Creation of separate forestry bank exclusively for afforestation scheme.
- Incentives for the investors as being offered in agricultural sector such as removal of licence fee, exemption from taxes, tax rebate scheme, facility of non-disclosure of source etc.
- Investor should have usufructory right reserving a certain proportion for the local people. Usufruct sharing should be directly proportionate to the investment made and other stakeholders should be given proper weightage.
- Direct lending by NABARD for the investment in afforestation schemes and should be available after the approval of their Board, e.g. RIDF.

GROUP 5: MANAGEMENT AND ADMINISTRATIVE ISSUES

Observations

- Financial institution to promote funding in the forestry sector.
- Reducing the burden of forest department.
- Decentralization of the process of sanctioning the management plans.
- Dispute resolution and arbitration among the different stakeholders.
- Direct and indirect investments and concessional credits for wasteland development.
- Creating data bank on the various attributes of available degraded forests/wastelands.

Recommendations

- As creation of a new and exclusive financial institution to promote funding of projects for the forestry sector may not be feasible in the near future, the group feels that the existing financial institutions should appreciate due importance of the forestry sector. They must accordingly earmark and allocate larger resources (both human and non-human) for investments in this sector.

- The financial institution should, not only operate from the supply side (by providing cheap and easy resources), but also take pro-active steps to promote demand for credit-worthy projects. As a part of this demand-creation exercise, governmental, non-governmental, as well as financial bodies should encourage undertaking of credit-worthy projects by self-help groups of stakeholders (inclusive of owners of non-forest lands).
- To facilitate and strengthen both the demand and supply sides of investible resources for wasteland development, IIFM should arrange regular discussions of all important stakeholders.
- As professional foresters working with government are already over-burdened with departmental works, there is a strong need to promote chartered foresters among qualified personnel to guide private wasteland development activities. IIFM and other competent management organizations should take a lead in training and promoting such chartered foresters. There is a strong recommendation specially to promote agro-forestry clinics.
- Territorial DFOs should be entrusted the job of according sanctions to management plans for non-forest lands.
- The group favours evolution of institutional framework for dispute resolution and arbitration in course of a continuous process of dialogue, discussion and negotiation among the relevant stakeholders. They should start the process of evolving contractual arrangements and safeguards for farmer-industry partenership for wasteland development.
- The group recommends for credit at concessional rates for the community based organizations engaged in development of wastelands under government ownership.
- For private wasteland development the same concessional credit should be provided even to individuals and their associations.

- Existing JFM organizations like FPCs (and similar bodies that may be created on non-forest lands) and their higher-tier bodies may be encouraged to receive funds from non-government sources (including individuals) as a reward/incentive for their ecological services of local as well as global importance.
- Governmental bodies must initiate the task of creating a strong database on the various attributes of available wasteland both inside and outside of forests. Private bodies (including farmer and industry associations) must at the same time be encouraged to create such data base in the interest of increasing their returns from use of such lands on project basis.

APPENDICES

APPENDIX I

LIST OF PARTICIPANTS IN THE WORKSHOP

1. **Mr. M. Hashim**
   Managing Director
   M.P. Van Vikas Nigam Ltd.
   Panchanan Building
   Malviya Nagar, Bhopal-462 003

2. **Prof. K.C. Malhotra**
   C/o IIFM, Bhopal

3. **Mr. Vijay Ratan Khare**
   M.P. Minor Forest Produce Federation Ltd.
   M.P. Nagar, Bhopal

4. **Mr. A.M. Paul**
   NABARD, E-5, Arera Colony
   P.O. Ravishankar Nagar
   Bhopal-462 016

5. **Dr. Dinesh Kumar Tyagi**
   DCF, Shahapur Forest Division
   Shahapur, Thane (MS)

6. **Mr. Muran Chandra Mishra**
   Deputy Conservator of Forests
   Chhatisgarh Forest Department (Development)
   Raipur-492 001

7. **Mr. S.N. Rao**
Dy. General Manager (Plantations)
ITC Ltd.
Paper Boards and Speciality Papers Divisions
Sarapaka, Dist. Khammam-507 128
Andhra Pradesh

8. **Mr. Vivek Sharma**
CARD, E-7/803, Arera Colony
Bhopal-462 016

9. **Mr. R.C. Saxena**
National Centre of Human Settlement and Environment (NCHSE)
Girish Kunj, Arera Colony
Bhopal

10. **Mr. Krishna Kumar Singh**
E-9, K/2, Char Imli
Bhopal

11. **Mr. D.P. Singh**
F-1, Janki Enclave
Kolar Road, Bhopal-462 016

12. **Mr. J.K. Jain**
Dy. General Manager
A.P. Paper Mills
Rajahmundry-533 105

13. **Mr. D.V. Negi**
CCF (Centre), MoEF
GOI, West Regional Office
Bhopal

14. **Mr. A.K. Bisaria**
CF, O/O PCCF
Satpura Bhawan
Bhopal

15. **Mr. Ajay Noiding**
Forest India Place
BILT, Gurgon-122 002

16. **Mr. Shailendra Bahadur**
CF (Working Plan)
Nagpur

17. **Mr. Debasis Bhattacharya**
CII, 23, INS Area
Lodi Road, New Delhi-110 023

18. **Mr. Samar K. Datta**
Centre for Management in Agriculture
Indian Institute of Management
Ahmedabad-380 015

19. **Dr. P.B. Gangopadhyay**
Addl. PCCF
Bhopal (M.P.)-462 004

20. **Mr. Milindo Chakrabarti**
CREATE, St. Joseph's College
PO: North Point
Darjeeling-734 104

21. **Mr. Akhilesh Bhargava**
FM Consultants India Pvt. Ltd.
11/3, Amaltas Complex, Shahpura
Bhopal-462 039

22. **Mr. Raj Chaurasia**
Ballarpur Industries Ltd.
First India Place, Tower C
Mehrauli Gurgaon Road
Guragaon-122 002

23. **Mr. R.R. Okhandiar**
DFO
Van Bhawan, Tulsi Nagar
Bhopal

24. **Mr. O.P. Khare**
CF (Coordination)
O/o PCCF
Bhopal

25. **Mr. Mangesh Kumar Tyagi**
CF Development (CCF) Dev.
Bhopal

26. **Dr. R.P. Singh**
M.P.R. Van Vikas Nigam Ltd.
Panchanan Building, 5th Floor
T.T. Nagar, Bhopal-462 003

27. **Dr. Nishi Naryani**
Foresters Welfare Society
156/A, Indrapuri, Bhopal

28. **Thriveen Sankar C.**
IIFM, Nehru Nagar
Bhopal

29. **Ms. Krishna Patra**
Research Associate
IIFM, Bhopal

APPENDIX II

WORKING GROUP'S REPORT ON THE PROSPECTS OF MAKING DEGRADED FORESTS AVAILABLE TO PRIVATE ENTREPRENEURS FOR PLANNING COMMISSION, GOVERNMENT OF INDIA

Composition of the Working Group

Dr. N.C. Saxena CHAIRMAN
Secretary, Department of Wastelands Development

Secretary
Ministry of Environment and Forests

Inspector General of Forests
Ministry of Environment and Forests

Shri P.C. Srivastava
Principal Chief Conservator of Forests
Government of Uttar Pradesh

Shri S.D. Mukherjee
Principal Chief Conservator of Forests
Government of Andhra Pradesh

Shri C.G. Mishra
Principal Chief Conservator of Forests
Government of Orissa

Shri R.D. Sharma
Additional Principal Chief Conservator of Forests
Government of Madhya Pradesh

Shri V.R. Chitrapu
Principal Chief Conservator of Forests
Government of Tamil Nadu

Shri R.D. Jakati
Chief Wildlife Warden
Government of Haryana

Shri K.N. Singh
Chief Conservator of Forests
Lucknow

Prof. Shekhar Singh
Indian Institute of Public Administration

Shri S. S. Rizvi
Consultant, WWF India

Dr. Anil Aggarwal
Director, Centre for Science and Environment

Shri Uma Shanker Singh
Member-Secretary
Deputy Adviser
Planning Commission.

Preface

Planning Commission constituted a Working Group to examine the prospects of leasing out Degraded Forest Lands to the private entrepreneurs/Forest Corporations for production of industrial raw material. Out of 14 members of the Working Group, we could not get the views on the draft report of Shri P.C. Srivastava. Amongst the remaining 13, 10 members have *fully* supported the findings of the Working Group that Degraded Forest Lands should *not* be leased out to private entrepreneurs; and that industries needing forest raw material should establish contact with farmers.

The remaining 3 members, Shri Vishwanath Anand, Secretary, Ministry of Environment and Forests, Shri C.P. Oberoi, Inspector General of Forests, and Shri S.C. Mohanty, Special Secretary, Forests and Environment Department, Government of Orissa have also given general support to the report. Shri Anand has stated that "I am not in favour of sale or lease of forest lands to the private sector for the purpose of raising timber". Shri Oberoi has stated that, "Regarding the participation of industries in afforestation, I agree that they should have durable tie-up with the private farmers for producing the raw material." Shri Mohanty has stated, " I would generally endorse the arguments developed in the report, both for and against leasing of degraded forest lands to raise industrial plantation."

In the light of these views and the full support of the other 10 members, the report should be taken as reflecting the unanimous consensus of the Group.

However, Shri Anand has raised other important policy issues which have a bearing on this subject. Therefore, his views are enclosed, in full, as Annexure-IV to the report. Similarly, Shri Oberai stressed the need to bring more investment in the forestry sector. His views have, therefore, been fully quoted in the report in the section "Views of the Ministry of Environment and Forests". Shri Mohanty has given arguments in favour of making certain categories of degraded forests available for industrial companies. His views have been reproduced, in full, in the section on "State Governments".

As Chairman of the Working Group, I take this opportunity to thank all the members of the Group, as also experts, who fully cooperated with the Group and who have helped in drafting the report. I also thank Shri Mahapatra, Addl. Secretary, Wastelands Development, and other officers of my Ministry who contributed their ideas to the report. It was, indeed, my pleasure to work with the best brains in the forestry sector and produce this report. I hope it would be found useful by the Planning Commission, Ministry of Environment and Forests, state governments, industry and the academic world.

N.C. Saxena
Chairman of the Working Group

October 16, 1998

Executive Summary

The Planning Commission set up a Working Group to examine the prospects of leasing out of degraded forests to the private entrepreneurs/ Forest Corporations. The terms of reference for the Group were:

- Economic, social and environmental feasibility of leasing or otherwise making degraded forest land to the private entrepreneurs/Forest Corporations.
- Implication of Ninth Plan, forest policy, forest law on this issue.

Since it is paper and pulp industry which has been demanding forest lands on lease either directly or via the Forest Development Corporations (FDCs), we essentially considered the merits of making degraded forests available to the paper industry for meeting their raw material needs. The concept *prima facie* is quite appealing; government has wastelands but lacks funds, industry has both capital and technology, and hence best suited to afforest degraded lands. In the process the poor get jobs. So why should one object?

Non-forest barren lands: The Group supported the involvement of industry in reclamation of non-forest wastelands, such as desert lands of Rajasthan, Bhal lands of Gujarat, ravines of M.P. and saline lands of U.P. The total area of such lands is estimated as 20 million hectares. In fact several state governments had in the past offered on lease such non-forest degraded lands, but industry showed no interest. The state laws relating to land ceiling in Rajasthan and Gujarat are liberal enough to grant exemption on such leases from Ceiling limits. Despite the initial high cost of reclamation these lands have the advantage of being available in contiguous patches and hence amenable to economies of scale. Besides, these degraded lands do not support the livelihood needs of the poor, and therefore, handing them over to industry does not exacerbate social tensions.

Degraded forests: The situation would be entirely different if village commons or degraded forests are leased to industry. Industry has argued that it would make 'unproductive' lands productive through captive plantations. However, if productivity is defined in terms of subsistence value, such forests are a vital source of living for the poor. Such lands may have a low tree density, but satisfy the fuelwood, fodder and livelihood needs of about 100 million poeple. In fact, these lands are degraded because they suffer from extreme biotic pressure, and require neither capital investment, nor higher technology, but protection and recuperation, which can be done only by working with the people, where industry has neither expertise nor patience. The West Bengal experience shows that about 2000 peoples' forest protection committees have regenerated more than 300,000 hectares of sal forests at little extra investment, simply by protection on the promise of sharing wood and non-wood products with them. If lands on which peoples' livelihoods are dependent are given to industry, they may have to employ muscle power to keep people at bay, thus escalating social tensions, which are already quite acute in several forest and park areas.

Even within the industrial sector, paper and other large industries consume just a fraction of forest products. 90 per cent of forest raw material is processed by 25,000 saw mills and a larger number of cottage units, who would also lay claims on forests, once their big brother is able to snatch concessions from government. Besides there would be claims from coffee, cashew and palm plantation industries. Like paper industry they will raise short-term and quick growing species in place of multi-layer mixed forests obtained through regeneration. Its ecological implications need to be taken into consideration. Using forests for growing raw material for industry will be setting the clock back to the 1960s, showing that we learnt nothing from the mistakes of the past 30 years of trying to create man-made forests, which were ecological disasters, besides completely alienating the people and leading to faster degradation.

According to industry's own admission, their requirement can easily be met from 2 million ha of degraded land. As against this there is 141 m ha of cultivated land and 20 m ha of farmer owned uncultivated wasteland. These lands have the potential of producing pulpwood, especially in view of the fact that both eucalyptus and bamboo are short rotation crops and eminently suitable for farm sector. In fact the bogey of raw material crunch is no longer valid, given the vast expansion in farm forestry programme. There is a surplus of eucalyptus wood today, as is evident by the sharp fall in eucalyptus prices all over the country, well documented in the World Bank reports on forestry in India. The problem is locational; industries got established close to government forests, where as raw material is available in green revolution areas. A practical solution would be to split the processing units; to establish a new pulp making plant close to farm forestry areas, and transport pulp to the paper mill. If pulp can be imported from as far as Canada, surely it can be moved from Punjab or Haryana, where a pulp producing mill can be located, to Central and eastern India, where paper mills are located.

If industry produces its own raw material, who would farmers sell to? Where is their market, if not industry? 60 per cent of farm land is owned by rich and affluent farmers, who are market oriented, and can be trusted to fulfil the requirements of industry. They are even prepared to produce teak wood, if government removes restrictions on the felling of teak trees from private lands and on its movement. Since the overall demand of the industry is limited, and if allowed to be met by leasing it would adversely affect the farm forestry programme, which is one of the cheapest and most sustainable method of producing wood.

The prosperity of the Indian farmers is the very basis for the prosperity of the country. The farmers are at a very great disadvantage as their crops are susceptible to great fluctuations in market prices, perishability of commodities and consequent urgency to dispose of the produce, adverse climatic conditions as well as pests and diseases. The only crop that is not susceptible to these adversities is tree farming, which can be taken up either in place of crops or along with them. Leasing of degraded forests to industries would deprive the farmers of even that opportunity of taking up tree farming.

Thus the claim of the industry that it would create additionality of production is not true, as any afforestation by them will be at the cost of tree planting efforts by farmers on privately owned degraded lands, tubewell enclosures, and homesteads, where the social cost of production is minimal, as these lands are of no use for cultivation. Thus the argument that farm forestry would compromise food security is irrelevant, in the face of empirical historical evidence of significant tree plantation by farmers on lands which were not suitable for agriculture. Farmers exploit their own family labour (which is unpaid), and therefore can produce wood cheaper than industry. Farmers harvest their trees during the lean agricultural season and thus are able to achieve further saving in costs by spreading family labour inputs more evenly throughout the year. In fact the government policy of subsidising bamboo production on forest lands for supplies to industry acts as a deterrent to cheaper production on homesteads. The present proposal of the industry means getting land almost free of capital cost, thus involving subsidy worth several thousand crores. In the light of new liberalisation policies of the government, such subsidies on non-merit goods are highly undesirable.

The proposal of the industry to allow them to use government forests for industrial plantations is thus against two groups of people; lakhs of farmers who would be deprived of a market for agroforestry, and millions of voiceless forest dwellers who would be denied access to NTFPs and other biomass that they gather.

The introduction of Panchayat Act in tribal areas, where most forests are located, will render leases to industries of forest lands illegal and unconstitutional, as the spirit of the Act is in favour of local ownership and control over natural resources in Schedule V areas. In any case, involving industry in afforestation of government forests is against the Forest Conservation Act and the Forest Policy, both on grounds of management and species choice. Even Forest Department is prohibited from doing plantations of cashew or rubber on forest lands, how can then private industry be allowed to raise monocultural plantations?

During the course of the enquiry by the Group, representatives from the industry clarified that they require forest lands with at least 1 metre soil depth. Such fertile lands, even when not having much tree cover, would regenerate on their own without much cost. Thus regeneration woud be a cheaper economic option than plantations. Besides, the industry has no credible plan to resolve the demands that local communities have over such lands. As such, any leasing of forest land is likely to result in hardships and oppression for the local communities who have historically depended on such lands for meeting their basic needs.

It is now being proposed by some that the management of degraded lands should remain with government corporations but private companies may invest their money in the plantations. If government banks have not found the activities of these corporations worthy of investment, why would private companies do so? Forestry being a long gestation activity, it requires stable tenure. If land, management, and source of funds is going to be with three different parties; government, FDCs, and private industry respectively, the risk for the investor is very high, unless there is unwritten and unstated understanding between the parties.

Lastly, degraded forest lands with crown density of 25 and 40 per cent are not likely to be available in contiguous patches, as these are likely to be interspersed with better quality forests.

Therefore, industry should be asked to establish links with farmers who will produce raw material if given a remunerative price, in ways similar to the linking of poplar growing farmers with a match factory in north U.P. This experiment shows that, with technological back-up, timber size trees suitable for sawing can be raised on farm lands within eight years. In fact, due to farmers' enthusiasm for growing poplar its enhanced supplies have led to the establishment of several plywood factories in that area, thus providing considerable downstream employment. It is a myth that industry cannot deal with farmers directly. For several crops like sugarcane, potato, rice, cotton etc. industry has been in touch with farmers for decades.

The Ministry of Environment and Forests has argued that the industry are interested not in degraded forests on which the poor depend but only in barren forests (approx. area 6 m ha, density less than 10%). If that is so, why don't they show interest in revenue wastelands which are available in many states? In any case the Group was unequivocally informed by the industry that they are interested only in lands with sufficient soil depth of at least one metre. Paper industry has to be internationally competitive which would need raising plantation on good quality lands available in one patch. According to the Group, such lands would necessarily be good quality forest lands away from habitations, or farm lands and homesteads. If industry is not to get good quality forests, they have no other alternative but to establish links with farmers.

In the ultimate analysis the question to be asked is, whether the claim of the industry over forest lands is based on transparency and sound economic rationale, or is it a seductive myth and against the interest of farmers and forest dwellers?

Introduction

Vide its letter dated 6th June, 1997 (at Annexure 1), the Planning Commission set up a Working Group (henceforth called the Group) under the chairmanship of Dr. N.C. Saxena, Secretary, Rural Development, Government of India, to examine the prospects of leasing out of degraded forests to the private entrepreneurs/Forest Corporations. The terms of reference for the Group were as follows:-

1. Economic, social and environmental feasibility of leasing or otherwise making degraded forest land to the private entrepreneurs/Forest Corporations.
2. Implication of Ninth Plan, forest policy, forest law on this issue.

Since it is the paper and pulp industry which has been demanding forest lands on lease either directly or via the Forest Development Corporations (FDCs), we shall essentially consider the merits of making degraded forests available to the paper industry.

In addition to holding meetings of the group, written views were solicited from all members and state governments. The representative of the Government of India suggested in the first meeting that the following, who have generally been representing the case of industries, should be invited and their views heard:

Shri Pyare Lal, ITC, Bhadrachalam

Shri A. K. Mukherjee, ex-IG, Forests

Shri N. R. Krishnan, ex-Secretary, MoEF

Dr. B. N. Ganguly, IFS (Retd.)

This was done and they (except Shri Krishnan, who could not attend the meeting) were heard at length. In addition, Mr. V. M. Thapar, Chairman, Joint Committee of the Paper Industry was also invited and given full opportunity to present his views. Some members of the Group and the Additional Secretary, Wastelands Development visited degraded forests in Andhra Pradesh which were considered suitable for leasing by the industry. Shri Pyare Lal accompanied the team. The report is at Annexure 2.

Case for Leasing

The case for leasing or otherwise making degraded forest land to the private entrepreneurs either directly or via the government Forest Corporations is based on the following premises:

1. There is a great deal of degraded forest land in India which is of no use to anyone. It is lying waste and should be made available to anyone who wishes to use it.
2. The main constraint in the way of restoring the productivity of such lands is lack of resources. The resources required for increasing the productivity of degraded forest lands through large scale technology based plantations are considerable and are not easily available in the public budget.
3. Industry is keen to be involved in the afforestation of degraded lands because it is an economically profitable venture.
4. Involving the industry in establishing high-yielding fast-growing plantations on degraded forest land will result in accelerated rehabilitation of these lands and thereby help in regreening India.
5. There is a shortage of raw material required by the paper and pulpwood industry; the involvement of capital and technology by the industry would amount to additionality of efforts towards afforestation, and therefore will be socially desirable. It will, without asking for public resources, create new jobs and thus will be a win-win strategy.
6. There are no serious social or environmental issues in putting degraded forests under industrial plantations.
7. Making industry self-reliant in their own raw material supply is cost-effective as more public resources could then be diverted for better protection and conservation of natural forests.
8. Farmers cannot be trusted to supply raw material to the industry, as farmers are essentially concerned with meeting their subsistence needs and have no interest in growing long-gestation crops. In any case relying on farm forestry alone will not give the industry the required cost-effectiveness necessary in today's reformed markets.

These issues will be discussed in this report in detail. We begin with the nature and extent of degraded lands in India. How much is the area of degraded lands in India, and what factors have led to its degradation? We would look at both, the non-forest and forest wastelands.

Area of Degraded Lands in India

One can estimate the area of degraded lands in India in three ways, based either on the figures of the State Revenue Departments or on the NRSA imagery, or on the FSI data.

According to the nine-fold land-use classification, out of 304 million ha of land in India for which records are available, roughly 40 million ha is considered unfit for vegetation as it is either under urban and other non-agricultural uses such as roads and rivers, or is under permanent snow, rocks and deserts. The break-up of the rest 264 million ha of land which has some potential for vegetation is as follows:

Table 1: Breakup of Culturable Lands in India (in m ha)

I.	Cultivated land	142
II.	Forest land	67
III.	Fallows/culturable wastes/pastures/groves	55
IV.	Total area of culturable lands	264

Land statistics are well developed for cultivated and forest lands, but not for the category III in the above table for non-forest and non-cultivated lands. Information needed for assessing their potential, like ownership, extent and type of degradation, and present and possible future uses, has never been collected on a systematic basis for the entire country. This makes it difficult to estimate how much of the 55 m ha of the non-forest and non-cultivated land is degraded.

The National Commission on Agriculture believed that the entire area of 55 m ha classified under groves (3 m ha), fallows (24 m ha), pastures (12 m ha) and culturable wastelands (16 m ha) was degraded. Had it not been so it would be under the plough! Not all this land can be degraded: land classified as "groves" would seem likely to have trees on it; not all fallows are necessarily degraded; and on some of these lands social forestry has been undertaken. In the absence of any other estimates, we make the arbitrary assumption that one-fifth of this land is not degraded, giving a total of 44 ha of uncultivated degraded land. Even this may not be entirely available for tree plantation, as a part may be required for schools, panchayat buildings, cattle resting place, wells and dwellings. In dry areas, the proportions available would be higher. We estimate the area of degraded non-forest lands available for tree growing as 75 per cent of 44 m ha, i.e. 33 m ha. Adding 27 m ha of degraded forests (out of a total of 67 m ha as shown in Table 1), the total comes to 60 m ha.

On the other hand, the NRSA has surveyed 241 districts in India, and they have estimated the area under different categories of wastelands as follows:

Table 2: Category of Wastelands

Category	*Area (m ha)*
Snow Covered/Glacial	6.992
Barren Rocky/Sheet Rock	6.251
Sands-inland/coastal	5.572
Salt affected	1.988
Gullied/or ravinous land	2.020
Upland with or without scrub	26.515
Water logged	1.220
Marshy/swampy	0.824
Shifting cultivation land	2.824
Mining/Industrial Wastelands	0.116
Degraded grazing land	3.105
Degraded forests	16.274
Forest Blank	1.814
Grand Total:	75.515 m ha

According to the Forest Survey of India (FSI, 1998), 48 per cent area notified as forests had in 1997 a crown density of more than 40 per cent, 34 per cent between 10 and 40 per cent, and the rest 18 per cent has less than 10 per cent or no tree cover at all. Of the area under forests, 37 per cent is tropical moist deciduous forest (where *Shorea robusta* is the main species), 29 per cent is tropical dry forest (where teak is the valued species), 8 per cent is tropical wet evergreen forest, and the rest 26 per cent is subtropical, temperate, alpine and other forests. Nearly 12 m ha has been set apart as protected area for wildlife, out of which 8 m ha has tree cover (World Bank, 1993). It may be useful to undertake actual state-wise ground surveys so as to establish the extent of degraded lands in the country.

The FSI estimated current productivity for the entire forests at 0.7 cu m (cubic metres) of wood per hectare per year, which includes both recorded and unrecorded removals from forests. These levels are dramatically lower than the potential, which has been estimated at 2 cu m per ha per year. Achieving this potential which is about three times the current productivity would bring considerable improvement in the economic and environmental well-being of India's land and people.

The annual changes in forest cover, as estimated by the FSI are given below.

Table 3: Changes in Forest Cover in India Annually as per Remote Sensing Data

Period	*Area gained/lost annually in ha*
1981-85	(-) 1,47,000
1987-89	(-) 47,500
1989-91	(+) 28,000
1991-93	(+) 2,200
1993-95	(-) 25,000
1995-97	(-) 2,75,000

Source : FSI, 1998.

These three sets of figures, though not comparable (the NRSA data is not from all districts and may include non-culturable wastelands too), give a rough idea of the area of degraded lands in India, though no firm data is available about the nature and extent of degradation. The moot question is: can these lands be commercially exploited and used for industrial plantations?

Leasing of Non-forest Wastelands

One of the most critical aspects of information about non-forest wastelands pertains to ownership. Not all may be owned by

Government. In addition to uncultivated lands which were historically part of the farmers' holding, especially in ryotwari semi-arid areas, many poor families have been allotted lands under various programmes in the last 20 years, although such lands may still be lying uncultivated. Figures from a few states given in Table 4 show that substantial culturable waste area has been privatised as a conscious policy outcome.

Table 4: Allotment of Government Wastelands to the Poor

Name of the State	*Area Distributed in lakh acres*
Andhra Pradesh	42.02
Assam	5.89
Bihar	9.75
Gujarat	13.81
Karnataka	13.72
Kerala	4.57
Madhya Pradesh	1.71
Maharashtra	10.23
Orissa	7.26
Tamil Nadu	2.07
Uttar Pradesh	24.89
West Bengal	4.32
Rajasthan	.93
Total (including others)	144.96

Source : MRAE, 1998.

A substantial part of this land may be suitable for growing trees, but not for agriculture. These lands are certainly not available for leasing to industry, but may promote farm forestry, and thus may help in meeting the raw material requirements of the industry. The constraints in meeting this objective are discussed later in the report.

As regards non-private and non-forest wastelands, government wastes close to the villages are used by the community for grazing, but lands distant from villages are lying barren, and are available in large chunks. Some state governments have in the past offered barren lands, such as desert lands of Rajasthan and saline lands of U.P. and

Gujarat, on lease to industry. The scheme by the Rajasthan government was advertised in the newspapers in 1990-91 and a minimum of 2000 acres was being offered on a long-term lease, whereas the scheme in U.P. was initiated in 1979, but given up after a few years for want of proposals. The Gujarat government is at present offering land upto 800 ha for reclamation, but again there has been no positive response from forest based industry. Details are described below.

Revenue Department of Gujarat evolved a policy on August 9, 1994 under which non-cultivable wasteland including sandy and saline land can be granted to NRI companies, cooperative societies, Indian industrial houses, public limited companies and other persons for developing the land or plantation of medicinal trees by Israel technology or other scientific methods. Taking into account the experience, expertise, access to technology, employment opportunities etc. land not less than 100 acres and not more than 2000 acres can be granted on lease of 20 years. The conditions of the lease will be as under :

- The lease will be for 20 years at the first instance which could be renewed for another 20 years.
- No lease rent will be charged for 20 years after which lease rent as prescribed can be charged which could also be *toptop* based on percentage of profit.
- If the activity could not be started within two years then the time limit can be extended by another year if there are reasonable grounds but after that the lease will be cancelled.
- The lease will be subject to the use of land for the purpose for which it has been granted otherwise the land will revert back to government.

So far one industry, M/s Vadilal Industries Ltd. has been granted 316 acres for the purpose of plantation.

Similar facilities for granting leases to industry beyond ceiling limits exist in Tamil Nadu. For instance, one industry, Galaxy Crystal has been given a lease over 300 acres of unsurveyed/umeasured land in village Thillaivilakam, Taluk Thirutthuraipundi, District Pattukotai.

While endorsing such efforts of the state governments, the Group would like to reiterate that the intention here is not to recommend leasing out of lands which are or can be used as pastures and grasslands by the poor. These lands are in any case highly dispersed as these are close to villages. We are recommending to lease such barren lands which are far away from habitations and are available in larger plots of 1000 ha or more, and are of no use to the villagers. These are desert or ravine or saline lands, which require huge investment before they can be made productive. This distinction must be borne in mind, because if grasslands get occupied with plantations then we end up just shifting problems from degraded forests to degraded commons. The experience of leasing of C and D lands in Karnataka for pulpwood plantations should not be forgotten. In 1986 the state government of Karnataka decided to lease out 30,000 ha of common lands to a paper company on a lease rent of 12 per cent of the produce. When agitation and petitions to government did not yield any result, a public interest writ petition was filed by several NGOs before the Supreme Court. The Government of India in this litigation decided to oppose the decision of the state government and side with the NGOs, with the result that the state government was forced to cancel the lease in 1991 (Hiremath *et al.*, 1994). This was celebrated as a big victory against the lobby of forest-based industry wanting to grab government lands.

The Group would, therefore, recommend to the private entrepreneurs to consider reclaiming barren non-forest lands remote from habitations through modern technology. There is an Investment Promotion Scheme of the Department of Wastelands Development, Government of India, where such enterprises would be entitled to 25 per cent subsidy. These lands have the advantage of being available in contiguous patches and hence amenable to economies of scale. However, there are two inhibiting factors; poor soil conditions and consequently high cost.

Mr. V.M. Thapar, Chairman, Joint Committee of the Paper Industry, explained that plantations could not be raised on rocky areas or inhospitable terrain. According to him, desert lands of Rajasthan may be thought of for certain type of low price vegetation, such as grasses and bushes, but for paper industry one required lands with good soil depth. Small and dispersed patches are economically not viable. He opined that non-forest wasteland may be useful for plywood industry, but not for the paper industry. The minimum pulp requirement for a viable industrial plant is 1,000 tonnes per day which may be obtained from 4,000 tonnes of wood. This would require about 10,000 ha of either contiguous or land in sufficiently big patches for plantation. Paper industry has to be internationally competitive which would need raising plantation on good quality lands available in one patch.

As regards cost, it is admitted that the initial cost of reclamation may be high. The Agriculture Finance Corporation in 1986 had identified large chunks of wastelands in the country, and prepared model projects for their development. The cost of reclamation is summarised as follows:

Table 5: Cost of Reclamation of Wastelands in 1987

	Name of District	*Cost per hectare in Rs.*
1.	Chittoor (Andhra Pradesh)	10,100
2.	Srikakulum (Andhra Pradesh)	9,600
3.	Surat (Gujarat)	8,600
4.	Sabarkantha (Gujarat)	8,900
5.	Hassan (Karnataka)	10,300
6.	Meerut (U.P.)	15,000
7.	Raibareli (U.P.)	13,500

Source : NWDB, 1988.

The above costs need to be compared with the cost of block plantations on farm lands, which were estimated to be only Rs. 2800 per ha by the World Bank in 1988 in U.P. Another study (Chatha, 1991) found the cost of raising trees on farm lands in Haryana as only Rs. 2100 per ha. The cost of raising eucalyptus on farm bunds was even less. One can appreciate the indifferent attitude of industry towards afforestation of barren lands which have little soil left, as the scheme is not economical when compared to the costs of obtaining the same raw material through other sources, such as government forests, imports and private lands.

Present Price Distortions

As the industry gets subsidised raw material from government forests, there is further no incentive to them to turn over to barren lands. Until very recently highly privileged prices have been the standard practice for industries. A report (CSE, 1985: 91) has mentioned that in Madhya Pradesh in 1981-82 industrialists paid the Forest Department 54 paise for a 4-metre bamboo, while forest dwellers paid a little over Rs. 2 a bamboo supplied by the Forest Department. The UP Forest Corporation had calculated that during 1983-84 the actual cost of raising eucalyptus in government forests was Rs. 220 per tonne, whereas it was supplied to the Saharanpur paper mill at Rs. 140 per tonne, and to the Nainital mill at Rs. 196. The market rate for equivalent quality of eucalyptus, as judged from the auction prices, was between Rs. 500-700 per tonne during the above period (information based on government records, and all prices are for dry and debarked wood). Although the element of subsidy has been greatly reduced lately, it still continues in many forms. Bhabbar grass is sold by the UP Forest Corporation at Rs. 72.50 per quintal to rope-makers, but to industry at Rs. 40 per quintal. In A.P., the price that F.D. got by selling bamboo in the open market ranged from 800 to 1200 Rs. a tonne in 1990-91, as against Rs. 550 at which bamboo was supplied to the industry.

It is obvious that the market for industrial raw material is totally distorted by the present system of committed supply from government forests at concessional rates. Its impact can be summed up as under:

(i) It leads to creation of over-capacity and inefficient use of material by industry. It increases sickness in the industry and prevents its technological upgradation.

(ii) Price difference in open market and concessional rates leads to undesirable practices. There are reported cases of concessional supplies being siphoned off to open market for quick money by some concessionaires.

(iii) Arbitrary fixation of supply/quotas and prices to various units in similar business encourages corruption within the forest department at the cost of industrial efficiency. At the same time the state loses valuable revenue.

(iv) It discourages development of private efforts at tree plantation because of depressed market prices and low market demand from consuming industries.

(v) Since there is a lagged response of forest product supplies to the current signal of price, and expectation of change in government policy (such as allotment of forest land to industry, or new partnership arrangements with FDCs), there is a tendency on the part of industry not to take any initiative, and 'wait and see' becomes the safest policy.

To sum up, so long as these price distortions are not corrected, there will be no incentive to the industry to turn its attention to barren lands.

Barren Lands vs. *Degraded Lands*

It may be recalled that it has often been asserted by the proponents of leasing that only barren forest lands are to be utilised for plantations. Even some of the statements of the Government of India have created this impression. In an answer to a Parliament Question on the 2nd of August, 1994 asked by Shri S.Vaghela and Shri P. Patel the Minister of State for Environment and Forests, Government of India disclosed that the government was considering to 'involve industry in afforestation of *severely degraded areas*'. The stand of the government as to the quality of land which will be offered to industries was further made clear in a letter dated 2nd August from the Minister to Shri M.S. Swaminathan, President, WWF India, in which he stated that, '*In this case the fact is that there is absolutely no proposal to alienate forest land, or take away the rights of local communities, or to establish mono-cultural plantations, or supplant in anyway the efforts of NGOs and cooperatives. The proposal under consideration of government is restricted to areas which are* ***almost entirely denuded of forest cover****, which in no way contribute fuel and fodder to local communities, and which are not under joint forest management schemes. ... a mix of species would be planted to avoid mono-culture, a significant proportion of the land would be planted with purely forest species, would not be harvested ... Far from abridging the rights of local communities, the proposed scheme will seek to enhance those rights, giving them content and meaning, by increasing biomass availability as well as employment.*' The Minister's letter hinted that such an arrangement was being considered primarily for paper and plywood industry.

Why has the paper/pulp industry been particularly keen to lease "barren" forest lands, and not other categories of equally degraded lands?

Uncultivated public lands can be put in two categories; where density of tree cover is less than 10 per cent, and where it is between 10 and 40 per cent. For the sake of clarity these would be called barren and degraded lands respectively. This distinction is vital, otherwise there is a danger that better quality lands may be put under industrial plantations under the garb of 'afforestation of wastelands'. Although there is no scientific basis for choosing 10 as a cut-off point, yet a limit has to set in order to distinguish lands which are almost devoid of vegetation and have almost ceased to be of importance to the local people from degraded lands which support local livelihoods and provide fuelwood and fodder to them. From the statement and letter of the Minister quoted above it is apparent that the proposal is restricted to barren lands and there is no move to involve the industry on lands with better density of present tree cover. This is also supported by para 9 of the draft guidelines (though not yet finalised, at Annexure 3) prepared by the Ministry on this subject some years back: 'First, the most severely degraded forests below 10 per cent density will be allotted and afterwards next category shall be offered for plantation. The land above 25 per cent density shall not be available for plantation.

It may be pertinent to mention here that despite the efforts of the Ministry of Environment and Forests, Government of India, to convince everyone that only lands denuded of forest cover will be offered to industry, neither the industry nor the state governments of M.P., Orissa, and Arunachal Pradesh, the states which have shown keenness to lease forest lands to industry, have categorically accepted this position. Industry has clearly stated about the nature of lands of their interest, that these will have to be forest lands with a high tree density, may be ranging between 25 and 40 per cent. In a meeting the District Magistrate of Chindwara, M.P. disclosed that the paper industry was offered 5000 ha of forest lands, but they refused as they thought that it was too degraded and laterite in nature. The industry for good reasons would like to invest their capital only on land which has a good soil texture and where the cost of plantations is less than the cost of alternative supplies. Such lands have to be forest lands of around 40 per cent crown cover, and not barren wastelands, whether inside forests or outside.

The Group was informed that the paper industry which has been in the forefront of putting pressure for leases on government in particular, required about 2 m ha of degraded forest land over a period of 10 years, with soil depth of at least one metre. Industry was not interested in barren lands with poor soil cover. Given the requirement of the industry of good forest land with at least one metre soil depth and an almost contiguous patch of a 10,000 hectares, a question arises whether such lands can ever be located within forest areas and still be classified as degraded forests?

To answer this question we have to look for the extent of forest degradation in India and factors that caused this degradation.

Nature and Causes of Forest Degradation in India

Before considering the viability of leasing of degraded forest lands it is important to analyse the reasons for degradation, and why previous efforts to make these lands productive through plantations had limited success.

Forests are subject to intense pressure from human beings, livestock and urban markets. The relative contribution to deforestation of the two categories of consumers, people and industry, has been a subject of controversy in India, which has blurred its objective analysis. FD officials argue that since commercial and industrial requirements are low as a proportion of the total demand for wood, at less than 20 per cent, peoples' demands put an unbearable burden on forests. The almost continual lopping for fuelwood and/or fodder as well as cattle and goat browsing, that occurs in many areas and prevents adequate regeneration, must play a major role in forest destruction. For instance, most of the Maharashtra soils contain a large clay proportion, compaction by trampling makes the surface impervious to water penetration and increases the soil density. The former leads to run-off, flooding and erosion during the monsoons and the latter reduces the tree growth significantly. The result is that most forest lands have no regeneration.

On the other hand it is argued that subsidised supplies to industries and giving them priority has reduced availability for the people and resulted in their further alienation from the forests turning it into an open access resource. A study showed that one-third of rapid deforestation in Himachal Pradesh was due to excessive exploitation by the forest-dwellers, and the rest due to commercial interests. Often the two processes of industrial extraction and use by the people follow each other. The selective logging of a few large trees creates openings in the crown cover leading to better grass production, which invites cattle and goats. Their browsing makes regeneration difficult, and then the area is invaded by exotic, non-palatable weed species.

Some authors make a distinction between use of fuelwood by rural people, which is largely twigs and branches and hence potentially sustainable, and by urban sector. The greater use of logs and larger branches in the towns means that reasonably-sized trees are sought after and cut, possibly in large patches thus having a more degrading effect on the forest than may be the case with village needs that can be obtained more often from pruning or pollarding branches of trees or even bushes in a limited area. Thus collection of fuelwood for sale in urban areas is the cause of much destruction and degradation of forests.

Most forests are located in regions of poor soil, where agriculture is not very productive, hence forests are the source of a major portion of income through sale of fuelwood, called headloading. A CSE study estimated that at least 3 to 4 million people were involved in headloading, making it India's biggest source of employment in the energy sector. It is a low paid and a high risk occupation, as pilfering wood from RF for sale is an offence (collecting wood for self-consumption from PF is permitted on paper, but frowned upon by the forest staff in actual practice). The study commented that it was ironic that tribals, who for centuries lived in harmony with forests were today forced to eke out a living by further destroying their forests.

Discussions with field officers indicate that during the last three decades two other processes led to fast deforestation. One, arising out of political populism, was to allow people to harvest in an unsustainable manner, and the other was pressure on officers to contribute more to revenues. It was not unusual for forest officers to receive letters right down at the operative level from the highest officials of the Forest Department instructing them to produce more revenue. The same pressure was repeatedly conveyed in the monthly meetings of Range

Officers, always in the same form: "Range X has contributed (a stated amount revenue) and Range Y is not contributing enough." Ironically, before an election, it was common for a Minister to order that the forests be opened for the people without penalty. But after the election, the same Minister was likely to demand more revenue.

Finally, deforestation has often been associated with sudden policy change or periods of uncertainty, like takeover of private forests, abolition of landlordism, setting up of Forest Corporations, and political unrest. Once large trees are felled, the old harmony between people's demands and supplies through twigs and branches gets disrupted, and Department's efforts towards replanting come to naught. Thus deforestation, rather than being a continuous phenomenon, could be interpreted as a one-shot operation often directed by governmental activity. It occurs as a result of not just local pressures on resources, but also 'any momentary disruption of the institutional framework responsible for resource protection and management'.

Aggravating the situation was the paucity of funds for the forest sector—forestry accounting for less than one per cent of the development budget as compared to agriculture which received between 20 to 24 per cent. Although the availability of funds has improved in absolute terms, the proportion of forestry budget to the overall plan budget has remained generally less than 1 per cent. However, if flows to this sector from the Ministry of Rural Areas and Employment are taken into account, the share of funds for forestry may well be above one per cent.

Deforestation is, however, not a post-independence phenomenon or linked to the increase in population of people and cattle. It is very important to highlight nearly 100 years of extensive destruction that the forests of this country have suffered under the colonial regime. The colonial regime was mainly interested in converting into money as much of the resources as was possible. In the southern parts of the country where British rule lasted little longer, destruction was for even a longer period. The forests were initially worked for producing high quality timbers for the British Navy as well as for export. Large areas of forests were also destroyed for the creation and operation of vast railway network. In several parts of the country which were far removed from the coal fields, firewood from the forests was the fuel for the railways. Very extensive tracts of forests were also destroyed for conversion into plantations of tea, rubber, coffee, cinchona and indigo. Further, these plantation crops destroyed forests not only for creating plantations but even for their operation due to supply of fuel for the labour and even for marketing the product. Every kilo of tea produced required at least two kilos of firewood. Working of forests for clearfelling as fuel coupes or for large scale plantations of mono-cultures of commercially important trees is another very widely practiced forestry operation which was totally alien in this country and is against all principles of conservation and sustainable management.

One more very significant omission about causes of forest degradation is forest fires. Forest fires have been causing havoc for a very long period of time and were encouraged and tolerated as these fires were not particularly destructive to commercially important species like teak.

To sum up, it is not possible to determine to what extent each of the above factors causes deforestation. It can occur through excessive depletion of young trees by households with even a limited demand for fuelwood under open access forest exploitation. Denying rights to the local people has led to the organised fellings in the Jharkhand area by the people themselves who have been damaging the forests. Local patterns of deforestation vary and it is 'never a simple matter of numbers outstripping environment'. On the whole, pressure from people do contribute a great deal towards forest degradation, which is compounded by subsidised supplies to industry. The alienation of forest lands from the people who need it for satisfying their needs, and consequently forests turning into open access lands has been one of the main cause for degradation as well as for increasing misery of the people. Till the mid-eighties the response of the government to this crisis of deforestation was to bring more area under non-browsable and market oriented single product timber trees in order to reduce pressure from local population and increase state revenues. This strategy turned counter-productive and hastened the process it was designed to prevent.

The correction came in the form of a new Forest Policy in 1988 followed by June 1990 guidelines recommending Joint Forest Management (JFM). Although the old thinking is still not dead, yet by now almost all major States have passed enabling resolutions to implement the programme. The Government of India and the donor agencies, appreciating the realistic foundation of the new concept, have decided to make the JFM the central point of its future forest development programmes.

Characteristics of Degraded Forests

Degraded forests have three characteristics that must be kept in mind when one is thinking of how best to make them productive. Firstly, degraded forest lands with crown density of 10 and 40 per cent are not likely to be available in contiguous patches, unless combined with good quality forests. According to Mr. Thapar, industrial plantations would require about 10,000 ha. of land in a few close patches for plantation. Paper industry has to be internationally competitive which would need raising plantation on good quality lands available in one patch. The industry would not be able to supervise plantation and protection over lands that are distributed over hundreds of villages. The experience of social forestry showed that there was no continuity in the management and control of thousands of scattered pieces of planted village lands creating enormous problems of protection.

Dependence on Forests

Secondly, forests are not spread evenly in India, but are concentrated in the Northeast, the Himalayas and Shivalik ranges, the central belt, strips along the Western Ghats and other hill areas, and in patches of coastal mangroves. More than 50 per cent of forest land is located in the central belt, which is the poorest region in India with heavy tribal concentration. India's forests have generally speaking not been uninhabited wildernesses. Even in the remote forests people have either been living traditionally or were brought by the Forest Department and settled there to ensure the availability of labour. Today, there are about 100 million forest dwellers in the country living in and around forest lands for whom forests have continued to be an important source of their livelihoods and means of survival (Lynch, 1992).

Besides fuelwood and other wood products, forests provide what were misleadingly termed "minor" forest products, and now better known as NTFPs. Most NTFPs come from forests, although some trees yielding NTFPs occur on private fields too, providing valuable assets and flows for subsistence and cash. Seventy per cent of NTFPs are collected from the five states of Maharashtra, Madhya Pradesh, Bihar, Orissa, and AP where live 65 per cent of the tribal population (Guha, 1983: 1890). Much collection is done in the lean agricultural months of March-July when other employment is not available.

As most degraded forests are close to habitations, they have extreme biotic pressure, and often have rights of the people recorded on such lands. In fact these lands are degraded primarily because of peoples' pressure. It will be impossible to extinguish such rights, or to reduce biotic pressure. Already, in many protected areas reserved for biodiversity and wildlife, government is finding it tough to deal with people's rights of access. If land is used by private entrepreneurs (which is a less legitimate cause than preservation of wildlife), such problems will multiply manyfold, and will attract attention of the social activists and press, who will get a readymade battle field to fight and criticise government. One should not forget the intensity of public feeling and agitations sparked by the Silent Valley Project in Kerala, The Blue Pine Project of M.P. and the Harihar Polyfibre Project of Karnataka. In all such cases proposal to use lands which had peoples' livelihoods depend on it, were strongly resisted by the people.

Thirdly, departmental plantations raised by Forest Department on degraded forest lands have generally been a failure because of pressure from cattle and people. In Orissa, a study by SIDA showed that survival of plantations after a few years was as low as 15 per cent, because peoples organisations were not consulted and their requirements were not taken into considerations. Degraded forest lands require not capital investment, nor even higher technology, but protection and recuperation, which can be done only by working with the people, where industry has neither expertise nor patience. The West Bengal experience shows that about 2000 peoples' forest protection committees have regenerated more than 300,000 acres of sal forests at no extra investment, simply by protection on the promise of sharing wood and non-wood products with them. People were willing to give their time and labour in looking after regenerating forests, because they identified with "their own forests" and were confident that they would benefit from such an effort. It was this sense of identification which was missing in governmental efforts in the past. Finding funds was not the main constraint. If lands on which peoples' livelihoods are dependent are given to the industry, they may have to employ mafia gangs to keep people out, thus completing the cycle; in addition to politician-mafia nexus and politician-industry nexus we shall witness the rise of industry-mafia nexus.

To sum up the three points made in this section, degraded forests with some soil depth are dispersed, they have extreme biotic pressure and therefore are ideally suited for regeneration provided one works with the people. Such forests are certainly not "useless", as they satisfy livelihood needs of about 100 million people.

Past Experience of Plantations

One may state here that there is nothing new in the present proposal of using forest lands for commercial plantations. In fact, as this was the practice for about a hundred years in India, a great deal of knowledge on environmental, economic and social impact of using forest lands for industrial plantations is available. What is perhaps new is the suggestion of bringing government forests under private management (although informal arrangements of industry's control over forest lands had existed in the past and still continues over bamboo forests in some states). Therefore, we will consider the likely impact of this suggestion under two heads, the impact due to commercialisation of forests, and the implication of privatising forests.

Environmental Impact

Till 1988, forest policy in India was to obtain sustained yield of timber from government forests by planting non-browsable and market oriented timber trees. Right from the First Plan in 1952, emphasis was laid on the conversion of "low" value mixed forests into "high" value plantations of commercial species like teak, and eucalyptus. Thus scientific forestry was equated with raising of man-made forests, designed to meet industrial requirements. A diverse forest eco-system was converted by government into a single species 'timber mine'. The foresters, who were to conserve the forest ecosystem, became the main agents of reducing the diversity of forest species.

The National Commission on Agriculture (NCA) unfortunately put its stamp of approval on this approach in the following terms, *'There should be a change over from the conservation oriented forestry to more dynamic programme of production forestry. The future production programme should concentrate on clear felling of mixed forests, mixed quality forest and inaccessible hard wood forests and planting these areas with suitable fast growing species yielding higher returns per unit area .... Production of industrial wood would have to be the* raison d'etre *for the existence of forests. It should be project-oriented and commercially feasible from the point of view of cost and return.'*

It is interesting to note that the NCA made a specific recommendation (see Para 42.3.18) against leasing forest lands to industry. It wanted government to allocate to them forest raw material from the enhanced programme of commercial forestry on forest lands. Out of 64 m ha of forest lands in India, the NCA recommended that 48 m ha should be dedicated to production forests, and the rest 16 m ha for biological diversity. Thus the NCA did not want any part of forests to be used for grazing or meeting other subsistence needs of the people.

For instance, in Madhya Pradesh, as in other states, the entire thrust of forestry was towards the high forest system, created after clear-felling and ruthless cutting back of all growth, except of the species chosen for dominance. The 6th Five Year Plan (1980-85) of Madhya Pradesh stated, *'to produce 25 million cm of industrial wood it would be necessary to subject 5.5 million ha of production forest lands to the intensive management, that is to clear-felling and planting. .... with the massive plantation programme being launched in the state, there would be extensive monocrops of teak in the forests... we should clear-fell and plant roughly one lakh hectare annually if we want production of industrial wood to keep pace with demand in future.'* As regards efforts made in the past to meet tribal demands for fruit, medicinal herbs etc. from forest lands, the same Plan document admitted, *'no special programme were taken, which could directly contribute to the upliftment of the tribal economy.*

For their part, plantations have usually been single species, or involving only a few species, equally entailing loss of diversity and access, and often on a large scale, and in practice hardly pursuing an objective of benefiting the local people, beyond wages. In Kerala, for example, since 1970, 1.5 lakh ha of area has been brought under man-made plantations replacing mixed forests (FAO, 1984:37). Although watershed protection was mentioned as one of the objectives in the working plans, no prescriptions existed and no steps were taken to enhance protective value of forests. The same FAO study observed that timber production and revenue maximisation got the highest priority, while watershed management was ignored.

Some of the basic roles to be performed by the forests which cannot be performed by any other land use are (i) regulation of the water cycle; and sustaining stream flows; (ii) bio-diversity conservation and (iii) amelioration of climate. While timber and pulpwood can be grown on any piece of land, these basic roles can be secured only by the well managed and well preserved natural forests.

Conventional forestry based on clear felling disrupts the annual circulation of nutrients, and increases soil erosion (Spurr and Barnes, 1980: 240). Monoculture plantation forestry is also prone to pest attack. Thus eucalyptus plantations in Kerala raised after clearfelling dense green and deciduous forests have been devastated by fungal diseases, and the consequent low productivity defeated the very purpose for which they had been raised (Nair, 1985). Similar was the experience of teak plantations in north Bastar, M.P. (Anderson and Huber, 1988: 63). On the other hand, mixed forests draw and give nutrients to the soil at different stages of their growth, and hence are ecologically far more beneficial than plantations. One of the main outputs from forests should be water, which is possible only when forests are considered more in the context of local rather than "national" needs.

Tropical forests support complex ecological chains while playing an essential and salutary role in the earth's climate and atmosphere. They can return as much as 75 per cent of the moisture they receive to the atmosphere. Thus they have a profound effect on rainfall. Yet these vast natural forests, surrounded by poor populations, were rapidly diminished as they were being converted into plantations to meet market demands.

Thus raising short-term and quick growing species in place of multi-layer mixed forests has serious ecological implications. For restoring ecological balance mixed species should be encouraged through protection and regeneration, and not plantation of mono-cultures. Using degraded forest lands for growing raw material for industry will be setting the clock back to the 1960s, showing that we learnt nothing from the mistakes of the past 30 years of trying to create man-made forests, which were ecological disasters, besides completely alienating the people and leading to faster degradation. The proponents of leasing to the private entrepreneurs have argued that the environmental risk would be minimised by the following measures:

- 20 per cent of identified areas will be developed as permanent forest.
- Balance 80 per cent area to be planted with (minimum) 3 species—no single species to cover more than 50 per cent area.
- Lops and tops to be made available to local communities free to meet fuelwood requirements.

However, looking at the past record of involving industry, it is very difficult to believe that the above measures would be carried out with sincerity. The industry may, for instance, turn around and claim that other species did not survive, or that the people themselves uprooted the seedlings of other species. Once the control over land is given to the industry, it is difficult to ensure safety clauses. As the above measures reduce the profitability from investment, it is but natural that industry would tend to ignore these, once it has physical control over land, and with money power has 'managed' those in power.

One may also bear in mind that two or three species do not make a forest. Besides these would be periodically harvested, thus leading to loss of biodiversity and risk of soil erosion as well as encroachment for agriculture. The needs of the local communities, who basically need small amounts of NTFPs, are in harmony with the basic objective of biodiversity conservation, as it does not involve clear-felling. Industry has no credible plan to meet the hardships which would be caused to the local communities, which have historically depended on these lands for meeting their basic needs.

Economic Impact

One of the main arguments given in favour of involvement of industries in afforestation is that government lacks funds, and with limited resources it is not possible to afforest degraded lands. It is conveniently forgotten that this same argument was given by the National Commission on Agriculture when it recommended in August, 1972 that Forest Development Corporations (FDCs) registered under the Companies Act should be created to attract institutional finance.

There are 26 FDCs in the country, out of which 11 have raised plantations over 12,36,487 ha, mostly in the states of AP, Maharashtra, Karnataka, MP, Tamil Nadu and Gujarat. A government evaluation of FDCs mentioned, 'FDCs have not achieved the desired improvement of productivity of forest lands as evidenced by low yield, poor growth and low survival of plantations created by them.'

These were set up in each state to augment state resources; logic was simple, for commercial activities banks will be too willing to lend money. Did the Corporations succeed in attracting bank funds? A simple answer is no. For instance, the Madhya Pradesh State Forest Department Corporation was registered in July, 1975, but its total paid up capital of Rs 16.30 crores is contributed entirely by the state government. The Corporation was created so as to attract bank funds, and create additionality of resources for the government. However, in the last ten years, the Corporation has been able to get bank funds in only one year. In other years, there was no additionality of resources. It depended entirely on government funds for its schemes. There were several reasons for the Corporations in not having succeeded in achieving even the limited economic objective of raising commercially viable industrial plantations.

Apart from poor management, most commercial species require the presence of mixed forests for adequate growth. Shri R.S. Mishra, IFS, in the working plan of South Betul Division for the period 1973-74 to 1987-88, has appropriately remarked that it is in the mixed forest that invasion of teak is rapid, whereas in blank and understocked areas teak makes slow gain. Both the Sal Expert Committee and Dr. Ram Prasad, ex-Director, S.F.R.I., Jabalpur, have commended that sal itself regenerated best when there is appropriate support vegetation. The position in relation to teak could not be better summed up than in R.S. Mishra's words, "Mixed species found on the archaen soils are naturally deep rooted. In doing so they extend the zone of underground drainage and soil aeration to a greater depth. It is this..... which is responsible not only for spreading but also for good health, vigour and quality of teak, which requires both good drainage and good soil aeration for its pure expression. The removal of mixed species very soon leads to surface compaction.... This is how mixed species are ecologically important in relation to teak in the high forests of Betul district." Provided that the mixed forests are worked for nistar in a scientific manner, there is no reason why the very biotic pressures which now operate in the forests should not be used as a means of creating the best growing conditions for teak and sal.

It is now being proposed by some that the management of degraded lands should remain with government corporations but private companies may invest their money in the plantations. If government banks have not found the activities of these corporations worthy of investment, why would private companies do so? Forestry being a long gestation activity, it requires stable tenure. If land, management, and source of funds is going to be with three different parties; government, FDCs, and private industry respectively, the risk for the investor is very high, unless there is unwritten and unstated understanding between the parties.

Another reason for failure of commercial plantations has been the acute pressure on such lands, as well as the activities of smugglers. Commercial plantations increase hostility between government and the local people. In some districts, such as Bastar and Rajnandgaon, tribals have organised themselves and resist planting of commercial species on forest lands. Instances are not wanting where people (not naxalites) uprooted commercial seedlings planted by the FD or Forest Corporations. The experience of tribal states, such as Andhra Pradesh, shows that awareness among the tribals is likely to increase in future (as it should), and that they can no longer be taken for granted in the matter of choice of species. Despite the popularity of teak in the Forest Department in A.P., it was difficult to protect teak in natural forests, as it was being smuggled and stolen at a very fast rate. The experience of Maharashtra and Gujarat is that after teak is ten years old, villagers cut it illegally and take it away. The future of pure teak plantations, which take 60 to 80 years, is therefore bleak. Thus planting market oriented species encourages smugglers or indisciplined behaviour.

To sum up, if public sector banks have not found the performance of FDCs inspiring enough to invest in them, it is difficult to believe that the private sector will lend money to the FDCs, unless of course industry wishes to revert back to the days of almost free supplies from government forests. The PCCF Andhra Pradesh has suggested that 'a review is necessary of the utilisation of forest lands given to FDCs, as they are serving the interest of the industry rather than the poor forest dwellers.' Industrial plantation whether by the private entrepreneurs or by the FDCs has had little chance of success so far in the face of acute human and cattle pressure.

Social Impact

Turning a complex forest into a genetically simplified plantation converts the local tribal into a second-class inhabitant. One-third of the total growing stock in Madhya Pradesh is in one district alone, Bastar, where there should be no shortage of fuelwood and other forest products, and yet it is ironic that foresters—tribal relationships are at its worst in Bastar, thus suggesting that the real issue is not shortages of fuelwood and fodder, but larger issues of control and power. The argument of the industry that only 2 m ha out of 67 m ha of government forests need be given to industry is of no consolation to those local forest dwellers from whom their life support system will be snatched.

No less a figure than Shri Dalvi, IG Forests, while addressing the 1981 International Conference on tropical forest management at Dehradun illustrated the inherent conflict arising out of forest plantations in the following terms:

> *'Let us consider another example of a natural forest predominantly of sal. This forest represents to poor forest-fringe-dwellers a source of livelihood yielding seeds for sale, branches and leaves for fuel and manure. The decision to convert this sal forest to industrially more valuable species like teak may satisfy the needs for higher revenues which may or may not be used for the welfare of these same people, but would certainly deprive them or an output from the forest which they were enjoying.'*

Other writers have been less charitable about the intentions of the government. An ex-Forest Secretary of Madhya Pradesh writes,

> *This (the policy of giving priority to industries and subsidising industrial raw material) is clearly discriminatory. The rights of a huge section of society cannot be wiped out in order to benefit a few industrialists. For instance, the Orient Paper Mills was promised a lakh (100,000) ton of bamboo per year from four districts of the state. This eliminated all bamboo from Rewa, Panna, Satna and Shahdol. When such a situation arises the Forest Department tells the villagers to fend for themselves because there is nothing in the forests for them (NCHSE, 1987;* iv).

The total number of cane, bamboo and basket weavers in 1981 was 8.2 lakhs, of which 6.9 were in rural areas. A common problem of all bamboo artisans is that of raw material shortages. The major cause for this is their diversion to paper and other industries, which often procure the raw material from state government at throw-away prices, much to the disadvantage of the artisans.

A plantation loses the old character of the forests. For instance, Mahua is of no significance to the Forest Department, nor any efforts have been made to increase its number in forests. No doubt, Mahua is also not felled by the department, but its significance for them is not the same as for tribals. Compared with the Forest Department, tribal involvement in Mahua is pervasive to a profound degree. In addition to collecting flowers and seeds for sale at the weekly market, or for exchange for salt or cloth, tribals use the wood to support the canopy at wedding celebrations, the dried flower to add bulk to their food or to feed their animals, the seeds and the flowers for preparing liquor and for religious ceremonies. Creation of such mixed forests is possible only when people are not alienated, and are at the centre stage of forest planning.

Transfer of Management of Forest Lands to Industry

As already stated, Ministry of Environment and Forest claims that there is no proposal to alienate forest land, and that forest land will continue to be with government. However, the proposals which have been drawn by the state governments and the industry assume that the management of forest lands will be transferred to them, though not ownership, through a MOU between the FDC and the industry. These proposals are being discussed below.

The Orissa Government has proposed that the industry should pay 25 per cent of the outlay of the project as interest free loan, and then it will appoint a Project Manager and become overall in-charge of the project, solely responsible for day to day working, execution of works and implementation of the Project till harvest. The industry will have the option to buy the produce from the plantations at the prevailing market rate, and the FDC can dispose of the produce only after the refusal of the industry in writing (which gives a veto power to the industry, and hence it will decide what is the market rate). Twenty-five per cent of the produce including lop-tops, bark and small timber will be earmarked for local people at rates to be decided by government in consultation with the Project authorities. It appears from the MOU between the FDC and the industry (which has already been signed in April 1993) that the rest of the 75 per cent of the expenses will come from banks, which is possible only when plantation is on fertile lands. In any case the MOU never states that only degraded forest lands will be taken up. Also, if land remains under the control of the industry, chances of a part of the production being siphoned off under the table cannot be ruled out.

Similar to Orissa, the Arunachal Pradesh Government has also proposed that industry will have the right of surface utilisation for the purpose of raising tree plantation. In other words, land will remain under the control of industry. They will also be allowed to raise temporary structures. One wonders if this is not alienation of forest lands or if this is not leasing, how else does one define leasing or subletting. The Arunachal Pradesh Government proposes that the gross revenues accrued from the plantation activity will be shared with the government in ratio of 75:25. In other words, the government is leasing out the forest lands by charging a rent of 25 per cent only. In addition, government would charge Rs. 2500 per ha. as 'opportunity cost' of the allotted land. The annual share of the department from the sixth year onwards should not be less than Rs. 10,000 per ha. The government has already identified 25,000 ha. which the wood-based industry proposes to regenerate over a period of 15 years on an investment of about Rs. 250 crores.

The draft conditions formulated by the Ministry of Environment and Forest in 1994 (at Annexure 3) reveal that the industry will get the right over land for undertaking afforestation and right over fixed percentage of forest produce, but land will not be leased to the industry. It is difficult to understand the fine distinction which is proposed in the guidelines between leasing and handing over possession to the industry. These guidelines suggest that the land above 25 per cent density shall not be available for plantation, although the proposals finalised by the State Governments do not restrict themselves to 25 per cent density.

In a letter dated 15th July,1994 addressed to the Minister, Enviornment and Forest, Government of India, Shri Thapar, Vice-Chairman, Joint Committee on Paper Industry has given two models for raising pulpwood plantation. Both envisage handing over of the possession of forest lands to industry. The proposal is silent over the question of parting any revenue or royalty to government. Establishment of social forestry on 15-20 per cent of land is seen as rent, in other words, almost free of charge. Purely from commercial point of view, it may be more profitable for government to auction forests!

There is no mention in either the Government of India guidelines or the State Government's proposals or those from industry as to how the rights and concessions which people enjoy over these lands will be affected. As degraded forest lands are most likely to be protected forests, the issue of peoples' rights cannot be swept under the carpet. It is significant that the Government of Arunachal Pradesh, which has the second highest area under forest after Madhya Pradesh, has identified reserved forests of Paya and Digru located in district Lohit to be leased to industry. On the one hand, the Forest Conservation Act specifically mentions that the reserved forest land will not be de-reserved and the new Forest Act (which is still in a draft stage) bans not only de-reservation but disallows any form of Joint Forest Management on reserved forests, on the other hand control and possession over reserved forests is being proposed to be handed over to private parties. It is feared that other States will also follow suit because industry will not like to get into legal hassles of dealing with rights and concessions of a large number of forest dwellers. Is it fair to surrender exclusive control over forests to the rich, and deny people's participation even on a joint basis? Politically too, it would be difficult to sell this idea in a country where there is so much of land hunger.

Sometimes industries, in order to maximise production use methods which are destructive to trees. An obvious example is extraction of resin from pine trees. Where industries held bamboo leases they utilise even the better quality bamboo for pulp, although according to rules only inferior quality bamboo should be used as pulp, and the better quality should be sold to artisans. Furer-Haimendorf (1985) describes how a particular paper mill in Andhra Pradesh exploited bamboo in a tribal region by bringing in hundreds of labourers from different states and used methods of extraction which endangered future regeneration.

Many countries such as Malaysia and Philippines have experience of leasing out forests to logging companies. The general experience has been that bifurcation of ownership and management leads to faster deforestation and the arrangement is not conducive to long-term investment by either parties. Moreover, political masters see such arrangements as patronage, and give concessions to the industry at a price. For instance, in many developing countries, licence fees is waived or its recovery stayed after the industry has paid "political subscription". Once possession over forest lands is transferred to the rich, such concessions appear justified in the interest of "sustained production". The suggestion that the management of forest lands should be given to private industry amounts to nothing but share cropping. A great deal of literature exists in agriculture on the evils of share cropping and leasing. In our own country, the abolition of intermediaries and zamindari system in the early 1950s was based on the rationale that it is only owner cultivation which can maximise production, and that share cropping is inimical to production. When compared with agriculture, forests require even more secure tenure, otherwise with diluted rights over forests, industry is likely to behave in a very irreponsible manner, and maximise short term gains at the cost of environmental considerations.

Implication of Leasing for Forest Policy and Law

Considering the heavy dependence of the people on forests, the Group has come to the conclusion that forests can be sustained only if the forest policy is geared to meeting the needs of the people, and involves them in its management. For almost a century and until 1988, the policy was to use these lands for commercial purposes, and to look upon the people as a liability. This led to deforestation, which increased rural poverty and unemployment on the one hand, and social unrest and conflict over rights on the other. A shift in policy occurred in 1988 as planners recognised the difficulty of excluding people from forests, although the policy instruments of improving the effectiveness of communities in local protection are still not very well understood.

The Forest Policy, 1952 declared that village communities should in no event be permitted to use forests at the cost of 'national interest', which was identified with defence, communications and vital industries. It wanted forests to be used to produce valuable timber for industry and other national purposes. The policy stated,

> *The accident of a village being situated close to a forest does not prejudice the right of the country as a whole to receive benefits of a national asset. The scientific conservation of a forest inevitably involves the regulation of rights and the restriction of the privilege of users depending upon the value and importance of the forest, however irksome such restraints may be to the neighbouring areas....... While, therefore, the needs of the local population must be met to a reasonable extent, national interests should not be sacrificed because they are not directly discernible, nor should the rights and interests of future generations be subordinated to the improvidence of the present generation.*

In budgetary allocations too, emphasis was laid on the conversion of 'low' value mixed forests into 'high' value plantations of commercial species such as teak and eucalyptus. Forestry at that time meant raising trees in order to get sustained yield of timber in perpetuity. Exotic species were introduced to create man-made forests. Between

1952 and 1980 over three million ha of plantations were established, the major proportion of which were to fulfil industrial needs (CSE, 1982). Out of Rs. 670 million spent on afforestation during 1966-74, roughly Rs. 560 million was on production forestry alone (GOI, 1981: 45). The use of bamboo for paper manufacture accelerated from a low of 58,000 tonnes at the end of the Second World War to over 5 million tonnes by 1987 (Hobley, 1996).

The National Commission on Agriculture (NCA) also supported the view that, '*Resources for industrial raw material, both for internal consumption and export, should be stepped up through large-scale industrial plantations on forest lands.*'

Thus the entire thrust of forestry during the first four decades after Independence was towards the production of a uniform industrial cropping system, created after clear-felling and ruthless cutting back of all growth, except of the species chosen for dominance. Far more emphasis was placed on plantation rather than on management of existing trees.

The 1988 Forest Policy

In all sectors of the economy, where both government and private sector work, it is generally the government sector which looks after the infrastructure or welfare needs of the people, whereas market needs are met from the private sector. Thus health, education, roads etc. which are non-commercial programmes and look after the welfare needs of the people are under the domain of government, whereas private sector looks after commercial production. It was strange that in forestry this distribution of responsibility was not being followed, and the reverse was being attempted so far and even endorsed by the NCA, as it suggested that forest lands were to meet the commercial needs of the economy and farm lands were to produce 'fuelwood and fodder'. As regards village lands, despite rhetoric to the contrary, these lands in actual practice produced commercial polewood or urban fuelwood, and did not meet subsistence needs of the poor. This conceptual weakness was perhaps one of the main reasons for the failure of the social forestry model.

This anomaly was corrected by the new forest policy announced in 1988 which is radically different from the two previous policies.

According to this, forests are not to be commercially exploited for industries, but they are to conserve soil and the environment, and meet the subsistence requirements of the local people. The policy gives higher priority to environmental stability than to earning revenue. Derivation of direct economic benefit from forests has been subordinated to the objective of ensuring environmental stability and maintenance of ecological balance. It discourages monocultures and prefers mixed forests. The focus has shifted from 'commerce', and 'investment' to ecology and satisfying minimum needs of the people, providing fuelwood and fodder, and strengthening the tribal-forest linkages.

Parra 2.2 of the new Policy defines the basic objective of forest management:

> *The principal aim of the forest policy must be to ensure environmental stability and maintenance of ecological balance including atmospheric equilibrium which are vital for sustenance of all life forms, human, animal and plant. The derivation of direct economic benefit must be subordinated to this principal aim.*

Para 4.3 of the new Policy reads as follows:

> *The life of tribals and other poor living within and near forest revolves around forests. The rights and concessions enjoyed by them should be fully protected. Their domestic requirements of fuelwood, fodder, minor forest produce, and construction timber should be the first charge on forest produce.*

Similarly Para 4.6 of the Policy lays down:

> *Having regard to the symbiotic relationship between the tribal people and forests, a primary task of all agencies responsible for forest management, including the forest development corporations should be to associate the tribal people intimately in the protection, regeneration and development of forests as well as to provide gainful employment to people living in and around the forest. While safeguarding the customary rights and*

interests of such people, forestry programmes should pay special attention to undertaking integrated area development programmes to meet the needs of the tribal economy in and around the forest areas, including the provision of alternative sources of domestic energy on a subsidised basis, to reduce the pressure on the existing forest areas.

The Policy states in para 3.5 that

'*Minor Forest Produce provides sustenance to tribal population and to other communities residing in and around the forests. Such produce should be protected, improved and their production enhanced with due regard to generation of employment and income*'.

As regards supplies to industry, the first part of Para 4.9 states:

As far as possible, forest based industry should raise the raw material needed for meeting its own requirements, preferably by establishment of a direct relationship between the factory and the individuals who can grow the raw material by supporting the individuals with inputs including credit, constant technical advice and finally harvesting and transport services.

It is also stated in the same para that

'*the practice of supply of forest produce to industry at concessional prices should cease. Industry should be encouraged to use alternative raw materials. Import of wood and wood products should be liberalised.*'

Two factors must have weighed upon the minds of policy planners in suggesting a diminished role for forest industries on forest lands in the new Forest Policy. First, the popularity of eucalyptus among farmers increased the availability of pulpwood at a cheap price for the paper industry. In some states, such as Gujarat, large farmers even took to teak plantations, a timber crop which takes 30 to 60 years to mature. And secondly, liberalised imports of pulpwood was permitted which eased the supply for the industry. With new sources of supply, it was no longer considered crucial for the industry to depend on government forests.

The implementation of the Policy was facilitated by the Government of India by issuing a resolution in June 1990 making it possible for the Forest Departments to involve people in the management of forests. In the states of Orissa, Andhra Pradesh, West Bengal, and Bihar there are a large number of groups protecting forests either on their own or initiated by the Forest Department, as Table 6 would indicate.

Table 6: Actual Forest Cover and Area Under JFM in Certain States (sq.km)

State/UT	*Forest Cover*	*Area Under JFM*	*JFM Area as per cent of Forest Cover*
Kerala	10,334	20	0.19
Himachal Pradesh	12,521	60	0.48
Jammu & Kashmir	20,440	141	0.69
Karnataka	32,403	339	1.05
Rajasthan	13,353	244	1.83
Gujarat	12,578	259	2.06
Madhya Pradesh	131,195	3,500	2.67
Orissa	46,941	2,960	6.31
Haryana	604	63	10.43
Andhra Pradesh	43,290	6,480	14.97
Bihar	26,524	7,103	26.78
West Bengal	8,349	4,493	53.81
Total	358,532	25,662	7.16

Source : SPWD, 1998.

Thus in the previous policies people and the environment were seen, all too often, as antagonistic. The forest-people interaction was conceptualised as a zero-sum game, in which both parties could not win. According to the JFM philosophy, the conflict model is neither necessary nor useful. To the contrary, ways can be sought in which the interests of people and of long-term sustainability are harmonised in a mutually supporting manner. This has also sharply reduced the rate of forest degradation, as stated in a recent World Bank report:

> *Forest canopy cover has held up surprisingly well under these depredations. Four biennial estimates of forest surface cover during the 1980s and early 1990s show little variation over roughly last ten years despite net removal before reforestation/ afforestation of about 3.3 m ha worth of wood annually. Trends in "Forestry and Logging" output and prices over the last 30 years suggest that there has been some curtailment of the exploitation of forest resources since the mid-1970s.*

The proposal of the industry would reverse the present happy trend and bring us back to the 1970s when the rate of deforestation was alarmingly high.

Forest Conservation Act

The other policy initiative taken by the Government of India is the Forest Conservation Act, 1980. This Act limits the powers of the state governments, and they cannot de-reserve Reserved Forests or divert forest lands for non-forest purposes unless permission of the Government of India has been obtained. The states cannot now lease out forests to the private entrepreneurs/Corporations not owned by government. The operative section 2 (iii) of the Act (after amendment in 1988) reads as follows:

> *'No State Government or other authority shall make, except with the prior approval of the Central Government, any order directing that any forest land or any portion thereof may be assigned by way of lease or otherwise to any private person or to any other authority, corporation, agency or any other organisation not owned, managed or controlled by government.'*

Despite the stringent law, some states, such as Orissa, have decided to assign bamboo forests to paper industry, who have been appointed sub-agents of the OFDC. An interesting issue of law arises whether the arrangement with the paper industry amounts to leasing of forests or not. The official position is that it is leasing of forest produce, and not of land, and hence the provisions of the Forest Conservation Act are not attracted. However, an arrangement would amount to leasing of land if the lessee pays a royalty to the owner of land, the payment to wage labourers is made by the lessee, and the lessee has the right to take the produce away. In the present case all the three conditions are satisfied. Therefore forest leases in favour of private parties in Orissa are against the Forest Conservation Act. As the permission of the GOI has not been taken for the assignment of bamboo forests to industry, the entire arrangement is unlawful. Officers are taking grave risk of prosecution in being a party to such illegal arrangements. Even if the GOI does not launch prosecution, NGOs are likely to drag government officials to court on this vexed issue.

In the late 1960s the system of appointing labour contractors was abolished all over India, as it had led to corruption, theft, and development of unhealthy nexus between bureaucracy/politicians and contractors. The present arrangement in Orissa not only revives the old contractor system, but is even worse, as at least in the older system there was some transparency and fair play, as only the highest bidder could become contractors. The order appointing industry as 'raw material procurer' goes to the extent of stating that 'other terms and conditions may be settled in consultation with the industry.' This must be the only case in government where lessees are appointed first and then are left free to lay down their own terms! Probity demands that one advertises and gets the offer of terms from the lessees, before deciding who should get the lease.

The staff of the OFDC, who were supposed to check malpractices by the industry, has been transferred to the paper industry, with the result there is dilution of checks on the number of trucks taken out, whether proper harvesting precautions are being followed, or whether minimum wages are being paid to the workers, etc. The industry has rights of harvesting but it makes no efforts to follow the prescribed silvicultural practices to ensure adequate regeneration. Technically, it is obligatory that the party which is felling bamboo should simultaneously undertake cleaning and other silvicultural operations to ensure adequate regeneration, but under the present scheme, the industry is under no obligation to do so, thereby endangering further production of bamboo.

It is interesting that for satisfying the needs of the common people, GOI has asked the paper industry to harvest 5000 tonnes of long bamboo. Thus, even the owner has to request the industry to meet the demands of the poorer section of society. Local enquiries showed that the industry was not meeting the demands of the poor on the plea that the poor do not turn up to get subsidised bamboo!

Despite the prescription in the new Forest Policy, 1988 that the needs of the forest dwellers will be the first charge on the forest produce, the poor in Orissa have to meet their demand for bamboo by stealing, while the industry gets subsidised bamboo and has the first charge. From one depot the disposal of bamboo was as follows:

To industry	33,60,000 pieces (roughly at 15 paise a piece)
Through open auction	27,275 pieces (Rs. 10 to 13 a piece)
Sent to other divisions	2,892 pieces
Local sale to cultivators (tenants)	350 pieces (at Rs. 4.30 per piece)
Sale to artisans	nil

Only those who own land and thus are tenants and pay cess are entitled to fuelwood and bamboo, that too after a lot of verification from several officials. There is no system by which the landless and artisans can get bamboo even at a price, and thus are forced to resort to illegal harvesting.

In pursuance of the National Policy of 1988, Government of India on 24 December, 1996 enacted a new Act, Provisions of Panchayats (Extension to the Scheduled Areas) Act, 1996. Section 4 (m) (ii) of this Act provides that -

> *'while endowing panchayats in the Scheduled Areas with such powers and authority as may be necessary to enable them to function as institutions of self-government, a State legislature shall ensure that the Panchayats at the appropriate level and the Gram Sabha are endowed specifically with the ownership of minor forest produce.'*

All States with Schedule V areas were requested to amend the State laws so as to bring them in conformity with the Government of India Act. In view of the legal rights of PRIs created over forests in the Schedule V areas (where most forests are located), it would now be illegal to lease out forests to the private entrepreneurs/Forest Corporations.

Views of the Ministry of Environment and Forests and State Governments

The latest stand of the Ministry of Environment and Forests, Government of India is best explained in a letter from Minister for Environment and Forests, dated 27th October, 1997 addressed to Prime Minister:

'In reacting to a similar demand, this Ministry had prepared a proposal in 1995 for the Cabinet that severely degraded forest lands may be leased out to industry for growing timber for meeting the needs of the wood pulp based industry. The main ingredients of this proposal were:

(i) Severely degraded forest areas (less than 10% density), which are preferably free of traditional and community rights, to be made available for afforestation to State Forest Development Corporations (FDCs) in financial and technical collaboration with private sector.

(ii) A pre-design spatial mix of least three species with no single species having more than 50 per cent population to ensure a minimum level of bio-diversity.

(iii) 20 per cent of the project area to be planted for non commercial use in consultation with local village communities for meeting their *bona fide* requirements of fuel, fodder and small timber.

(iv) Installation of a management committee with equal representation of the State, FDC, industry and village community/ panchayat to be formed for overall technical supervision.

(v) 5 per cent of the quantities of the forest produce or its value from thinnings and final felling will go to FDCs as its share.

(vi) The scheme will also be open to registered cooperative societies consisting of local landless unemployed tribal educated youth, ex-service personnel etc.

The then Cabinet considered this proposal on 3rd July 1995 and decided that a Group of Ministers should consider this proposal, after which the proposal should be brought back to the Cabinet.

Subsequently, the Planning Commission set up a Committee on 6th June, 1997 headed by Shri N. C. Saxena, Secretary, Department of Rural Development, to examine the issue of leasing forest land to industry for the purpose under consideration. This Committee held one meeting on 16th June, 1997 and has yet to arrive at a conclusion.

This Ministry organised its Annual Conference of State Forest Secretaries and PCCFs on 26th and 27th September, 1997. Some States like Andhra Pradesh, Madhya Pradesh raised a demand that State Governments should be allowed to lease out severely degraded forest lands to industry for raising commercial species of timber. As the matter was one that permitted two strongly opposing view points, it was decided to set up a Working Group to go into this issue and give its report within 8 weeks.

Prior to this, the Supreme Court, while dealing with the issue of how to prevent large-scale illicit felling in the North Eastern States and other areas had, *inter alia*, observed that the Government should not take any steps to dilute the existing forest Policy. Further, the definition of forests was also expanded to cover what is understood to be a forest in the dictionary sense of the meaning and as is understood in common parlance.

Keeping in view these sensitivities and the somewhat complicated nature of the issues involved, I feel we should wait for the existing Committees to examine the matter in depth after which we can proceed further.

In view of these reasons, I thought that I would bring to your notice that we will require some more time to examine the issues involved and to come out with a proposal within 4 weeks may not be possible. As soon as the Committees have examined the various issues, my Ministry will examine their recommendations for further necessary action.

In response to the draft report of the Group, Shri Oberoi, the IG Forests sent the following comments:

> 'In this regard, I would like to state that we should attempt to bring more and more investment in the forestry sector to ensure achieving the national target of bringing one-third of land mass under tree/forest cover. Regeneration and improving the productivity of degraded forest is a national imperative. To achieve this, we must try to involve the people as well as other sectors in afforestation activities. Regarding the participation of

industries in afforestation, I agree that they should have durable tie-up with the private farmers for producing the raw material. Regarding greening the degraded forest areas, I would like to state that 15 million ha. of marginally degraded forests where adequate root stock exists, would be tackled through aided natural regeneration, gap filling and enrichment planting under the JFM Programme. Apart from other arrangements to green another 15 million ha. degraded forests devoid of root stock and self-regenerative capacity, attempts should also be made to enhance and utilize the institutional capacity of the State Forest Corporations. For this purpose, the corporations should take up forest plantations with adequate inputs and suitable genetic material propagated through most appropriate technology. This will ensure bringing vegetation cover on the completely barren lands. Then only, it would be possible to meet the requirements of various sectors of the society and to keep the pressure of the well stocked forests, which are the bio-diversity repositories and serve diverse ecological roles.'

State Governments

Several states expressed their views either in writing or in the meetings. These are described below.

Tripura

Massive plantation of industrial species as captive plantation would result in monoculture, which is ecologically undesirable. Insisting on mixture and imposing restrictions on monoculture would not be encouraging to the industries and they may back out. Though the forests are degraded, they may not be totally depleted of biodiversity values. Minor Forest produce may be available and many herbs, shrubs and other lower vegetation of food, medicinal and ecological values may be existing in those areas. Large-scale plantation (monoculture) would wipe them out.

Most such degraded lands need eco-restoration through soil conservation, which would need several ameliorative measures. Since such measures are costly, industries may not be motivated enough to take up such operations. Furthermore, Agroforestry and Farm forestry are likely to be adversely affected.

Ecological status of any forest land fully stocked or degraded, cannot be decided on the basis of tree density alone. Total importance of land is to be judged on the basis of aesthetic, ecological and biological parameters. Therefore even if areas of less than 10 per cent tree density are selected, biodiversity conservation aspect will not receive due care.

It has not been clearly established whether there has been actual shortage of raw materials particularly, in pulp and paper industries. In many cases it has been shown that price of some raw materials like eucalyptus poles in certain states registered a decline in real terms. It can hardly be said that there is actual shortage of such raw materials.

It is a fact that industry would like to invest their capital only where the cost of plantation is less than the cost of alternative supply. Thus only good forest land can meet this requirement. Incidentally, these forest lands are also the areas on which local poor depend highly for subsistence. It is also a point of controversy whether industry would come forward to invest in degraded forests, which are quite poor in innate productive capacity.

Regarding augmentation of financial resources inviting corporate investment for afforestation, it can be said that substantial external funding is now available for such work in the country and such funding is expected to grow in magnitude in future. The proposal is not going to effectively attract institutional finance for the SFDCS, either.

The expectation underlying the proposal that such lease would help in combating encroachment may not be achieved because the scheme if ultimately materialises, would be implemented over actually a very small percentage of degraded forest land and it would hardly cover the extensive balance degraded area (30.8 million ha.) of the country. On the contrary, the biotic pressure may be diverted to the adjoining better quality forests. Also the problem of encroachment is very complex and such lease is not going to address anyone of them. Therefore, this will be of little significance.

It is felt that leasing of degraded forest and will amount to snatching away of life support system of local tribals and poor fringe dwellers. Most degraded forests have extreme biotic pressure and often have rights of the people recorded on such land. It will be virtually impossible to extinguish such rights or reduce biotic pressure. People's rights to access is already a chronic problem in protected area network of the country. If such land is used by industries who will generally be perceived as alien to the area, will accentuate to the problem to an unmanageable limit. The past experience of granting lease to industries have not been always good. It is reasonable to expect that industry would maximise profit of their venture and may use methods, which may not be very conducive to sustained conservation and development of forests. Exercising strict control and vigilance every time may not be possible. Bifurcation of ownership and management may result in complications leading to faster deforestation. Furthermore, it is also not known how many industries/ entrepreneurs demonstrated their keenness to afforest degraded forest lands and if so, with what terms.

State specifities in Tripura : Referring to the proposal under Tripura scenario, it is to state that total degraded forest land in the state is 85,100 ha. (58,400 ha. in erstwhile P.F. and 26,700 in R.F.). Out of this area, 8620 ha. are encroached. Due to lack of infrastructure, industries and employment opportunities, a large section of rural population depend on forests for livelihood. It will be almost impossible to extinguish such rights or reduce biotic pressure on such lands. Even in the 4 sanctuaries of the state reserved for biodiversity and wildlife conservation, much difficulty is being experienced to deal with people's right to access and protecting the resources there.

The *bona fide* householder and cultivators who are inhabitants of villages entirely surrounded by Reserved Forests in Tripura are eligible under Notification No. 1 dated 29-4-1952 of Forest Department to have the forest produce free from such Reserved Forests free of royalty up to a limit as indicated therein for their own personal use but not for barter or sale. Such concessions were also extended from protected forests which are now unclassified government forests by Notification No. 3 dated 29-4-1952. Besides, other concessions had also been allowed like grazing of cattle free of charge, shifting cultivation jhuming) by hillmen jhumias etc. The hillmen jhumias also enjoy several other concessions from these forests. Thus, the seriousness of the conflict likely to generate as a result of such lease can easily be comprehended.

Ecologically the state is situated in the maximum potential productivity zone. Many degraded forest areas of the state do not require capital investment, not even very advanced technology but proper protection and regeneration methods. Existing rootstock in degraded forest areas, if appropriately protected, would effectively afforest such areas.

Furthermore, there are no large-scale wood based industries in the State and in the past no industry so far came forward with any request or proposal to afforest in degraded forest on lease. The State Government has undertaken a large-scale project of economic resettlement of tribals. There have been serious constraints of availability of suitable land out side the forest area. In conformity with the recommendation of Sukia Commission, the State Government is preparing a proposal for utilisation of degraded forest land for resttlement of tribals.

In view of the details presented in the aforesaid paragraphs, the State Government is not in a position to recommend leasing out of forest land for raising industrial plantations.

Madhya Pradesh

The MP Rajya Van Vikas Ltd. invited tenders from private entrepreneurs in June 1996 to undertake plantations on a contract basis in four tribal districts, Seoni, Rajnandgaon, Chindwara, and Betul. This was objected to by the GOI on the grounds that it was not permissible under the Forest (Conservation) Act, 1980. There were also problems related to technical and financial bids, therefore all the tenders called for were rejected.

Sikkim

All forest lands in Sikkim are located in slopes which are fragile and plantation by private entrepreneur will not be environmentally sound, besides there is no Forest Corporation.

Pondicherry

Degraded forest lands should be brought under social forestry by involving local people, and should not be leased to the private entrepreneurs/Forest Corporations.

Tamil Nadu

Mr. Chitrapu, PCCF, Tamil Nadu opined that increasing prices of wood especially firewood and pulp motivated the farmers to produce wood on their farm land. Use of farmland and degraded private land for tree planting is helpful in preserving the biodiversity of the country. Leasing of degraded forest land to industries may be beneficial to industry but from social and environmental point of view we might be losers. Moreover, the paper industry is a very polluting industry. Citing the example of Indonesia where the forests have been privatised, he further said that the present problem of "smog" is a fall out of such privatisation. He, therefore, did not favour leasing of forest land to industry.

Andhra Pradesh

Shri S. D. Mukherjee, PCCF, Andhra Pradesh, informed that the Andhra Pradesh Government had embarked upon an ambitious programme of joint forest management in which people's committees have been given a very important role in deciding the choice of species and in forest management. If industry wanted particular species to be raised on such forests, they would have to persuade people's committees. He had sent a proposal to this effect to the paper industry, who have not responded so far. The industry should also promise a guaranteed minimum price, so that communities can put some area under eucalyptus and bamboo. He calculated that even if 5 per cent of the area under JFM is put under commercial species, it would produce sufficient wood for industry. According to Shri Mukherji, the Bhadrachalam Paper Mill spends Rs. 35,000 per ha in producing eucalyptus on its private farm, whereas similar quality can be produced by the communities at 1/3rd the price.

When this issue was raised in the meeting in the presence of the industry representatives, they said that it amounted to communal farm forestry which is worse than farm forestry on private lands, as far as hassles of supply to industry are concerned. The industries could raise sample clonal plantations on 1/2 to 1 ha of private plots taking the villagers in confidence. The industry is therefore in no mood to negotiate with the communities. Despite this reluctance of the industry to liaise with community groups, the PCCF would welcome a tripartite agreement among people, industry and Government but stated that management and control cannot be given to industry. The field notes given at Annexure 2 throw more light on this proposal. Representatives from industry were not in favour of such an arrangement, as they wanted outright control over land without paying any price for it.

Haryana

Mr. R. D. Jagati, Chief Wild Life Warden, Haryana, expressed his fears that leasing out of degraded forest land to private entrepreneurs may strangulate the flourishing agro-forestry sector of the State. The Yamunanagar Paper Mill gets 75 per cent of its raw material from farmers and thus supports farm forestry in a big way. Haryana also exports farm forestry wood to other States. This was confirmed by representatives of industry who informed that wood is transported up to 500 kms for supplying the needs of industry. He further added that degraded forest lands are of two types— (1) barren lands where no economically viable plantations could be raised and (2) marginally productive degraded lands which form bulk of the wastelands. Degraded forest lands are generally in the vicinity of good forests and access of the industry to degraded forest lands would put immense pressure on them. He also expressed his doubts about the machinery identifying the "suitable degraded forest land". It is quite likely that the field machinery may be pressurised in declaring good quality forests as degraded. He further added that improved clones and techniques require good soil and inputs and are not suitable for barren lands.

Orissa

Responding to the draft report of this Group, Orissa Government sent detailed comments disagreeing with some of the recommendations of this Group. These are quoted below in full.

"The main argument against leasing out degraded forest land for the above purpose is that good and not-so-degraded forest lands may be diverted for captive industrial plantation, and in the process, forest-dependent livelihoods of the local communities may be adversely affected. This could be resolved by entrusting the task of selection of degraded forest areas to independent committees (of both Government officials and non-officials and village spokespersons) which would confine strictly to scrub lands with less than 10 per cent standing tree cover, where peoples' livelihood dependence is minimum. In Orissa, there are extensive tracts of such degraded lands recorded as forest, which do not provide any real livelihood support to the local communities, as there is hardly any vegetation or biomass output from these lands. These lands have suffered this fate on account of various unsustainable practices, the most important of which is Podu (shifting) cultivation. Some of these lands are in reasonably large compact chunks, and are quite suitable for growing Eucalyptus and similar other quick-rotation pulpwood species. There is very little viable root-stock left on these lands which could not be regenerated even with appropriate involvement of the local communities. Plantation is the answer to rehabilitation of such lands, and if industries are willing to come forward to invest funds in raising plantation on these lands, it should be welcome.

Such lands could be selected on widely dispersed locations so as to minimise the socio-economic costs to the local communities. Also, the mix of species to be raised, model of plantation, harvesting plant and sharing of the produce could be decided in each case, separately as local issues, by integrating the stake-holding interests of the local communities into the detailed design of any such plantation scheme. Difficulties would arise only when any single blueprint of plantation model is sought to be implemented at all such places. The proposed joint venture between the State-owned Forest Development Corporation and the Private Industry could carefully work out an appropriate mechanism to adequately address the site-specific social and ecological issues in grounding any such plantation.

One of the express objectives of allowing some industries a direct role in raising their raw material resources on forest lands will be to introduce the genetic improvement technologies so as to achieve a quantum jump in productivity from all such plantations. It is possible that the raw material now being supplied to the paper industries in Orissa by working the bamboo forest coupes could be met alternatively by raising highly productive plantations over about only 25000 ha. of degraded forest land, and if this is feasible, it would be a positive and pragmatic intervention to help relieve the pressure on the forests on account of bamboo exploitation.

The debate between 'farm forestry' and 'captive industrial plantation on degraded forest lands' need not be an either-or situation. There is space for both in our kind of situation where large scale farm forestry is not feasible in the immediately foreseeable future as a source of raw material for the industrial requirement. An industry-led farm forestry or contract farming, with appropriate liberalisation of transit and other restrictions and price-support intervention is certainly worth pursuing vigorously as a long term objective.

The market requirement of commercial and industrial wood being far in excess of the present availability, there need not be any apprehension that farmers will lose their motivation to take to farm forestry when industries are allowed to produce part of their raw material requirement from forest lands.

The question of equity on access to forest lands as between the local communities and the private industries will have to be resolved by carefully selecting the lands to exclude those with substantial user-dependence of the local people, and also by working out the appropriate royalties to be paid by the industry for the produce of the captive plantation, which may be based on the social cost and opportunity cost of making the degraded forest lands available for such plantation. There need not be any subsidy of any kind to the industry in the proposed arrangement. The whole arrangement would be based on the willingness of any industry accept the appropriate conditions.

Coffee, cashew and palm plantation are non-forestry activities, and there is a legal prohibition on such activities on forest lands. So, there is no question of entertaining any claim of such industries to degraded forest land on the analogy of paper industries."

Andamans and Nicobar

Andamans and Nicobar opposed leasing on the ground that there are no degraded forests in the UT.

Meeting Demand for Pulpwood through Farm Forestry

Since the private entrepreneurs/Forest Corporations would use land for commercial production of wood, it is relevant to discuss present demand/supply scenario for timber and pulpwood, which are the two main commercial output from trees. Out of these, timber used by the construction industry, mining, transport and communication, etc. is processed in small and cottage sector by a large number of decentralised sawmills. Its number is estimated to be around 25,000 in the country. They have made no demand for allotment of land, also because timber takes about 40-60 years to grow.

The main demand for leasing of forests has come from the paper industry, which uses bamboo and eucalyptus, which can be grown in about 5-8 years. The current utilisation of wood and bamboo by the paper industry is 3.2 million tonnes (mT), whereas the demand has been assessed as 6.4 mT by the industry. Mr. Hari Shanker Singhania, the then Chairman of the Development Council for Paper and Pulp industry, prepared a report in June 1990 according to which the demand for pulpwood was likely to increase from 4 to 8 million tonnes between 2000 and 2015, whereas production would increase only from 2.5 to 3.3 million tonnes. It also suggested that the entire requirement for paper industry could be met by allocating 2 million ha of forest lands. This looks exaggerated as the industry claims to be able to produce 20 cum/ha/yr by using clonal technology.

Timber prices after 1987 seem to be stabilising because of liberal imports of logs. The government has been encouraging import of logs and pulpwood by providing relief in custom tariffs. Timber (in log or sawn form) and pulp have been included under Open General Licence (OGL). On the whole the rate of customs duty is low. Since import of timber is under OGL, private entrepreneurs have been making their own arrangements for import. Hardwoods are mostly imported from the tropical countries of Malaysia, Burma, Indonesia, Brazil, Papua New Guinea, Singapore and Vietnam. Import of coniferous wood, though in small quantities, has also taken place from temperate countries like USA, USSR, Canada, France, Austria and Finland. Pulp is being imported from Canada, Finland, Norway, Portugal, Sweden, USA, USSR, Brazil, Indonesia, Thailand, Chile and Australia. As imported timber is comparatively cheaper than locally grown timber, liberalised imports have helped to conserve forest resources and check prices in the timber market.

Imported timber is mostly used by the building and construction industry, plywood industries and by the railways for making sleepers. The imported timber has also benefited a large number of saw mills, particularly in urban centres and in the vicinity of sea ports. There has been a quantum jump in recent years in the import of timber. In 1989-90, the total import was estimated at 1.5 mcum (Singh, Ashbendu 1992: 72). The total value of imports in 1995-96 had touched Rs. 5000 crores, and the quantity imported may well be 50 per cent of recorded timber production from forest lands. In addition, 1 to 1.5 mcum of newspaper grade pulp, which is almost 50 per cent of the present consumption, is imported. The Group felt that although timber imports may continue, the desirability of importing pulpwood in the face of its surplus indigenous production may be examined again.

Shortage of Raw Material and Farm Forestry

As already stated, paper industry has been demanding 2 m ha for growing hardwoods such as eucalyptus for pulp. As against this there is 141 m ha of cultivated land and 20 m ha of farmer owned uncultivated degraded lands. These lands have the potential of producing pulpwood, especially in view of the fact that both eucalyptus and bamboo are short rotation crops, eminently suitable for the farm sector. In fact the bogey of raw material crunch is no longer valid, given the vast expansion in farm forestry programme in the last twenty years.

Trees on farm lands before 1975 : In regions of abundant forests, farmers rarely plant trees on farm lands, as gathering from forests is considered more cost effective, but even in these regions trees occur naturally on farm lands. Trees can be good investments as savings banks for the poor, enabling them to accumulate capital. But trees would become poor people's banks only when their right to ownership of trees is not disputed and impeded by law or bureaucratic regulations. Quite often this was not the case. Movement of tree logs and tree products obtained from private lands was restricted during the colonial period, in order to prevent theft of similar products from FD lands. A widespread impression prevailed in the villages that all trees belong to the Government, and so does land on which such trees exist. In contrast to Africa, where trees are often planted to establish tenure rights, in India they are often removed to demonstrate claims to land.

These laws were continued and often made more stringent even after the British left. There are three different sets of laws: a farmer cannot harvest trees without permission; a permit is required to transport them; and lastly, some trees can be sold only to the FD. In most States people can sell trees only after going through a laborious process of getting a permit from designated minor functionaries - with consequences easy to imagine. These forest laws leave villagers in doubt whether any wood they produce will belong to them. Perversely, the immediate impact of such legislation is always more destruction of private trees as people wish to cash their assets before the government machinery to enforce laws is set up. Besides, such controls act as self-fulfilling prophecies: they deter protection and promote irresponsible felling.

Such laws have contributed to a decline in tree density on homesteads, boundaries and crop lands. The All-India Rural Credit Survey Report studied 75 districts from all over the country in 1951-52, and noted that in most districts either no expenditure was reported or less than one per cent of the families reported expenditure on the laying of new orchards and planting of trees. Farmers' indifference to planting of trees was also due to market distortions caused by government selling timber and pulpwood at a highly subsidised price to industries during the period 1950-1980. Another factor was the spread of irrigation, because of which trees on crop lands were not needed, and were felled or not protected. The area under privately owned tree crops and groves in the country was 2.77 per cent of the total reporting area in 1951-52, but fell to 1.15 per cent by 1980-81, showing a substantial reduction in the area under farm trees.

The 'Hog Cycle' of Farm Forestry

The trend of decline in the density of trees was reversed in the late 1970s when wood prices shot up and farmers responded to price signals and market opportunities by planting trees, especially eucalyptus on their lands.

The salient features of the farm forestry programme (USAID, 1988; World Bank, 1990; SIDA 1990) are:

- More trees were planted in commercialised and surplus producing agrarian regions than in subsistence oriented eastern states, despite the fact that rainfall and soil conditions were more favourable to trees in the east rather than in the low rainfall (but irrigated) NW India.

- Eucalyptus was the most favoured tree with the farmers as it grew straight, had a small crown, which allowed more trees to be planted per unit of area, and caused little shading when planted on field boundaries. It did not attract birds, was non-browsable, hence easy to protect, and yielded straight poles which were perceived to have a good market.
- Eucalyptus was planted more for sale as small timber, poles or pulpwood, than for use as fuelwood, although because of glut conditions it was often sold as fuel to brick kilns and fuelwood depots.
- And finally, farmers' enthusiasm to plant eucalyptus declined after 1988, as the tree failed to generate the kind of returns farmers were expecting from its sale. In contrast to other agricultural innovations (e.g., high yielding variety of seeds) which, after a fast initial growth, stabilised at a high level of adoption, the popularity of eucalyptus farming shot up quickly during 1981-86 but declined equally fast afterwards.

The disenchantment with eucalyptus can be attributed to the following three factors:

1. Production problems : In most farm forestry plantations, farmers resorted to dense plantations of over 4000 eucalyptus seedlings to a hectare, or the distance between two adjacent trees in case of bund plantation was just about a metre, which led to a poor quality of produce good enough only for fuelwood (Athreya, 1989; IMRB 1989). Second, due to rapid increase in demand for eucalyptus seedlings by farmers in the early eighties, seed collection was not done properly by the Forest Department and private nurseries, and poor quality seedlings were allowed to be planted (World Bank, 1988). Third, many farmers had planted eucalyptus on farm bunds, hoping to get a good income after six years. They were not advised that trees could cause loss of agricultural production. And lastly, the genetic status and composition of eucalyptus hybrid is less than optimal, due to gradual and continuous genetic deterioration this material has undergone over almost 150 years in India. This leads to poor yield. Further performance could be improved if the species now used was replaced by pure strains of suitable eucalyptus species and provenances selected to match specific site conditions. And last, under the climatic and soil conditions prevailing in many areas of the country intensive weeding and soil working were essential, which was frequently neglected by the farmers (ODA, 1989).

2. *Market imperfections :* Compared with markets for other agricultural products, wood markets are still not fully geared to receiving farm production. Farmers rarely bring their produce to the markets, they rather wait for the buyers to come to them. There are legal restrictions on the transport of wood. These are designed to prevent illicit felling from government forests, but they act against the interests of producers, as these rigid laws act as a barrier between the producer and the market, and bring uncertainty in the operation of sale transactions. These laws have acted as a single most inhibiting factor behind the unpopularity of farm forestry programmes. Even when some laws were liberalised, these were not publicised, with the result that harassment continued as usual. The main thrust of government's intervention should be to remove uncertainty and reduce risk, which farmers face while planting trees.

Unlike agricultural products which are supplied only from private farms, government was a big supplier of eucalyptus logs and poles to the market. Traders and paper mills preferred eucalyptus from government depots to collecting it from farms as it was of a superior quality, cheaper and available in bulk. Farmers' awareness about buyers, the prevailing market price, and government rules is weak. Due to these market imperfections, middlemen's margins are sometimes large, leading to high price differentials between what producers get and what consumers pay. Producers are thus deprived of the true market potential.

3. *Lack of demand :* Due to the above factors eucalyptus grown by farmers remained thin, and had little use as timber, which required trees of bigger dimensions and higher density than available from farm lands in a short rotation period of 5-7 years. The output was suitable for pulping, but paper mills, based on wood, are supplied subsidised wood from the State Forest Departments. Although lately many of them have started collecting eucalyptus from the farmers too,

yet there are organisational and locational problems in buying wood from scattered farmers and transporting it over a long distance. The eucalyptus crop from farms, when it reached the market in the mid-eighties, was initially sought to be used as poles for scaffolding. But the demand for poles is far less compared with demand for pulpwood or timber, and the market was not able to absorb the substantial production of eucalyptus poles. Even when trees gained girth timber from eucalyptus was found to be far inferior in quality and durability than other timbers, and hence fetched a very low price. Ultimately, much of eucalyptus wood is now being sold as fuelwood at less than the expected price. Farmers' produce as fuelwood competes with wood supplies from government, and with coal and petroleum products. These commodities have administered prices. Thus farmers could not get the expected price for their output.

There could be other factors too that have prevented farm forestry sector from realising its potential. We need to further energise the private farm forestry sector through other means also. The Ministry of Agriculture could be requested to examine why full potential has not been achieved here. Issues such as greater flexibility in institutional financing, involvement of the poor in large scale nursery raising, technical support from the Forest Department and elimination of legal restrictions, which hamper the growth in this sector need to be reviewed. The Ministries of Environment and Forests, Finance, Industry and other stakeholders could play a useful role in assisting the Ministry of Agriculture to draw up a dynamic package of incentives.

Impact of Tree Farming on the Availability and Prices of Pulpwood

Thus despite farmers enthusiasm for planting eucalyptus prices started falling, as the demand from other sectors for farm wood was not enough and industry preferred imports or subsidised supplies from government. In Haryana the total consumption of poles for construction by households in 1985-86 was 66,090 cum, which is only 2.3 per cent of the total supply of 3 million cu m wood from farm forestry (NCAER 1987). According to a study (Athreya, 1989), the

Haryana paper mill was buying eucalyptus from the farmers at Rs. 440-460 a tonne at the factory gate in 1986, but not only the price came down to Rs. 330 in 1988, also the factory started imposing a quality cut of 10 per cent, reducing the effective price to Rs. 300 only. If inflation of 8 per cent is taken into account the fall in prices during 1986-88 amounts to almost 40 per cent.

A fall in pole prices has been noticed in several regions, where eucalyptus planting by farmers had been popular. In Gujarat, the price of an eucalyptus pole of 35-40 cm girth and 3-4 metre length has fallen from Rs. 39 to 34 during 1985-89 at current prices, as quoted in the World Bank Report on India (1993):

Table 7: Price of Eucalyptus Poles in Major Market in Gujarat

(in Rs./piece)

Year	*Ahmedabad*	*Bharuch*	*Mehsana*	*Rajkot*	*Kutch*	*State*
1985	35	35	46	41	37	39
1986	31	30	39	41	35	35
1987	24	37	41	40	35	35
1988	23	39	39	29	35	33
1989	24	46	34	25	40	34

A World Bank/USAID team assessed the retail price of eucalyptus poles in February-March, 1988 in the north Indian markets at Rs. 400 a ton (USAID, 1988), a fall from the earlier price of Rs. 500 a ton. In Chandrakona (West Bengal), one of the largest eucalyptus market in eastern India, the price of eucalyptus pole of 30 cm girth fell from Rs. 25 to 18 during 1988-90 (IIM 1991). In the same area, the farm gate price for a similar product fell from Rs. 26.70 in 1985 to Rs. 10.20 in 1989 (Singh and Bhattacharjee 1991). In the Punjab, poles used to sell for Rs. 200-300 per piece in 1984-85 but the price fell to Rs. 40-45 per piece after 1988 (Khare and Rao, 1991). If inflation of eight per cent per annum is taken into account, the price in 1988 was only 15 per cent of the price in 1984.

The decline in the price of eucalyptus poles in West Bengal in nominal terms (if inflation is taken into account, the fall in prices would be even more drastic) has been reported by the World Bank report (1993) to be as follows :

Table 8: Changes in the Price of Eucalyptus Poles in West Bengal

Girth of tree (16 ft. long) in inch	*Diameter of pole (16 ft. long) in inch*	*Market Price (Rs. per pole)*				
		1985	*1986*	*1987*	*1988*	*1989*
9	3	15	15	13.5	12	12
12	4	28	28	25	24	22
15	5	58	58	52	50	48
18	6	100	105	96	90	82
21	7	125	125	110	105	100
24	8	160	128	125	115	115

These figures clearly show farmers' keenenss and capacity to supply eucalyptus to industry. It is also learnt from the industry sources that the profitability of paper mills has considerably improved in the last ten years due to fall in prices. According to the Labour Bureau, Shimla, fuelwood prices in 20 towns of India increased in absolute terms by 73 per cent between 1989-96, whereas the rise in wholesale price index during the same period was 93 per cent, thus confirming increasde availability of wood.

Due to extensive fuelwood plantations outside the forests in Tamil Nadu working of natural forests for meeting fuel requirements has been banned. Contrary to the expectations of fuelwood famine consequent on the closure of coupes, many of the drier tracts of Tamil Nadu now produce enough quantities of casuarina wood to meet not only the fuelwood requirements of the towns and cities but also pulpwood requirement of some of the industries located within Tamil Nadu as well as the adjoining States. Extensive teak plantations have been taken up on the canal banks in the Cauvery delta in Thanjavur district and significant quantities of timber are becoming available from these plantations. It is also relevant to point out that Tamil Nadu receives hardly 800 mm. average annual rainfall and if these things are possible in a State so poorly endowed this can be replicated in a much more striking manner in the rest of the country.

It was unfortunate that in many farm forestry programmes, like in Orissa and AP, where bamboo could grow quite well on homesteads and other such lands with good moisture (such as tubewell enclosures), no special emphasis was given to bamboo. The area under bamboo, both on forest and private lands, must have gone down in the last 30 years. On the other hand, the area under eucalyptus has considerably increased. In several states natural forests were clear-felled and eucalyptus planted in their place during 1960-80, as encouragement to man-made forests was the priority at that time. Seventeen per cent of the geographical area of UP is declared as government forests, and more than half of the total timber output from government forests consisted of eucalyptus in 1987-88, despite the fact that before 1960 its area in UP was negligible (UPFC, 1990).

The success of farm forestry seems to have influenced the general price of fuelwood too throughout the country.

Although the supply of eucalyptus to markets has gone down a great deal in the last five years, yet the price of pulpwood has not increased, except in northern region, as the following Table supplied to the Group by Mr. Thapar shows:

Table 9: Prices of Pulpable Bamboo, Wood and Grasses
Rs/air dry MT (1% Moisture)

Year	*Eastern India*		*Western India*		*Northern India*		
	Bamboo	*Wood*	*Bamboo*	*Wood*	*Eucalyptus*	*Bhabar*	*Khair*
1993-94	1630	1280	1310	1670	1425	2170	985
1994-95	1700	1285	1320	1575	1725	2370	1025
1995-96	1935	1325	1465	1725	2240	2650	1120
1996-97	2450	1420	1610	1800	2375	4400	1645
1997-98	2470	1540	1715	1920	2820	4500	1710

Considering the rise in general price index over the same period, the rise in price of pulp material does not indicate "acute shortage". For Southern region, the following figures from Tamil Nadu show that administered prices were increased drastically only recently.

Table 10: Details of Prices for Pulpwood and Wattle Bark

Year	*Blue-gum*	*Eucalyptus grandis*	*Debarked Wattle wood*	*Pine wood*	*Eucalyptus hybrid*	*Bamboos*	*Wattle Bark*
1990-91	965	830	680	—	510	380	2200
1991-92	1065	920	690	—	520	380	2300
1992-93	1175	1015	760	840	575	420	2530
1993-94	1295	1120	840	925	635	465	2785
1994-95	1425	1235	925	1020	700	515	3065
1995-96	1570	1360	1020	1125	770	570	3375
1996-97	2748	2380	1785	1969	1348	1140	4725
1997-98	—	—	1785	—	1348	—	4135

If industry produces its own raw material, who would farmers sell to? Where is their market, if not industry? Sixty per cent of farm land is owned by rich and affluent farmers, who are market oriented, and can be trusted to fulfil the requirements of industry. They are even prepared to produce teak wood, if government removes restrictions on the felling of teak trees from private lands and on its movement. Since the overall demand of the industry is limited, and if allowed to be met by leasing it would necessarily reduce the size of the market for farmers. Leasing of forest lands will thus adversely affect the farm forestry programme, which is one of the cheapest and most sustainable method of producing wood.

Thus the claim of the industry that it would create additionality of efforts and funds is not true, as any afforestation by them will be at the cost of tree planting efforts by farmers on privately owned degraded lands, tubewell enclosures, and homesteads, where the social cost of production is minimal, as these lands are of little use for cultivation. Farmers exploit their own family labour (which is unpaid), and therefore can produce wood cheaper than industry. Farmers harvest their trees during the lean agricultural season and thus are able to achieve further saving in costs by spreading family labour inputs more evenly throughout the year. Quality of soil on farm lands (which is the most crucial factor in cost of production) is far better than the "degraded forests" which could be made available to the industry, and hence despite the constraint of absence of marketing infrastructure, farm wood would always be cheaper than wood produced on degraded forests.

Group's Suggestions

If degraded forests are not to be leased to industry, two questions need to be answered. First, what is the best strategy for halting degradation of forests and putting them back to greenery? Second, how should industry meet its demands for raw material? These issues are discussed below.

Forest Lands

The two main components of afforestation, farm forestry and afforestation of Forest lands, should have different objectives and approach. Farm forestry and agroforestry should aim at maximising sustained economic returns from land, whereas public forestry should aim at maximising welfare through production of such commodities like fuelwood, fodder, NTFPs, etc., which are needed by the people. The choice of species, though subject to agroclimatic and technical considerations, would also be different for the two programmes. Short duration exotics, which give high market value would be suitable for farm forestry, whereas species which have their value in the crown, and not in the stem would be suitable for public forestry. Therefore, as mandated by the Forest Policy, 1988, government forests should conserve soil and the environment, and meet the subsistence requirements of the local people. Timber is a product of the dead tree, whereas people collect branches for fuel from living trees allowing the stem to perform its various environmental functions. Moreover, gathering is more labour intensive than mechanised clear-felling. Hence in a poor country, livelihood based forestry will be more sustainable than timber-based forestry.

This would be reversing the priorities as were traditionally understood between what is the main product and what is the by-product from forests. Choice of species and management should be radically changed to suit the new policy. From forest lands, leaves, twigs, and dead wood should become the main intended products, and timber would be a by-product from large multipurpose trees. For quick benefits to the poor, long gestation trees should be supplemented with an understorey of bushes and shrubs so as to satisfy their immediate needs. The proposed changes are explained in brief in Table 11:

Table 11: Technical Options

	Traditional	*Suggested options*
objective	reduce people's dependence on forest lands	increase supply of goods desired by people
production goal	high stem biomass	high crown biomass
client	market and industry	forest dwellers and local people
timber	main product	by-product
silviculture	conversion to uniform	selective felling and protection
species	exotics and commercial	grasses, bushes, shrubs and NTFPs
production through	planting	mainly natural regeneration
usage through	harvesting	gathering

Regeneration vs. Planting on Degraded Lands

The success of JFM is generally seen as proving the superiority of natural regeneration over plantation as a technique for improving productivity and biodiversity in forests. While this is no doubt true, there are several circumstances where plantation cannot be avoided. Three such situations could be: creating a fuelwood reserve before beginning protection by the community; planting on lands incapable of regeneration; and thirdly, where desired species do not come up as a result of protection. These are discussed below.

Mere protection of a not-so-degraded area may transfer human and cattle pressure to some other area, as people have to meet their daily requirement of fuelwood somehow or the other. Therefore, production of biomass through quick growing shrubs, bushes and grasses must be undertaken on degraded and barren lands before the beginning of community protection, so that peoples' demands are met in a sustained manner from these bushes and shrubs, while people protect forest lands in anticipation of more valuable NTFPs and forest products. The issue of how to meet the economic needs of the people for the first few years, during which they have to reduce their dependence on the protected land must be faced squarely. Although the success of many JFM experiments is generally alluded to leadership or peoples' efforts, it is seen that in almost all such cases there was an alternative source of fuel available to them. In Southwest Bengal, the task of peoples' protection of degraded forest lands became easier because the farm forestry programme in that area had been highly successful, increasing fuelwood supplies and incomes even for the poor. In Eklingpura, Udaipur, where community protection has been highly successful, plenty of prosopis shrubs in and around the village provide fuelwood to everyone almost at zero opportunity cost. On the other hand, in another village of the same district, Shyampura, which had no prosopis in its vicinity, a local NGO was struggling to promote protection, but was finding it difficult to prevent unauthorised removals from the area (Choudhury, 1993). These examples illustrate the importance of creating a fuelwood reserve before expecting people to start protection.

The other situation warranting planting is where land is so degraded that regeneration is slow, or root stock is absent because of which regeneration is not possible. There may be other barriers to natural recovery such as the presence of weeds, unfavourable soil and climatic factors, a low presence of fertile trees and a lack of symbiotic microbial associations necessary for seedling establishment (Perera *et al.*, 1995). In extreme situations where soil erosion has reached conditions characterised by gullies and ravines with little or no vegetation, natural regeneration alone may have very little or no impact on improving vegetative cover. In such a situation, there may not be sufficient incentive for the people to give their time and labour for protection in lieu of the intermediate and final products, which may be available after inordinate amount of delay and waiting. Intensive soil working is required in such cases. Natural regeneration can also be enhanced or accelerated by soil and water conservation measures like contour trenching, vegetative bunding, and small check dams. In denuded areas where severe over-exploitation has reduced possibilities for rapid natural regeneration, nurseries and plantations will be needed to provide employment, and so would be fodder and fuelwood in the quickest possible time. The whole idea is consider how a continuous flow of forest products can be ensured to the communities.

The third situation where planting may be necessary is where due to protection, species which come up do not coppice well. As peoples' demands cannot be curbed for a long period, some amount of harvesting becomes unavoidable after a few years of patient waiting. In case the species do not coppice well, harvesting leads to a non-sustainable situation, and land may become denuded again. A similar situation will be when the root stock is already quite degraded, and species likely to come up are not valuable in the perception of the people, and therefore, they may be reluctant to contribute their labour for protection. In such cases the strategy of natural regeneration alone may not be enough to enforce the necessary discipline. Maybe, if plantation of species desired by the people is taken up the perception of the tract's value may increase, and everyone may cooperate. Enrichment planting could also increase the supply of raw material for the local craft or artisan based activity. An active strategy of forest restoration through management is likely to be more successful than simply abandoning lands and hoping that the regenerating forest will survive the numerous threats to it. Such a strategy would include accelerating by simple silvicultural methods, such as trenching, ploughing, weeding and thinning. However planting should be done in such a way that existing root stocks or advanced regeneration is encouraged to grow as part of the mixed stand.

Choice of species : Having thus established the need for artificial planting in many situations, the question arises which species should be given priority. Species have been so far based on convenience of staff rather than needs of people. People have not been asked what trees they prefer, least of all the poor. Socially useful species producing fruit, fodder and other NTFPs have had little place. Felling for timber may produce income to government but village consumption of tree products is little increased. Technology for subsistence goods has to be different from large scale plantations for markets.

Therefore even on barren forest lands, where plantation is required, one should change the nature of species from commercial species to planting of usufruct based trees. These should be supplemented with grasses, legumes, shrubs and bushes to yield fuelwood and fodder in the shortest possible time. An immediate identification of quick growing shrubs with high calorific value, with their retention in the forest to serve fuel requirements, the development of pastures, and the development of massive fuelwood plantations around centres of high consumption and encouragement of silviculturally sensible exploitation of fuelwood species would also be important components of the new policy. This would strengthen tribals' access to forests, and therefore, benefits would be directly appropriated by them.

Foresters and foreign experts who advise the Government of India and the donor agencies, because of their training and experience, have looked upon trees as timber, to be obtained after felling. Therefore, even in the social forestry programmes market oriented species were planted. The traditional Indian way of looking at trees has, however, been different. As opposed to trees for timber, Indian villagers for centuries have depended on trees for their sustenance. There has been little felling. Instead, trees have been valued for the intermediate products they provide, which sustain and secure the livelihoods of the people.

The difference can be understood by comparing how fuelwood species are viewed in the two perspectives. As per received wisdom, fuelwood is obtained by felling trees having a high calorific value, or as a by-product from lops and tops of timber trees. Casuarina and eucalyptus therefore seem perfectly justified on public lands. But the poor tribals obtain fuelwood from twigs and branches of living trees, and not by felling trees, and often get little from the felling of so-called fuelwood trees. Casuarina and eucalyptus may be justified on farm lands, if they improve farm incomes on a sustainable basis. But these hardly serve the poor, when raised on public lands.

Given the inefficiency of administration and 'soft' character of the political system, one could generalise that out of a tree on public lands the stem goes to the rich and the towns, whereas branches, leaves and twigs belong to the poor. Therefore the strategy should be to opt for species which have high proportions of branches and twigs relative to stem wood.

This requires a complete reversal of the recommendations of the National Commission on Agriculture, 1976 which favoured commercial plantations on forest land, and trees for consumption and subsistence on private land (Saxena, 1990). "Scientific" forestry should therefore mean that wild fruits, nuts, NTFPs, grasses, leaves and twigs become the main intended products from forest lands and timber a by-product from large trees like mahua and sal. The reverse has been the policy for the last 100 years. Although after the advent of the new forest policy in 1988 there has been a lot effort to involve forest communities in management, sufficient thought has not been given to make necessary changes in the technology which will be suitable to the changed objectives.

At the same time JFM should not be seen as a panacea for forest degradation. Not all social groups are hit equally by the decision to protect and keep livestock out of the proposed area to be protected in JFM. For low income rural families to participate, it is important that benefits start flowing as early as possible, either in the form of gatherable biomass or new opportunities of employment. This will require policy shifts and afforestation programmes on other categories of land - farm lands, village lands and FD lands remote from village habitation. Simultaneous development of all categories of land in the same region will also provide short-term benefits to the most needy.

The distinguishing features of these programmes are described in Table 12.

Table 12: Recommended Features of Afforestation Programmes on Different Categories of Lands

	Farm lands and homesteads	*Village lands*	*FD lands close to village*	*FD lands distant ant from village*
Who protects?	Farmer	Village group	FPC via JFM	FD
Management by	Farmer	Village group	FPC and FD	FD
Purpose	Enhanced production and profits	Village subsistence and village incomes	Subsistence and income needs of the village, esp. the poorest	Production of gatherable biomass and NTFPs
Technology	plantation and agroforestry	plantation	natural regeneration supplemented by plantation on barren lands	mostly natural regeneration

One practical implication of including programmes on different types of land in the same project would be that these would have to be implemented in the same Range or group of villages, rather than being implemented, as happens currently, in different Ranges, which loses complementarity.

As already stated, barren lands under JFM would need artificial plantation. Should commercial species be banned on such lands for all times to come? On this vexed issue we have an open mind. In the initial phase, communities are at a low level of awareness and preparedness, therefore, FD has an upper hand in decision making. During this phase subsistence and livelihood needs must be given topmost priority. However, with greater control and confidence, the communities become more democratic and less inequitous, and may also like to use such lands for income generating trees, including horticultural and pulpwood trees. When control over forest lands is truly decentralised to the poor and needy in the village, their desire for even pulpwood trees should be viewed sympathetically. At this stage the industry could help the community with technology and extension, just as they would do so for a private farmer.

In the village visited by AS (WD) it appears that the villagers were confident enough to deal with the industry on their own terms. The recommendation of the committee should, therefore, be seen as not permitting in all cases pulpwood species on forest lands, but responding favorably to villagers' desire in order to strengthen village control over forest lands. The relevant para of the report is quoted below:

> *'the team is of the opinion that since industries would like to have control over the management and implementation of the JFM projects to ensure required return on their capital, their association in VSSs* per se *is not feasible. However, their intervention by supply of cloned plantation material to the villagers, the VSSs and general buy back guarantee is welcome. The suggestion of PCCF, Andhra Pradesh for constitution of JFMDA is quite reasonable and workable. This agency can provide corpus and some areas in the degraded forest can be planted with high yielding cloned seedlings keeping in view the existing bio diversity and suitable inter mix of species.'*

One of the banes of planning in India has been that at a particular time there is one fashionable idea and it is sought to be imposed over the entire state/country ignoring agro-ecological diversities and location-specificities. Thus the earliest forestry projects emphasised fuelwood and fodder, and it was believed that these are the real priorities everywhere of the rural poor, although later research showed that the problem was more complex, and required varied approach. Some years later farm forestry became name of the game, and every district was given ambitious targets in seedling distribution, ignoring constraints imposed by the specificity of farming systems. This resulted in reporting of bogus figures to show unrealisable targets. Now Joint Forest Management rules the roost. Perhaps after five years it would be realised that structural conditions do not suit the JFM approach everywhere in the country, unless radical changes are introduced in the pattern of Forest Settlement. The disadvantage of this kind of uniformity in policy prescription are many and well-known. Even today there would be areas where social forestry concept is relevant or where the cheapest way of producing wood would be via the farm forestry. However, these programmes get a low priority by way of international funding when compared to Joint Forest Management for which high targets are set in all the villages ignoring both, the capacity of the Forest Department to build up and respond to local institutions, and the need to simultaneously introduce several other tenurial and silvicultural reforms so essential for the success of the JFM approach.

Farm Forestry and Forest-based Industry

For farmers, fuelwood would remain a by-product of wood, and the main product would be more remunerative products, such as poles or pulpwood. Therefore, they would be too keen to plant income generating trees, if the right conditions are created. To boost production of pulpwood market constraints should be removed. First, a great source of market imperfections in wood markets is the legal and procedural framework which makes cutting and selling privately owned trees difficult, irksome and complicated, besides unremunerative (Chambers *et al.*, 1989). Therefore these restrictive laws could be abolished. Second, government should stop import of pulpwood and subsidies on government supply of wood to industries, thereby inducing industry to buy from the farmers at a remunerative price. In fact, government should shift to non-commercial species on forest lands, as producing eucalyptus on public lands adversely affects the farm forestry programme, which is one of the cheapest and most sustainable method of producing wood.

Third, it could also initiate schemes for linking farmers with industries, in ways similar to the linking of poplar growing farmers with a Swedish match factory in north India (Chandra, 1986; Ghosh, 1994). This experiment shows that, with technological backup, timber size trees suitable for sawing can be raised on farm lands within 8 years. In fact, due to farmers' enthusiasm for growing poplar its enhanced supplies have led to establishment of several plywood factories in that area, thus providing considerable downstream employment. Improvement in technology and extension is required for all farm forestry species, so as to result in production of thicker logs suitable for sawing. WIMCO plans to sell 1.5 million seedlings to farmers a year.

Similar tie-up with farmers is being tried by the ITC-run Bhadrachalam Paper Mills in AP. The industry produces improved seeds, grows the seedlings in their nurseries, and gives them to the farmers for planting. Farmers get crop loans from the banks, and extension service from the industry. This example shows that improved planting material can improve productivity from 7 to 20 cum/ha/year. A minimum price is guaranteed to the farmers by the industry, although farmers are free to sell their produce to anyone they like. NABARD has rightly insisted on free choice for farmers, although industry would like the farmers to be bonded to them. The farm forestry project was formally approved by NABARD in June 1989 and re-finance assistance was sanctioned. The project envisages planting 1500 ha of marginal agricultural land (owned by individual farmers) with fast growing tree species (such as Eucalyptus and Casuarina) in eight districts. Trees are planted along field bunds, boundaries and irrigation channels in rows, and as blocks combined with intercrops. ITC Bhadrachalam provides important services to farmers within the project. These include the supply of high quality seedlings (from improved stock), extension services, a buy-back guarantee for pulpwood, assistance in loan procurement, and research and development support. This has increased the supply of pulpwood to the ITC Bhadrachalam and has provided incomes and fuelwood (from lops and tops) to farmers.

ITC Bhadrachalam Paperboards Ltd. of Andhra Pradesh increased its output from 42,000 tonnes in 1979 to 94,000 tonnes in 1995 without any increase in supply from natural forests. This was made possible entirely through wood obtained from farmers. The Company is also selling clonal plants at Rs. 6 a clonal seedling, and found that those farmers who paid a higher price for their seedlings took better care of their plants, and survival was as high as 95 per cent.

Between 1988-94 South India Viscose, Tamil Nadu (SIV) sold 20 million seedlings to farmers. Its sale of eucalyptus wood from farmers has been increasing every year as follows:-

Table 13: Eucalyptus Procured from Farmers by SIV in Tonnes

Year	*Qty in tonnes*
1993	5,000
1994	25,000
1995	40,000
1996	60,000

The recent upsurge in the number of companies offering private teak plantations to the urban rich is well known. This may be the only example in India where urban private capital is getting invested in rural areas; generally the flight of capital has so far been only from rural to urban sector, and therefore such schemes need not be discouraged. They have been able to get around the problem of ceiling on land by buying the land in the name of each investor. Paper and plywood industry, if desirous of acquiring large chunks of degraded forest land to take the advantage of economy of scale, should learn from the experience of such private sector initiatives.

Even small farmers can be benefited if they are linked with industry for markets. Whether an activity would be economically viable for small and medium holdings or not depends on two factors; divisibility of inputs and scale of economies. Tree planting requires divisible inputs of seeds, fertilisers, water and labour. It is quite feasible to plant just a few trees on a small piece of land. Secondly, economies of scale favour small scale production, as it requires family labour in off-season (for harvesting at least), uses land with little opportunity cost, and can be taken up along with agriculture in appropriate agroforestry models. The only disadvantage with small farmers is that of risk, which can be overcome by offering to the farmers proven technology and extension.

It is a myth that industry cannot deal with farmers directly. For several crops like sugarcane, potato, rice, cotton etc. industry has been in touch with the farmers for decades. No industry imposes a condition that farmers are bound to sell to that industry, as is being demanded by the paper industry in India. Such a restriction would mean exploitation of the farmers, and must be opposed.

It is not being denied here that there could be transitional problems in building up communications by industry with farmers. These were noticed in the case of supplies of eucalyptus from farms to paper mills. First, many mills are designed for bamboo, and not for eucalyptus. Because of a shortage of bamboo, the mills are closed, or running at low capacity. It was unfortunate that in many farm forestry programmes, like in Orissa and AP, no special emphasis is given to bamboo. Second, many others requiring wood are located in the east and south, where forest lands are located and where eucalyptus plantations were first started on forest lands. For them to transport wood over a distance of more than 200 km from the north-west would be uneconomic. Hence the paradox of abundant availability of raw material in the north and west part of the country, and low capacity utilisation of mills in the east and south of the country continues. A practical solution would be to split the processing units; to establish a new pulp making plant close to farm forestry areas, and transport pulp to the paper mill. Third, buying small lots from a large number of dispersed farmers requires the setting up of a new marketing infrastructure, whereas paper mills like to get large-scale consignments from forest depots. And last, it is not easy to obtain government permission to move wood bought from private sources, as restrictions exist on transport of wood in many states.

But these are temporary problems and can be sorted out by mills with the help of government. Wherever paper mills are active in buying from farmers, it is observed that they are able to buy almost the entire marketed farm eucalyptus from the region. In districts Kolar and Bangalore of Karnataka, where the interest of farmers in growing eucalyptus continued for a longer period than in north-west India, a study shows that most of farm eucalyptus is being bought by paper mills. In Kolar, 97.5 per cent of privately grown eucalyptus was marketed, of which 97 per cent went to the Harihar Polyfibres; in Bangalore 92 per cent was marketed, all to the same paper mill. Similar preference in favour of the paper mill was noticed in the village Bagwala (about 20 kms from a paper mill in district Nainital, U.P.) in early 1991, as almost the entire produce from the village was reaching the mill gate. Of the total wood arrivals in Lalkuan wood market, 7 km from the paper mill of Nainital, almost 90 per cent was being sold to the mills (Saxena, 1994). This was perhaps because traders were able to achieve a large turn-over if they supplied to the mills.

It has also been suggested by industry that farmers should produce fuelwood for their own consumption, leaving industrial wood to be produced by government or by industry. The experience of the last 15 years shows that farmers in commercial regions are more interested in producing for the market. While planning for wood production on farms one must keep in mind the likely requirement of marketed wood. A World Bank report (1991), quoting FAO figures, states that the total extraction of wood in India in 1988 was 264 million cum, of which 240 million cum was consumed as fuelwood.

Thus, of the total wood consumption in the country, only 10 per cent is industrial wood. Most fuelwood is collected, both for consumption and sale. The gatherers can always beat the producers over the pricing of fuelwood; the producers would be price-takers, rather than price-makers. This means that the market price of fuelwood would always be lower than its social cost for replacement of growing stock through investments in plantations. Therefore, the market price of fuelwood does not make its production on farms an attractive financial proposition in countries with large open access lands and vast poverty (French, 1985). Further, the entire arable land in green revolution areas is devoted to high cost commercial farming. Thus the opportunity cost of diverting land to tree crops is very high, which is not likely to be compensated by returns from growing fuelwood. Farmers would always prefer to use their lands for high value output, such as industrial raw material. Thus fuelwood which has to be gathered by the vast millions of people has to come from public lands, and pulpwood and other cash producing raw material from farm lands. Forest lands should continue to be used for preservation of environment and bio-diversity, and for meeting the subsistence needs of the poor, as envisaged in the Forest Policy. This should be done mostly through regeneration, which is cost effective provided peoples participation is ensured.

Summing up

Of all the forest based industries, paper and other large industries consume just a fraction of forest products. Ninety per cent of forest raw material is processed by 23,000 saw mills and a larger number of cottage units. The policy so far has been to provide subsidised raw material to large industry whereas small and cottage scale industry have to depend on market forces. In addition, now the large scale industry is using its political clout to get possession over fertile lands. The proposal, to say the least, is grossly unjust. It deprives forest dwellers the use of forest lands, which they are legally and morally entitled to use, and it deprives farmers from selling their produce to the only market that exists, which is large industry.

In a Parliament Question answered sometime in mid 1991, detailed reasons were put forward by the Ministry of Environment and Forests as to why land should not be allotted to industry. These were as follows:

1. Villagers have traditional rights over forest lands. Assigning forest lands to industry would create resistance in the local village communities.
2. Degraded forest lands should be regenerated to meet local needs as a first charge.
3. For restoring ecological balance mixed species plantation should be done, and not mono-cultures.
4. Industry should establish direct contacts with farmers, as provided in the New Forest Policy.
5. If paper industry is given land, it will lead to similar demands from tea, rubber and spices industry.

It is hoped that the Ministry will keep the above factors in mind before re-opening this issue. If government does not settle this controversy quickly and keeps the hopes of industry of getting free access to forests alive, investment by the industry in setting up of a new marketing infrastructure for buying from farmers would never be forthcoming. In the interest of viable farm forestry, this appears to be the only option. The claim of the industry over forest lands is not based on sound economic rationale, it is a seductive myth and a ploy to grab the better quality forest lands capable of regeneration free of cost. Handing over possession of valuable forests for captive plantations is against all principles of liberalisation and open markets. On the other hand, supply from farmers would be at the market price, taking into account all costs. Market should function in all cases where 'safety nets' are not called for (Chopra, 1995). Thus the proposal is not only against the ecological policies of government - the 1988 Forest Policy in particular - but also its economic policies.

To sum up, leasing of forests to the private entrepreneurs either directly or indirectly through Forest Corporations needs to be opposed for the following reasons:

- It will be against the interest of farmers who wish to supply wood to industry.

- It will not create any additionality of production. On the other hand, cheaper production on farm lands will be substituted by socially more costly production on forest lands. Government would be forced to give forest lands free to the industry, thus reverting to the old days of supplying heavily subsidised raw material to the paper industry.
- Degraded forests satisfy the fuelwood and fodder needs of a large populace. In fact, these lands are degraded because they suffer from extreme biotic pressure, and require neither capital investment, nor higher technology, but protection and recuperation, which can be done only by working with the people, where industry has neither expertise nor patience.
- The present market for pulpwood is distorted because of continuing subsidies to industry, especially for bamboo. The proposal of the industry to get free access to forest lands without paying its market price in cash will further distort and create imperfections in the market. Heavy subsidies will then make the supplies to industry totally dependent on the whims of bureaucracy, and thus will increase corruption. Subsidising the rich at the cost of tribals and forest dependent communities will attract public criticism, and may even give rise to militant movements.
- Paper and other large industries consume just a fraction of forest products. 90 per cent of forest raw material is processed by 25,000 saw mills and a larger number of cottage units, who would also lay claims on forests, once the large industry is able to snatch concessions from government.
- Besides there would be claims from coffee, cashew and palm plantation industries. Like paper industry they will raise short-term and quick growing species in place of multi-layer mixed forests obtained through regeneration. Its ecological implications need to be taken into consideration.
- Using forests for growing raw material for industry will be setting the clock back to the 1960s, showing that we learnt nothing from the mistakes of the past 30 years of trying to create man-made forests, which were ecological disasters, besides completely alienating the people and leading to faster degradation.

- It is not possible to find degraded forests in a contiguous patch of, say, 2000 ha suitable for effecting economies of scale. Such patches are found only in Reserve Forests of good quality.
- The proposal would be against the Forest Policy, the Forest Conservation Act, and the Provisions of Panchayats (Extension to the Scheduled Areas) Act, 1996.
- The soil quality (at least 1 m depth) demanded by industry is available in India only on best forests or farm lands. If good forests are not to be used for industrial plantations, the industry is left with no options but to establish contacts with farmers. Even if degraded forests with 1 m soil depth are there at all, they would easily regenerate on their own without much costs, if people are willing to cooperate. Such an arrangement will be via the JFM route.
- Industry has shown no interest in leasing-in non-forest wastelands, and therefore, their plans to operate on equally degraded barren forest lands is highly suspect.

ANNEXURE 1

No. M-12016/1/97-EandF
Planning Commission
(Environment and Forests Unit)

Yojana Bhavan
Sansad Marg
New Delhi-110001.
6th June 1997

ORDER

Sub: Constitution of Working Group on the prospects of Leasing out Degraded Forest Lands to the Private Entrepreneurs/ Forest Corporations

In pursuance of the decision taken in the meeting of Ninth Plan proposal of Ministry of Environment and Forests held on 5th May, 1997 under the Chairmanship of Principal Adviser (E&F), a Working

Group on the prospects of leasing out the degraded Forest Lands to the Private Entrepreneurs/Forest Corporations has been constituted under the Chairmanship of Dr. N. C. Saxena, Secretary, Department of Wasteland Development. The first meeting of the Working Group is scheduled for 16th June 1997 at 11.00 a.m. in Room No. 247-A, "A" Wing, Nirman Bhavan, New Delhi. The composition of the Working Group is as follows:

Dr. N. C. Saxena CHAIRMAN Secretary
Department of Wastelands Development
Nirman Bhavan, New Delhi.

Secretary Member
Ministry of Environment and Forests
Paryavaran Bhavan
CGO Complex, Lodi Road
New Delhi-110003.

Inspector General of Forests Member
Ministry of Environment and Forests
Paryavaran Bhavan
CGO Complex, Lodi Road
New Delhi-110003.

Shri P. C. Srivastava Member
Principal Chief Conservator of Forests
Government of Uttar Pradesh
17 Rana Pratap Marg
Lucknow.

Shri S. D. Mukherjee Member
Principal Chief Conservator of Forests
Government of Andhra Pradesh
Hyderabad.

Shri C. G. Mishra Member
Principal Chief Conservator of Forests
Government of Orissa
BHUBANESWAR.

Shri R. D. Sharma Member
Additional Chief Conservator of Forests
Government of Madhya Pradesh
Bhopal.

Shri V. R. Chitrapu Member
Principal Chief Conservator of Forests
Government of Tamil Nadu
Chennai.

Shri R. D. Jakati Member
Chief Wildlife Warden
Government of Haryana
Chandigarh.

Shri K. N. Singh Member
Chief Conservator of Forests
17 Rana Pratap Marg
Lucknow.

Prof. Shekhar Singh Member
Indian Institute of Public Administration
I..P. Estate
Ring Road
New Delhi-110002.

Shri S. S. Rizvi Member
Consultant
WWF India
172-B Lodi Estate
Max Mueller Marg
New Delhi-110003.

Dr. Anil Aggarwal Member
Director
Centre for Science and Environment
41, Tuglakabad Institutional Area
New Delhi-110062

Shri Uma Shanker Singh Member-Secretary
Deputy Adviser (Environment and Forests)
Planning Commission.

The terms of reference of the Working Group is as follows:

1. Economic, Social and Environmental feasibility of leasing or otherwise making degraded forest land to private entrepreneurs/forest corporations.
2. Implication of Ninth Plan approach, Forest Policy, forest law on this issue.

Sd/-
(UMA SHANKER SINGH)
Deputy Adviser(E&F)

Copy forwarded to:

All Members of the Working Group.

Copy also to:

PPS to Member-Secretary.

PS to Adviser (Agri. and E&F).

Joint Adviser (E&F).

Sd/-
(UMA SHANKER SINGH)
Deputy Adviser (E&F)

ANNEXURE 2

GOVERNMENT OF INDIA
MINISTRY OF RURAL AREAS AND EMPLOYMENT
DEPARTMENT OF WASTELANDS DEVELOPMENT

*****

REPORT OF THE CENTRAL TEAM ON FIELD VISIT TO ANDHRA PRADESH REGARDING PROSPECTS OF LEASING OUT OF DEGRADED FOREST LAND TO PRIVATE ENTREPRENEURS (5TH-7TH MAY, 1998).

A working group on the prospects of leasing out degraded forest lands to the private entrepreneurs/forest corporations has been constituted under the Chairmanship of Dr. N. C. Saxena, Secretary (RD-WD). The PCCF Andhra Pradesh, *vide* his letter dated 18.10.1997 had given a suggestion to associate the wood based industries/ entrepreneurs in the Joint Forest Management (JFM) programme to make best use of the degraded forest lands that will give maximum return to the people on sustainable basis and simultaneously provide raw material to the wood-based industries. Secretary (RD-WD) desired that Additional Secretary (WD) should visit some of the proposed areas along with officers of Ministry of Environment and Forests, nominee of IIPA. Accordingly, a team consisting of Shri S.B. Mahapatra, AS(WD), J.D. Sharma, DIG(DoWD), Shri Munindra, AIG(MoEF), Ms. Gopa Pandey, Professor, IGNFA, Dehra Dun (nominee of IIPA) had undertaken field visits in Khammam, Forest Circle and also had consultation with the villagers, executive committee members of Van Samrakshan Samiti (VSS); Vice President, ITC Bhadrachalam and PCCF, Andhra Pradesh. The CCF(JFM), A.P. and Conservator of Forests, Khammam Circle and other field officials also accompanied the team during the visit.

2. Originally, the proposal of PCCF was to allow the private entrepreneurs to invest money in degraded forest lands through the Joint Forest Management (JFM) committees so that the entrepreneurs can get back their investment by repurchase of surplus forest produce and the villagers may also get small timber etc. Later on, the PCCF informed that now the position stands changed and the industries are not willing to invest their money through VSSs for want of adequate security of their investment and return. It was suggested by the PCCF that an autonomous body (Joint Forest Management and Development Agency) as suggested by the Expert Committee set up by MoEF on 'Review of Afforestation Policies and Rehabilitation of Wastelands' and 'to Review the Implementation of National Forest Policy - 1998' be established at State level to implement JFM activities. This JFMDA, being an autonomous body, can get the funds from Ministry of Rural Areas and Employment, i.e. EAS, IWDP, DPAP, and MoEF/ NAEB, Ministry of Agriculture and Foreign Donor Agencies.

The apex body of JFMDA can have the representatives of State Forest Development Corporation, Wood based Industries, reputed Non-Government Organisations including Mahila Sangathans and Tribal Development Department. Funds from this agency can flow to the DFOs of the areas who will act as PIAs and JFM committees consisting of local people can execute the action plan.

3. Presently, in some of the divisions, the VSSs are assisted for development of these areas under the World Bank aided project. Further, the Commissioner(RD), A.P. has agreed to give an assistance of Rs.31.11 crores to A.P. Forest Department under EAS in DPAP areas which will be utilised for assisting the VSSs in the non-WB project areas in pursuance of D.O. No. 120742-2(3)95-E&F, dated 1st January 1998 of Adviser, Planning Commission. Still, a sizeable number of villagers have degraded areas to be regenerated.

4. During the field visits, the degraded forest areas being protected by Gommukoyagudem village in Bhadrachalam Mandal of Khamam circle were inspected. This VSS has been formed on 21.3.96 after detailed PRA exercise. An area of 150 ha. of degraded forest has been earmarked and handed over to the VSS for protection and treatment. During 1997, an area of 20 ha. has been treated, out of which clonal plantation of eucalyptus has been raised over 1.75 hectare. In the rest of 18.25 ha. area, teak stumps have been planted at the espacement of 3mt. × 2 mt. Water absorption trenches have been dug and in the mid space species like the *Sapindus emarginatus* (Soap nut), *Feronia elephantum* (Velga) and *Emblica officinalis* (amla) etc. have been planted. For clonal plantation of eucalyptus, the expenditure is about Rs.34,800.00 per ha. The cost is being met from the World Bank aided project. The villagers were quite enthusiastic and taking keen interest in protection and management of forest. Two NGOs - DARTA and JESS were associated with the villagers in this area. Bank account has been opened where the money available for the treatment of the area from the project funds was deposited and withdrawals are made jointly by the local person (Secretary) along with Forest staff. The villagers were happy about growth of clonal plantation. However, they were of the view that whole area should not be taken up under plantation and some area should remain earmarked for raising fuelwood and fodder. They were not prepared for giving management and control to industries or any other outside agency. ITC Bhadrachalam has been associated with this VSS in providing high quality clonal seedlings on 50 per cent subsidy, i.e. out of Rs. 7 per plant, Rs. 3.50 are reimbursed and they also stand committed for buying back any surplus produce from the private individuals as well as VSS on the rate applicable for the year.

5. The team also interacted with Mr. P. Lal, Vice President, ITC, Bhadrachalam and Mr. S.N. Rao, Dy. G.M. ITC, Bhadrachalam on issue of associating the industries and VSSs in management of degraded forest lands. During the discussions it emerges that the ITC, Bhadrachalam has an annual requirement of 2.6 lakh tonnes of pulp wood for producing 65000 tonnes of pulp for which at 7 years rotation, an area of 2600 ha. is needed annually and 18200 on gross basis for plantation. However, the ITC is also contemplating to enhance its capacity to 2 lakh tonnes of pulp per annum for which requirement of pulp wood will be 8 lakh tonnes per year needing an annual planting area of 8000 ha. Consequently, a gross area of 56000 ha. will be required for raising clonal plantation giving high yield. Apart from ITC Bhadrachalam, there are other wood based industries like A.P. Rayons, Sirpur Paper Mills, A.P. Paper Mills and Novapan. The current requirement of bamboo for these wood based industries is 13,70,000 tonnes per year. Assuming the productivity of 100 cum/ha to clonal plantation, the annual planting and harvesting area comes to 13,700 ha. and on gross basis 95,900 ha. for seven years harvest cycle. It was also informed by ITC that A.P. has nearly 36.8 lakh ha. of degraded land and this requirement comes to 2.6 per cent of such degraded forest lands.

6. Coming to the modalities for associating wood based industries in development of degraded forest lands, the ITC, Bhadrachalam expressed:

(i) The degraded lands require high input in the order of Rs. 32,000 per ha. An industry from its own resources cannot contribute this large amount. At the best, it can contribute 25 per cent funds from own resources and balance from the Government subsidy and institutional finance.

(ii) Since the industries are answerable to their shareholders and money borrowed from financial institutions has to be refunded along with accrued interest, they would like firm assurance that the investment along with interest will be available to them on maturity in cash or kind. To ensure this, the industry would like to have control in management and implementation of the programme.

(iii) The VSSs, State Forest Department and Government are not in position to provide assurance to such proposals due to typical nature of the venture.

(iv) The industry is also prepared to go for joint venture in this regard with the help of Andhra Pradesh Forest Development Corporation, however, its final admissibility and feasibility is debatable.

(v) The industry feels that if the Government is keen to attract large scale financial and technological investments by the wood based industries in support of JFM , innovative policy changes, legally valid institutional frame work defining specific role and responsibilities and rights of all partners in Joint Forest Management including the industrial units, need to be firmed up.

(vi) ITC Bhadrachalam is motivating the farmers to raise clonal plantations of eucalyptus on their marginal lands and providing buy back guarantee to the farmers. However, the farmers are free to sell the eucalyptus in open market if they so like.

Conclusion

7. The constraints of funding are paramount problem in forestry sector. The JFM movement remains a philosophy without assured financial support. However, in Andhra Pradesh, the JFM programme has been planned in very systematic manner. The State has identified 6806 villages for the purpose and 1665 VSSs have been covered in the World Bank aided project. For 3738 VSSs, the Rural Development Department of Andhra Pradesh has provided an assistance of Rs. 31.11 crores. However, financial resources are required for covering balance 1403 villages and other villages where VSSs will be formed in near future.

8. After considering the view points expressed by the VSS members, Forest Department officials and representatives of industries, the team is of the opinion that since industries would like to have control over the management and impiementation of the JFM projects to ensure required return on their capital, their association in VSSs *per se* is not feasible. However, their intervention by supply of cloned

plantation material to the villagers, the VSSs and general buy back guarantee is welcome. The suggestion of PCCF, Andhra Pradesh for constitution of JFMDA is quite reasonable and workable. This agency can provide corpus and some areas in the degraded forest can be planted with high yielding cloned seedlings keeping in view the existing bio diversity and suitable inter-mix of species. Instead of pure large plantation of one species, strip mixture/mixture of small blocks with interspersed natural growth can be thought of. About 25 per cent of degraded forest area can be kept for theuse by the villagers while in 75 per cent area the blanks can be afforested with quick growing species as suggested above. Since the VSSs will get good return from industries on sale of their surplus produce, high investment in these pockets by JFMDA can be considered. Constitution of JFMDA and its active role in rejuvenating the degraded forest lands will also help to chalk out long term perspective plan ans strategy for development of degraded forest/non-forest wasteland involving rural people.

ANNEXURE 3

Draft guidelines drawn up by the Government of India in 1994 for the afforestation of degraded forests through involvement of forest based industries.

1. Land will be leased to Forest Development Corporation (FDC) who, in turn, will enter into proper MOU with the user agency without leasing the land to them. This MOU will give right to the user agency for undertaking afforestation and right over fixed percentage of forest produce at the time of harvesting.
2. FDC will be represented in the Board of Directors of the user agency.
3. FDC will be entrusted with the responsibility of overall supervision of the plantations, e.g. species selection, number of trees to be planted, soil treatment etc. Actual operation will be done by the user agency.
4. Funds for afforestation will be provided by the user agency.

5. 25 per cent of the area will be planted with species for the purpose of supplying fuel, fodder etc. Cost of plantation and maintenance will be borne by the user agency. Entire produce out of such plantation will vest with the State Government.
6. At the time of harvesting 12.5 per cent of the forest produce from the balance 75 per cent area shall be made available to the State Government which the State Government will be free to sell at market price. However, the first preference shall be given to the user agency. This together with 25 per cent planted area will be treated as opportunity cost of the land.
7. Grass and lops and tops below 5 cm diameter will be made available free of cost to the local population.
8. Standing trees above 20 cm diameter in the area applied for will not be felled and shall continue to be the property of the Government.
9. First, the most severely degraded forests below 10 per cent density will be allotted and afterwards next category shall be offered for plantation. The land above 25 per cent density shall not be available for plantation.
10. Entire area will not be clear felled at a time which can be ensured by proper selection of species with different rotation. Projects with strong research base and advanced techniques of plantation will be given preference.
11. User agency will prepare a detailed project report giving details of land identified, choice of species, pattern of plantation, rotation period, expected yield, financial inputs, source of funding, proposed MOUs etc. after suitable lands are identified.
12. Each proposal will be examined under Forest (Conservation) Act, 1980 as well as from policy point of view and decision be taken after its detailed examination.
13. Only existing industries which already have access to wood from natural forests will be eligible. New industries which come up later will not be eligible.
14. Such industries will have to meet at least 40 per cent of their requirement from farm forestry, balance 60 per cent will also include imports, supply from natural forests and their plantations.

15. Formal proposals will be sent only after publishing about the proposed plantation and inviting objections from local community and also getting clearance from Gram Sabha.

ANNEXURE 4

D. O. No. PS/Secy(E&F)/10/98 October 8, 1998

Sub: **USE OF DEGRADED FOREST LANDS FOR GROWING INDUSTRIAL RAW MATERIAL**

Dear N. C.,

Please excuse me in having taken time in responding to the draft report sent by you on the above subject. Let me first compliment you on the quality of the report prepared by you, which certainly reflects your commitment to the issues that have been raised. I am giving below some suggestions/comments, which I hope you may find relevant.

1. The introduction section of the report mentions about a visit by some members of the Working Group to degraded forests in Andhra Pradesh, and their report has also been annexed. The conclusions of this team are at variance with the Group's final suggestions. Therefore, it is not clear whether or not the report reflects the consensus views of the Group as a whole.
2. The report goes to establish the extent, nature and category of wastelands, but there is no reference to FSI data on the subject. In any case, rather than rely on macro-level estimations alone, it would be helpful to carry out actual ground surveys so as to establish the extent of land available for this purpose. I understand that some States have already carried out such assessments.
3. In my view, we should arrive at a State-wise estimation of degraded forestlands and selectively spot check this. It is only after this that we should proceed towards the finalisation of views or a policy on the subject.
4. The draft report appears to suggest that the needs of the paper sector should be met entirely from the private farm forestry sector. The rationale perhaps being that private farm forestry sector will only be motivated if there is no other player in the market. While there is some merit in such a view, this is not the only factor that has prevented this sector from realising its

potential. We need to further energise the private farm forestry sector through other means also. The Ministry of Agriculture could be requested to examine why full potential has not been achieved here. Issues such as greater flexibility in institutional financing, involvement of the poor in large scale nursery raising, technical support from the Forest Department and elimination of legal restrictions, which hamper the growth in this sector need to be reviewed. The Ministries of Environment and Forests, Finance, Industry and other stakeholders could play a useful role in assisting the Ministry of Agriculture to draw up a dynamic package of incentives.

5. I am not in favour of sale or lease of forestlands to the private sector for the purpose of raising timber. It is, however, felt that forestlands do have a part to play in meeting some of the demand for commercial species of timber. It is also felt that the roles of the private farm forestry sector and State Forest Corporations are not mutually exclusive. It is correct that in the past displacement of eco supportive tree species by commercial species of timber on forestlands has had a negative impact on the biological diversity of an area. However, the Forest Policy of 1988 clearly recognizes this and the lessons we have learnt from our past experience. Even then, it can be ensured that Government enterprise does not repeat such mistakes. Similarly, the economic and social consequences of governmental initiatives in plantation activity should be consistent with principles of socio-economic justice, which keep in view the rights and needs of communities that are dependent on forestlands or meeting their requirements. It should not be difficult for these concerns to be woven into a set of guidelines, which would govern operations of the State Forest Corporations in this sector. After all, guidelines under the Forest Conservation Act, 1980 have been successfully implemented over two decades to strictly regulate the use of forestland for non-forestry purposes.

6. The Ministry of Agriculture would also similarly need to keep in view concerns arising from diversion of private agricultural rich lands for farm forestry, impact on soil nutrients and effect on water table. The importance of food security is something which will need to be given primacy. While forests have a very important role to play in maintaining ecological stability, conservation of biological diversity is no less important in non-

forest areas. Besides, plantation forestry is a legitimate component of forest management. And although the ecological and biodiversity values of plantations cannot be comparable with those of natural forests, bringing degraded forestlands under tree cover is in itself an adequate environmental compensation. The choice and mix of species to be planted can also be regulated to meet multiple objectives, including the needs of local communities.

7. As regards the proposal to ban import of paper pulp before such a recommendation is made, it would be advisable to project scenarios in event of significant shortfalls in supplies of timber on import prices of paper, cost to the consumer, effect on foreign exchange outflows and economic impacts on domestic industry. This would need a separate in-depth examination based on data and cost pricing which is not available to the present study and would perhaps be a matter more fit for examination by the BICP.
8. If some performers in the Public Sector, which include State Forest Corporations, have not measured up to the best levels of efficiency, we need to examine the reasons. It should not be forgotten that invariably they do not operate on a level playing field. Rather than see them as being managed by self-motivated interests, we need to see how they can be made to perform more efficiently.
9. The draft talks about a hidden agenda/statements which suggest a sell-out to industry, or which reflect on the administrative ability of members of the Forest Service to successfully perform their duties in an impartial manner. This is a reference which is based on a pre-supposition, which I find difficult to accept. If there have been some aberrations at some levels in the implementation machinery, this cannot lead us to a much wider conclusion which suggests a systematic failure on the part of the Forest Department.

With warm regards,

Yours sincerely,
(Vishwanath Anand)

Dr. N. C. Saxena
Secretary
Department of Rural Development

APPENDIX III

GUIDELINES FOR PARTICIPATION OF PRIVATE SECTOR THROUGH INVOLVEMENT OF NGOS AND FOREST DEPARTMENT IN AFFORESTATION/ REHABILITATION OF DEGRADED FOREST

Government of India, Ministry of Environment and Forests (F.C. Division) No. 8-21/96-FC, Dated 7th June, 1999

Preamble

Our National Forest Policy of 1988 envisages that one-third of the geographical area of the country should be under forest/tree cover. The total recorded forest area in the country is about 23% of the geographical area. According to the State of Forest Report, 1997, the actual forest cover in the country is only about 19.27%. It has been further estimated by Forest Survey of India that out of this 19.27% only 11.17% is dense forest i.e. having density of more than 40%. About 7.95% (2,62,310 sq.km.) of the geographic area is open forest i.e. having density less than 10%. This degraded forest needs urgent attention and sufficient monetary input so that it is rehabilitated and fully covered. These forest areas were endowed with rich biological diversity and should be our immediate concern to rejuvenate at the earliest.

The Ministry is greatly concerned over the slow rate of afforestation owing to financial and other constraints. The afforestation has gone down from about 89,000 sq. km in the VII Five-Year Plan to about 70,000 sq. km in the VIII Plan. Even if it is presumed that there will be no further degradation of forest areas it will take more than 25 years to restock the degraded forest areas. Even if we take a conservative norm of Rs. 20,000 per hectare cost it will require a budgetary allocation of more than Rs. 600 billions to rehabilitate them.

It is, therefore, clear that Ministry is not in a position even to restore the existing degraded forests from the available resources—leaving aside the goal of increasing the forest cover to 33% of the geographical area by afforestation.

It, therefore, requires some innovative policy decision wherein without compromising the basic provisions of our forest policy, we can attain the goal. One of the proposals received from the Government of Maharashtra under the Forest (Conservation) Act, 1980 appears to be aimed in that direction. It has been proposed by the Government that about 100 ha. of degraded forest land near Thane will be afforested/rehabilitated by BAIF, and NGO through funding from NOCIL in consultation/guidance of the local Forest Department. It will be done through an agreement signed between BAIF-NOCIL and the State Govt. BAIF-NOCIL will have no rights over the land or the usufruct at anytime thereafter whatsoever.

It has been observed that a large number of like minded industrial houses are willing to join hands in this green movement through NGOs and under the supervision and guidance of the forest department voluntarily without any rights whatsoever on the forest land or the usufruct. The executing NGO will work over the land for a time till the plantations/tree cover gets established.

Guidelines

The Ministry, therefore, after due consideration has taken a decision to encourage this joint participation of Industrial houses, NGOs and Forest Department in order to rehabilitate/reforest the vast degraded forest in the country. This will not only result in greening of the nation but will also result in employment generation. However, in order to have a clarity on the method and procedures, Ministry is issuing following guidelines :

1. There must be tri-partite agreement between the industrial house, an acceptable NGO of repute and the local Forest Department.

2. The agreement must have the commitment of the industrial house to bear the cost of the scheme; of the NGO to faithfully implement the scheme under the supervision and direction of the Forest Department.
3. The industrial house and the NGO will not claim any right whatsoever over the forest land or the produce anytime thereafter.
4. The rights of local people, if any, over the forest land will not be affected by this agreement.
5. The agreement will be for a period of say 5 to 7 years during which the area is likely to be tackled including maintenance.
6. The scheme should be prepared in such a way so that naturally occurring species should be given priority in the rehabilitation and even plantations should be of local native species.
7. Only such forest lands that are less than 0.4 density and that cannot be regenerated with natural regeneration should be taken up under this scheme.
8. Any other condition that State Forest Departments consider essential to fulfil objectives and preamble of these guidelines.

APPENDIX IV

गैर वन पड़त भूमि विकास में निजी पूंजी-निवेश हेतु राज्य की नीति

मध्यप्रदेश शासन, राजस्व विभाग, मंत्रालय, वल्लभ भवन, भोपाल, क्रमांक एफ 4-7/96/सात/2ए, दिनांक 22-11-97

प्रदेश में गैर वन पड़त भूमि काफी बड़े आकार में उपलब्ध है और इसे उपयोगी बनाने के लिये विशेषकर अतिरिक्त वनोपज पर आधारित उद्योगों के लिये कच्चे माल की पूर्ति के उद्देश्य से गैर वन पड़त भूमि का विकास किया जाना अत्यन्त आवश्यक है। राज्य शासन ने निर्णय लिया है कि निजी कम्पनी, पंजीकृत समितियों, संस्थाओं व उद्योगों को ऐसी पड़त भूमि के विकास के लिये आकर्षित कर पूंजी निवेश करने हेतु प्रोत्साहित किया जाय। राज्य शासन ने इस दिशा में निम्नानुसार नीति निर्धारित की है-

1. प्रदेश में उपलब्ध गैर वन पड़त भूमि वानिकी, उद्यानिकी, रेशम उत्पादन तथा अन्य वृक्षारोपण कार्यक्रमों के प्रयोजन के लिये निजी कम्पनियों, पंजीकृत समितियों और संस्थाओं को 30 वर्ष की निर्धारित कालावधि के लिये पट्टे दिये जायेंगे।
2. भूमि का बंटन इस हेतु गठित मंत्रि-परिषद की उप समिति द्वारा किया जायेगा। जो प्रत्येक प्रकरण पर गुण-दोष के आधार पर नीति अनुसार भूमि आबंटन का निर्णय लेगी।
3. गैर वन भूमि बंटन के पूर्व संबंधित ग्राम सभा से सहमति प्राप्त की जायेगी।

4. भूमि की प्रब्याजि, जिस वर्ष भूमि दी जाय उस वर्ष की गाईड लाईन के आधार पर निर्धारित की जाएगी। यदि एक ही भूमि के लिए दो या अधिक आवेदक हैं तो आबंटन नीलामी द्वारा उच्चतम मूल्य के आधार पर किया जायेगा।
5. पट्टे के लिये निर्धारित प्रब्याजि (प्रीमियम) की राशि निजी क्षेत्र (कम्पनी), पंजीकृत समिति और संस्था से पांच समान वार्षिक किश्तों में ली जाएगी।
6. पट्टे के लिये देय वार्षिक भू-भाटक का निर्धारण बढ़ते क्रम में (अर्थात प्रारंभिक वर्ष में कम तथा भूमि के वनीकरण के लाभ आने के संभावित वर्ष तक क्रमश: बढ़ते क्रम में और उसके पश्चात् निश्चित राशि के रूप में) किया जाय। अर्थात् पट्टे की अवधि के प्रथम वर्ष के लिये प्रब्याजि राशि का एक प्रतिशत, द्वितीय वर्ष के लिये दो प्रतिशत, तृतीय वर्ष के लिये तीन प्रतिशत, चतुर्थ वर्ष के लिये चार प्रतिशत, पंचम वर्ष के लिये पांच प्रतिशत, षष्ठम वर्ष के लिये छ: प्रतिशत, सप्तम वर्ष के लिये सात प्रतिशत और उसके पश्चात् आगामी प्रत्येक वर्ष के लिये प्रब्याजि राशि का साढ़े सात प्रतिशत वार्षिक भू-भाटक लिया जायेगा।
7. वार्षिक भू-भाटक की राशि संबंधित ग्राम पंचायत की "ग्राम पंचायत निधि" में जमा की जायेगी।
8. आबंटिती कम्पनी, पंजीकृत समितियों या संस्थाओं को ऐसी भूमि में किसी भी प्रकार की अंतर फसलों (इन्टर क्रॉपिंग) की सुविधा नहीं होगी।
9. राजस्व पुस्तक परिपत्र या अन्य शासकीय निर्देशों में उल्लेखित सामान्य शर्तों के अतिरिक्त गैर वन पड़त भूमि के पट्टों के लिये निम्न विशेष शर्तें भी रखी जायेंगी–

    (1) स्थानीय निस्तार अधिकारों को पूर्ण संरक्षण दिलाने के उद्देश्य से आवश्यक शर्तें अधिरोपित की जायेगी।
    (2) निजी कम्पनी, पंजीकृत समिति तथा संस्था पट्टे के प्रयोजन की पूर्ति हेतु आवश्यक मजदूर स्थानीय क्षेत्र से ही नियोजित करेगी।
    (3) भूमि के आबंटन/आधिपत्य पूर्ण होने के दो वर्ष के भीतर भूमि के विकास हेतु यदि स्थल पर कोई प्रभावी कार्य नहीं किया गया तो पट्टा स्वमेव निरस्त माना जायेगा।

10. मध्यप्रदेश कृषि जोत उच्चतम सीमा अधिनियम, 1960 की धारा 3(एच) के अधीन जारी अधिसूचना दिनांक 2-12-94 में यथा आवश्यक संशोधन किया

जाकर ऐसा प्रावधान लाया जाएगा कि ऐसी भूमि (गैर वन पड़त), जो राज्य की गैर वन पड़त भूमि विकास योजना के अन्तर्गत निजी कम्पनी, पंजीकृत समिति अथवा संस्थाओं को पट्टे पर आवंटित की जाए, अधिनियम के उपबन्धों से विमुक्त मानी जाये।

11. राज्य शासन की इन नीति के अन्तर्गत किसी भी ऐसे व्यक्ति की कोई भी भूमि जो उसके कब्जे पर पूर्व से है इस आशय के लिये आवंटित नहीं की जायेगी।
12. यह स्वीकृति वित्त विभाग पृ.क्र. 753-सीआर 2768/चार/ब-5/97 दिनांक 21-11-97 के द्वारा महालेखाकार, मध्यप्रदेश, ग्वालियर को पृष्ठांकित की गई है।

APPENDIX V

Tender Notice for Inviting Private Entrepreneurs for Afforestation on Degraded Forest Land of Madhya Pradesh on Work Contract Basis*

1. Introduction

Where the M.P. Rajya Van Vikas Nigam Limited, a Company incorporated under the Companies Act, 1956 being a State Government undertaking having its registered office at 'Panchanan' 5th Floor, Malviya Nagar, Bhopal-462 003 in Madhya Pradesh (herein after referred to as Nigam/State Government) invites sealed tenders from Private Entrepreneurs for afforestation on degraded forest land in Madhya Pradesh on work contract basis.

2. Definitions and General/Other Terms and Conditions of Tender and Instructions for Tenders

Annexure-I General/Other terms and Conditions : The Definitions of different words and expressions which have been used in this Notice including its Annexures and Schedule shall unless the context otherwise requires be as are given in the "General/Other Terms and conditions of the tender and instructions for the tenderers" contained in Annexure-I. This "General/Other Terms and Conditions of Tender" shall form part and parcel of this tender notice, and shall be constructed as included in this notice for all purposes.

* Used by Madhya Pradesh Rajya Van Vikas Nigam, Bhopal to invite private sector participation.

3. Contract Period

Schedule-I (identified areas for afforestation) : Tenders are invited for carrying out afforestation on degraded forest land of M.P. as shown in the schedule-I appended to this notice, for contract period of 30 years ending on ....

4. Tenders Form etc.

Annexure-II and III (Tender Form) :

(i) The tenderer shall submit his tender in the prescribed form (enclosed herewith as Annexure-II and III). The Earnest Money Deposit stipulated in condition 6 of "conditions of Tender" shall be paid either in Cash or in the form of Demand Draft. The tender form along with tenderer's agreement can be obtained from the O/o the Managing Director, M.P.R.V.V.N. on any working day during office hours from .... on words on payment of Rs. 5000/- or US$ 150 (Rupees Five Thousand only or US$ One Hundred Fifty only—Non-refundable) in the form of Cash or Bank Draft drawn on a Scheduled Bank, payable to Managing Director, M.P.R.V.V.N. Ltd., Bhopal.

(ii) Tender shall submit tender in two separate parts in different sealed envelopes. Part-I will contain a detailed Project Report for afforestation, maintenance and harvesting for each site as per terms and conditions given in Annexure-I. The Part-II shall contain the financial bid and shall be submitted in a separate sealed envelope.

Annexure-III (Tenderer's Agreement)

(iii) The tenderer shall personally verify and ensure that he has received the tenderer's agreement along with the tender form, as it is compulsory to enclose the tenderer's agreement duly signed along with the tender.

5. Submission of Tenders

(i) The tender complete in all respects shall be placed in a sealed envelope bearing the following words and address :

"TENDER INVITING PRIVATE ENTREPRENEURS TO CARRY OUT PLANTATION ON DEGRADED FOREST LAND ON WORK CONTRACT"

To

Managing Director,
M.P.R.V.V.N. Ltd.,
Panchanan, 5th Floor,
Malviya Nagar,
Bhopal-462 003 (M.P.)

(ii) The envelope containing the tender form should be presented to The Office of Managing Director either in person or through Registered Post, acknowledgement due in such manner so as to reach him before ...... Hours on .... Tenders will be received on this date, even if it is declared a public holiday.

6. Opening of Tender

(i) The Part-I envelope will be opened first on .... from ...... Hours in the presence of tenderers and the committee of experts constituted by the Government will examine the project technically. Only those financial bids will be opened which are found technically suitable. The decision of the Committee would be final.

(ii) The Part-II of the tender will be opened at Bhopal by a Commitee consisting of following officers on .... from .... Hours.

Members of the Committee

1. Executive Director, M.P.R.V.V.N., Ltd. or his nomminee
2. Conservator of Forests (Development), O/o PCCF or his nominee.
3. Conservator of Forests (Land Management), O/o PCCF or his nominee.

7. Execution of Contractor's Agreement

Annexure-IV Contractor's Agreement:

(i) The offer of the successful tenderer will be accepted by issuing telegram and/or letter and on the issue of such acceptance, the contract regarding afforestation on degraded forest land between the tenderer and Nigam will be deemed to have come into force.

Annexure-IV

(ii) The successful tenderer shall have to deposit the prescribed security and execute an agreement in the form given in the Annexure-IV in respect of site/sites for which the bid has been accepted within 30 days of the issue of the acceptance of the offer by the Nigam/State Government.

(iii) In the event of non-execution of the contractor's agreement the offer shall be liable to be cancelled and on such cancellation the Earnest Money Deposit/Security Deposit shall stand forfieted.

8. Annexures

Annexure-I to V and Schedule-I

Annexures-1 to V and the Schedule-I will be supplied with the tender form.

9. Acceptance of the Terms and Conditions

The act of submitting the tender shall be deemed to be unconditional acceptance of the terms and conditions of the tender.

10. English Version to Be Authorised Text

The English version of this notice including its schedule and annexures shall be deemed to be the authorised text for all purposes.

For and on behalf of the
Governor of Madhya Pradesh

Managing Director
M.P. Rajya Van Vikas Nigam Limited

ANNEXURE—I

General/Other Terms and Conditions of Tender and Instructions for the Tenderers Forming Part of the Tender Notice No. .... Dt. ....

The following are general/other Terms and Conditions of the Tender, instructions for the tenderers and the definitions of the different words and expression which have been used in the Tender Notice including its Schedule and Annexures. These shall form part and parcel of the Tender documents.

1. Definitions

In the tender notice, including its Annexures, unless the context otherwise requires :

(i) "Act" means Indian Forest Act, 1927 and the Forest (Conservation) Act, 1980.

(ii) "ADHINIYAM" means M.P. Van Upaj Vyapar (Viniyaman) Adhiniyam, 1969 and the Rules made thereunder.

(iii) "ANNEXURE" means an annexure to the Tender Notice.

(iv) "ARREARS" means any amount outstanding against tenderer which is due to the Government or Nigam and intimitation of which has been given to him by the Forest Department or Nigam or their officer, by registered post at least 30 days prior to the date of submission of the tender.

(v) "CONTRACTOR" means the person in whose favour the work contract has been executed.

(vi) "DEGRADED FOREST" means any forest area having a crop density less than 0.4 as adjudged by the Forest Department/ Nigam.

(vii) "FOREST PRODUCE" means as defined in the Indian Forest Department/Nigarm.

(viii) "GOVERNMENT" means Government of Madhya Pradesh.

(ix) "HARVEST" means any forest produce obtained from the crop planted by the contractor in the area specified.

(x) "MONOCULTURE" means planting of single species covering more than 50% of the given area.

(xi) "NATURAL GROWTH" means all the trees, other than the planted crop, growing and naturally regenerating in specified area before and during work contract period.

(xii) "NIGAM" means M.P. Rajya Van Vikas Nigam Limited, Bhopal.

(xiii) "NISTAR" means as defined in M.P. Land Revenue Code, 1959.

(xiv) "NON-FORESTRY ACTIVITY" means as defined in the Forest (Conservation) Act, 1980.

(xv) "PLANTED CROP" means the plantation undertaken by the contractor under the work contract as per approved project report.

(xvi) "PROJECT REPORT" means a document having among others a detailed plan of plantation along with investment and harvesting pattern.

(xvii) "PRODUCE" means forest produce obtained from the plantation crop which includes all the parts of trees above the stump.

(xviii) "WEEDING" means removal of any unwanted plant that interfere or tends to interfere with the growth of individuals of planted species usually within sphere of one metre diametre of planted seedling.

(xix) "ROTATION" means the planned number of years between the formation of a crop and its final felling.

(xx) "TENDERER" means a person or a Registered Firm or Legal Company who offers tender for carrying out plantation in accordance with the terms and conditions contained herein which expression shall include his heirs, successors, representatives and assignees.

(xxi) "TREE" means as defined in the Indian Forest Act, 1927.

(xxii) "WASTELAND" means as defined by The National Wastelands Development Board (NWDB).

(xxiii) Words and expressions which have not been defined above but which are defined in the Act or Adhiniyam, will have the same meaning assigned to them in the said Act or Adhiniyam.

2. Details of Sites

Schedule I : Details of sites (Comptt. No., Range, Division and Districts) on which plantation has to be done are given in the Schedule-I attached with the tender notice.

3. Provisions of Acts, Adhiniyam etc. Applicable

All the provisions of the Act and Adhiniyam, for the time being in force in so far as they are applicable to such contracts, shall specifically form part and parcel of the terms and conditions of the tender notice and the contract agreement.

4. Sites Availability on "As is Where is Basis"

(i) Subject to the condition 4(ii) below the site will be made available on "as is where is basis". The tenderers are advised to inspect the sites of plantation for which they intend to submit tender and satisfy themselves about the site quality and capability of the soil on which plantations are to be carried out. No dispute regarding site quality and their fitness for carrying out the plantation will be entertained after submission of tender at any stage nor will the Government/Nigam be responsible for any deterioration in quality after acceptance of offer of the tenderer.

(ii) The contract will be for carrying out plantation on the given sites notified in the given Schedule-I on the work contract basis. The contractor shall be responsible for making all financial arrangement for the work.

5. Persons Authorised to Submit Tender etc.

(i) Person or persons signing the tender form shall state in what capacity he or she or they are signing the tender form e.g. as sole Proprietor of the firm concerned or as Managing Director or Director or Secretary of Limited Company. In the case of Partnership firm the names of all the partners should be recorded and the tender form should be signed by all the partners or their duly constituted attorney to bind all partners in all matters pertaining to the contract as recorded in the power of attorney or in the partnership deed. True copy of the Registered "Partnership Deed" should be furnished along with the tender form failing which the tender shall be rejected. It shall be obligatory on the part of every partner of the firm, which enters into agreement to fulfil the terms and conditions of the agreement during the currency of the contract thereof, notwithstanding the dissolution of the partnership in the meantime. In the case of a Limited Company, the tender form shall be signed by a person empowered to do so by the Company, copy of Memorandum and Articles of Association of the Company and the letter authorising the person signing the tender documents shall be attached to the tender form failing which the tender shall be rejected.

(ii) The person signing the tender form on behalf of another or on behalf of a firm shall enclose with the tender form power of attorney or deed duly executed in his favour or the partnership deed giving him such power showing that, he has the authority to bind such other person or the firm, as the case may be in all matters pertaining to the contract. If the person so signing the tender form fails to enclose the said power of Attorney or partnership deed, his tender shall be liable for summary rejection. The power of attorney should be signed by the Proprietor in case of a Proprietory concern and by the person who by his signature can bind company in the case of limited company.

(iii) In case of a person submitting a tender in individual's capacity he should enclose certified copy of Income Tax return for last three years.

(iv) Tenders submitted by such persons who are minors, or insolvent or who have been black listed, shall be treated as invalid.

(v) A tenderer, who is in arrear, may pay the outstanding amount by Bank Draft/Demand Draft of any scheduled Bank along with his tender payable to the Managing Director of Nigam, but if he fails to do so, the tender submitted by such tenderer will be treated as invalid, and the E.M.D. shall be adjusted against the arreares.

(vi) The tender of any person/registered Firm/Legal company, who indulges in misconduct or disturbs peace during the opening of the tenders at the venue fixed for the purpose, shall be declared as invalid and any loss suffered by the Nigam on account of declaration of such tender as invalid shall be recovered from EMD/security deposit and in addition such person/registered Frim/Legal company shall be liable to be blacklisted with the permission of Government/Nigam for a period which may extend to three years.

(vii) Successful tenderer shall submit evidence to the satisfaction of Nigam, of having a registered branch office in Bhopal (M.P.) or at the district, in which site is located, before signing the agreement.

6. Earnest Money Deposit and Security Deposit (E.M.D. and S.D.)

(i) Every tender shall be accompanied by an Earnest Money Deposit of Rs. 1,00,000/- or US$ 3000 in the form of cash or bank draft drawn on any scheduled bank payable at Bhopal in favour of Managing Director of M.P. Rajya Van Vikas Nigam Ltd., Bhopal. Tenders accompanied by E.M.D. in any other form shall be summarily rejected.

(ii) The E.M.D. of unsuccessful tenderers shall be returned to them after the tender results are declared. However, in the case of the successful tenderers, E.M.D. shall be retained and adjusted towards security deposit.

(iii) An amount of Rs. One Lakh or US $ 3000 or 5% of total project cost, whichever is more, shall be deposited by the successful tenderer before signing the agreement as security deposit, which shall be returned after complation of agreement period. The security deposit may be in the form of bank guaranttee or bank draft or call deposit. The responsibility for renewal of Bank Guarantee before six month of its expiry shall rest with the tenderer.

(iv) No interest shall be paid on the amount of E.M.D. and S.D.

7. Manner of Filling Tender

(i) A tenderer can submit only one tender for one site. No tender of a tenderer will be considered if more than one tender is submitted by him for one site.

(ii) Tenders can be submitted only on the serially numbered form having seal of issuing office and signature of authorised officer. Tenders submitted otherwise will be treated as invalid.

(iii) The tenderer shall submit separate offer in the standard form for each site. The tender shall be submitted in two parts separately. Part-I will be the site specific project and the Part-II will be the financial bid for site. The financial bid shall be indicated in term of money along with payment schedule to be given to GOMP/ Nigam, in each harvest including cleaning, thinning and any other activities from which any forest produce is obtained. Submission of Part-I and Part-II without clear and authentic bases of arriving at different projections of production, sale value and financial return, shall render the tender as invalid.

OR

The tenderer shall submit separate offer in the standard form for each site. The tender shall be submitted in two parts separately. Part-I will be the site specific project and the Part-II will be the financial bid for site. The financial bid shall be indicated in terms of money with payment schedule tenderer expects from Nigam, in consideration of work executed by him. It is clarified that above mentioned money has to come from the project incomes proposed to the implemented by the tenderer. The Nigam takes no responsibility for payment in case of failure of tenderer or otherwise to yield/generate the projected income from the project.

(iv) The part-I containing the project report will be opened first. If the Project report is approved by the Committee constituted by the Government then only, Part-II i.e. the financial bid will be opened. In those cases where the project report is not approved by the committee constituted by the Government, the financial bid will be summarily rejected without opening the envelope.

(v) If in the tender submitted by a tenderer, the offer for any site is not clear, that is for which specific site or what financial bid it is or there is a mistake regarding identity of the site and if there is discrepency in word and figure regarding the amount then such offers shall not be considered, at all.

(vi) Tenderer shall mention his correct and complete postal address in his tender form. Communication sent to him by registered post on this address will be presumed to have been received by him. Responsibility for receiving all communications addressed to him will be that of the tenderer. If the address mentioned by the tenderer is found to be incorrect he will be liable to be blacklisted.

(vii) The tenderer should fill up and sign every page of the tender form, and enclose all necessary documents and Tenderer's agreement as duly executed with it and submit it as shown in clause 5 of tender notice. Failure to enclose Tenderer's agreement duly executed along with the Tender form and other documents will result in the tender not being considered at all.

8. Withdrawal of Offers etc.

(i) A tenderer may withdraw his offer for any site/sites before commencement of the opening of part-I of tenders, by submitting an application for withdrawal of offer to the committee constituted for opening of tenders under clause-6 of the tender notice.

(ii) A tenderer, shall not withdraw his offer for any site/sites after commencement of oepning of part-I tenders and shall be bound by his offer and by the terms and conditions of the tender notice including those contained herein till issue of communication of the acceptance or rejection of his offer by the Nigam. In case of breach of this condition the Earnest Money Deposit shall be forfeited and the tenderer may also be blacklisted for a period which may extend upto three years.

9. Acceptance of Tenders

(i) The Government/Nigam reserves the right to accept or reject offer(s) of all or any of the sites mentioned in Tender form without assigning any reason.

(ii) If for a particular site same financial bid is offered by more than one tenderer, the allotment of site shall be decided by draw of lots.

(iii) The tenderer will be bound to accept site/sites for which his offer(s) is/or are accepted.

10. Site Specific Project Report

The tenderer shall submit a site specific project of plantation, maintenance and harvesting for each site. The Project should also contain the percentage combination of forest species to be planted, harvesting rotation, plantation techniques to be used and man-power involved. The provision of fertilizers and insecticides should be based on soil analysis report and irrigation, area of irrigation should be based on ground water test. The clear and authentic basis of projection of production, sale value and financial returns shall also be mentioned in project. Brief guidelines for preparation of project report is given in Annexure-V.

11. Breach of Adhiniyam etc.

A tenderer who commits any breach of provisions of Act or Adhiniyam or any condition of the contractors agreement as a result of which his agreement is terminated, shall be liable to be blacklisted for three years in addition to forfeiture of his S.D. and E.M.D.

12. Transfer of Agreement

A tenderer can not assign or transfer his agreement to another person/ registered Firm/Legal company without prior written permission of Government. Such agreement can be transferred with the approval of the Government to another person/registered Firm/Legal company on payment of a transfer fee of 5% of total project cost in advance by him in form of bank/demand-draft of any scheduled bank payable at any branch of Bank at the head quarter of Managing Director of M.P. Rajya Van Vikas Nigam.

ANNEXURE—II

Tender Form for Inviting Private Entrepreneure for Afforestation of Degraded Forest Land on Work Contract Basis

[Annexure to Tender Notice No. PF (_______) Dt. ....]

M.P. Rajya Van Vikas Nigam Limited

Panchanan, 5th Floor, Malviya Nagar

Bhopal

1. I/We..(name and full address) hereby agree to carry out plantation on the forest land shown in schedule (one) on terms and conditions as specified in the Tender Notice and Tenderer's Agreement.
2. I/We .... shall pay a sum of Rs. .... (in words) .... only to Nigam as per schedule given at Page 12.

   Or

   I/We .... shall charge a sum of Rs. .... (in words) .... only from Nigam as per schedule given at Page 12.
3. I/We hereby declare that I/We have read and understood all the provisions of Indian Forest Act, 1927 and M.P. Van Upaj Vyapar (Viniyaman) Adhiniyam, 1969 & rules made thereunder, the conditions of Tender Notice No. .... Dt .... and the conditions of Tenderer's agreement appended to Tender Notice and I/We have personally inspected the plantation site, for which we have tendered.

4. As required under terms and conditions of the Tender Notice I/We enclose the following :

S.No.	*Particular of document*	*Details Tender From*	*Page No. from*	
			From	*To*
1.	(a) Evidence of payment of amount of Rs. 5000 or US $ 150 for tender form (Tender .... No. ....)	Office .... Receipt No. .... (Photocopy is attached)		
	(b) Evidence of payment of amount of Earnest Money Deposit of Rs. 1,00,000 or US $ 3000 (Tender .... No. ....)	Draft No. .... Dt. .... Payable in the .... Bank at Bhopal.		
2.	(a) True copy of the Certificate of registration of Firm/Company deed of partnership/ article of association of firm/company audited balance-sheet in case of Firm/Company and income tax return in case of individual of last three years (condition 5) of General/Other terms and conditions/contained in Annexure-I	(Please give details of the documents enclosed)		
	(b) Proof of technical experience			

5. I/We hereby declare that I/We shall be bound by my/our offer and terms and conditions of the tender notice, till the orders accepting or rejecting my/our offers are received or other person or registered Firm/Legal company is/are appointed as contractor for the above indicated plantation site.

6. Names and addresses of the individual or Directors (in the case of a Company) or Partners (in the case of Partnership firm) of the firm are as under :—

S.No.	*Name*	*Full Address and Phone*	*Nationality*	*Signature*
1.				
2.				
3.				
4.				
5.				
6.				

I/We desire and agree to carry out plantation as per the terms and conditions of tenderers and contract agreement attached in Annexure-III and IV.

Signature of the Tenderer
With Full Postal Address

____
____
____

Place : __________ Tel. No. __________
Date : __________ Fax No. __________
e-mail __________

Schedule of Payment of Nigam

Year	*Amount*	
	In Figure	*In Words*
1	2	3

OR

Schedule of Payment To Be Charged From Nigam

Year	*Amount*	
	In Figure	*In Words*
1	2	3

ANNEXURE—III

Tenderer's Agreement

[Condition 4(i) of Tender Notice]

(Annexure to Tender Notice No. PF (____________) dt. ......

This agreement is made .... day of .... between M.P. Rajya Van Vikas Nigam Limited (hereinafter called 'Nigam', which expression shall, where the context so admits, include its successors/representatives and assignees in office) on the first part and M/s .... District .... (hereinafter called the Tenderer(s) which expression shall include his heirs, successors, representatives and assignees) on the second part.

WHEREAS the Government of Madhya Pradesh, in the Forest Department, under its policy and existing laws desires to make available the wasteland/degraded forest lands in the Forest areas of Madhya Pradesh to the Nigam for carrying out plantations on such lands.

WHEREAS the 'Nigam' is desirous of inviting private entrepreneurs for afforestation of degraded forest land on work contract basis.

And WHEREAS the contractor is willing to undertake the afforestation on work contract basis under the supervision of the Nigam as per the project report approved by the competent authority.

AND WHEREAS it is agreed that the site offered for executing the plantation on work contract would remain in the possession of the Nigam and the contractor is only entitled to enter upon the site and execute the work as agreement upon. The Nigam will have the ownership right on existing stock and natural regeneration that comes up on the site.

AND WHEREAS it is agreed that the mode of payment for the work to be executed by the contractor would be in terms of money as per schedule mentioned under clause 2 of tender form.

AND WHEREAS it is agreed that the contractor shall start execution of plantation work within four months from the date of issue of work order, failing which the contractor shall be penalised by award of a fine of 1% of project cost of first year. If the contractor does not start execution within six months from the date of issue of work order, then the agreement shall be terminated and S.D. shall be forfieted. The contractor may be black listed as well.

AND WHEREAS the Nistar rights of the villagers shall remain unaffected.

AND WHEREAS the contractor is desirous to associate himself financially with the said project and has agreed to incur all necessary expenses for execution of the project including the supervision charges of the Nigam.

AND WHEREAS the parties after mutual discussions have agreed to implement the project for raising plantations of suitable basket of species, without adopting monoculture.

NOW those present witness, we registered firm/company (............................) hereby agree as follows :

1. I/We .... hereby declare that I/We have read and understoód all the provisions of the Indian Forest Act, 1927, Forest (Conservation) Act, 1980 and M.P. Van Upaj Vyapar (Viniyam) Adhiniyam, 1969 and the rules made thereunder; the conditions of the tender notice referred to above General/Other Terms and Conditions of Tender, etc. contained in Annexure-I of the Tender Notice and the conditions of the Contractor's agreement appended to the Tender Notice and agree to abide by the same.
2. I/We .... hereby declare that I/We shall not withdraw my/our offer after commencement of opening of tenders. I/We further declare that I/We shall be bound by my/our offer and by the terms and conditions of the tender notice till orders of competent authority, accepting/rejecting my/our offer, are passed or another person or party is appointed as contractor for the sites for which I/We have submitted the tender.
3. In the event of my/our failure to abide by the conditions of this agreement I/We .... shall be liable to pay such penalty, as may be leviable under the terms and conditions of the tender notice and also termination of the contract.
4. This agreement shall be deemed and always be deemed to be subject to the provisions of Indian Forest Act, 1927 and other Adhiniyam, the rules made thereunder and the orders issued from time to time under the said Act and Adhiniyam and other relevant rules and of the terms and conditions of Tender Notice

No. .... dt .... all of which shall form part of and shall be deemed to have become part of this agreement and shall be construed to have been specifically provided for in this agreement.

5. I/We .... hereby declare that neither any Forest Department/Nigam dues are outstanding against me/us in Madhya Pradesh nor have I/We been blacklisted by the Govt./ Nigam.

In witness whereof the person/registered firm/legal company hereto have put their signature on the day and year written first above.

In the presence of witnesses :

	Name	Father's Name	Address	Signature
1.				
2.				

In the presence of witnesses — Signature of Tenderer

1.

2.

For and on behalf of the
Governor of Madhya Pradesh

Managing Director
M.P. Rajya Van Vikas Nigarm Ltd.
Bhopal

ANNEXURE—IV
Agreement for Wasteland/Degraded Forest Land Development

Articles of agreement made this .... day of .... (month) One thousand nine hundred ninety eight between Madhya Pradesh Rajya Van Vikas Nigam Limited, a company incorporated under the Companies Act, 1956, being a State Government undertaking, having its registered office at 'Panchanan' 5th floor, Malviya Nagar, Bhopal-462 003 in Madhya Pradesh (hereinafter referred to as 'Nigam') on the one part, and M/s. .... a company/firm registered under the Companies Act, 1956 and having its registered office at the other part.

WHEREAS the Government of Madhya Pradesh, in the forest department, under its policy and existing laws desires to make available the wastelands/degraded forest land in the forest areas of Madhya Pradesh to the Nigam for carrying out plantations on such lands.

WHEREAS the Nigam is desirous of inviting private entrepreneurs for afforestation of wasteland/degraded forest land on "work contract basis".

AND WHEREAS the contractor is willing to undertake the afforestation on "work contract basis" under the supervision of the Nigam as per the project report approved by the competent authority.

AND WHEREAS it is agreed that the site offered for executing the plantation on work contract would remain in the possession of the Nigam and the contractor is only entitled to enter upon the site and execute the work as agreed upon.

AND WHEREAS it is agreed that the mode of payment for the work to be executed by the contractor would be in terms of money as per the given payment schedule.

AND WHEREAS it is agreed that the contractor shall start execution of plantation work within four months from the date of issue of work order, failing which the contractor shall be penalised by award of a fine of 1% of project cost of first year. If the contractor does not start execution within six month from the date of issue of work order, then the agreement shall be terminated and S.D. shall be forfieted. The contractor may be black listed as well.

AND WHEREAS it is agreed that the Nistar rights of the villagers shall remain unaffected.

AND WHEREAS the contractor is desirous to associate himself financially with the said project and has agreed to incur all necessary expenses from execution of the Project including supervision charges of the Nigam.

AND WHEREAS the parties after mutual discussions have agreed to implement the project for raising plantations of suitable basket of species, without adopting monoculture.

AND WHEREAS the parties are now desirous of recording the terms of agreement between them for execution of the project therefore it is hereby agreed by and between the parties hereto as follows:

1. The Project shall be set up in .... hact, the areas located at .... as per the details mentioned in schedule attached to this agreement. The period of the project shall be 30 years in the first instance which is subject to renewal by mutual agreement.
2. Period of Agreement
   This agreement shall commence from .... and shall remain in force upto .... (Date) .... (Month) ....(Year) unless terminated earlier, under the Terms and Conditions of this Agreement.
3. Parts of the Agreement
   This agreement shall always be deemed to be subject to the provisions of Indian Forest Act, 1927, Forest (Conservation) Act, 1980, and M.P. Van Upaj Vyapar (Viniyaman) Adhiniyam 1969, the rules made thereafter and the orders and notifications issued from time to time under the said Acts and Adhiniyam and the rules and Terms and Conditions of tender notice .... dated .... including the general/other Terms and Conditions of tender and instructions for the tender contained in Annexure—...... of the tender notice, all of which shall form part of and shall be deemed to have become parts of this agreement and which shall be construed to have been specifically provided for in those present.
4. The Nigam shall allow the use of the land for afforestation purposes only to the Contractor for 30 years and the absolute ownership of the land vests in the Government of Madhya Pradesh, the Nigam is only an agent. The possession of the site shall be exclusively of the Nigam. The contractor is only entitled to enter upon the land and carry out all the operations of planting as prescribed in the approved Project Report. The contractor can install a borewell or any such similar facility for irrigating the plantation. The contractor shall be bound to render every assistance to the Forest Department to extinguish fire and to prevent illicit felling, if any, in the adjoining forest

areas and shall also give immediate information of such ocurrence to the nearest forest, revenue or police authority.

5. The site made available to the contractor in phases, according to plantation programme, for executing the work shall be used exclusively for raising plantations for forest species as per the approved project report.
6. It is entirely the responsibility of the contractor to arrange funds for executing the project and the plantation shall be raised by him as per the approved Project Report. The Nigam shall not in anyway be responsible for arrangements of funds and will only provide technical guidance and supervision for raising the plantations.
7. The seedlings/plants required for plantation may be purchased by the contractor from the Nigam on payment of the cost of plants to be mutually decided.
8. The contractor shall as far as possible employ the latest technology for achieving optimum productivity.
9. The contractor shall not be allowed to clear any natural growth on the land on which the plantation are to be raised. However where felling is required before raising the plantation, the felling or clearance of area shall be done by the Nigam. The total expenditure on this work shall be borne by the contractor.
10. The cultivation of agriculture crop, horticultured crops oil-bearing plants and medicinal plants (except indigenous) are prohibited as per existing provisions contained in Forest Conservation Act, 1980.
11. The contractor shall be responsible for timely and proper execution of the plantation schedule and completion of various operations mentioned in the "approved project" report. If any delay or deviation is observed by the Nigam, the contractor shall be penalised as decided by the Nigam.
12. The Nigam shall undertake periodical or desired inspection of the plantation work done by the contractor to monitor and supervise the follow-up of approved project report and agreement.
13. The contractor shall provide periodic reports regarding the status of the plantation, survival and expenditure to the Nigam as per the proforma mutually agreed between the parties.

14. The rotation age for harvesting the produce shall be as per approved project report, however, no harvesting shall be done without prior intimation of 15 days to the Nigam. If at anytime, harvesting is done without the knowledge of the Nigam, the same shall be considered as illegal felling and all such harvest shall be seized by the Nigam without prejudice to the contractor being subjected to penal action under different Acts, Rules and Regulations governing raising of plantation/afforestation. The contractor shall not be permitted to harvest the crop after the expiry of the contract period. The harvesting pattern of the crop shall be indicated in the Project Report. There shall be provision of 5% of the intermediate harvest for Nistar to local villagers at the rates approved by the local Conservator of Forests as per the prevailing Nistar Policy.
15. The contractor shall leave 10% of planted crop spread over whole area at the time of final harvest as future crop to avoid soil erosion etc.
16. The contractor agrees to make payment of a sum of Rs............... (in words Rupees ....) to the Nigam as per yearwise details given below:

    OR

    The Nigam agrees to make a payment of a sum of Rs. .... (in words Rupees ....) to contractor in lieu of works executed by him from sale proceeds of produce generated by approved project, however the amount of payment shall be in proportion to the projected income from project. The yearwise detail of money to be paid by Nigam is given below :

Year	*Amount (Rs.)*	
	In Figure	*In Words*
1	2	3

17. All existing growth on the forest land given for afforestation will form part of the natural growth and the contractor shall not have any right whatsoever upon the produce obtainable from

the said natural growth. The natural growth will be enumerated and the record of such growth will be prepared prior to offering such land to the contractor to enter upon and carry out plantation. This natural growth and the forest produce will be the absolute property of the Government. The existing growth shall be specifically identified before handing over of the area for plantation work.

18. The contractor shall make all arrangements and be responsible for the security and protection of the plantations raised.

19. The Nigam shall not be responsible for any shortage or loss of production for any reason whatsoever nor shall the contractor have any right to claim any compensation from the Nigam/ Government on this count.

20. Any notice required to be given by one party to the other under this agreement shall be deemed to have been given if sent by registered post, addressed to the Registered Office of the Nigam/the contractor to whom such notice is given and such notice shall be deemed to have been received, when in the normal course, it should have been received by the party to whom it is given.

21. In the event of any dispute or difference arising between the parties to this agreement on any matter with respect to this agreement or any of the provisions herein contained or anything arising hereto, except in respect of any matter on which the decision is herein before declared to be final and binding it shall be referred to the Secretary to Government of M.P. Forest Department, Bhopal whose decision therein shall be final and binding on the parties.

22. In the event of breach of any condition of this agreement by the contractor or his authorised representative, the Nigam shall have the right to terminate the agreement after giving notice to the contractor. On such termination of the agreement the penalty to be imposed on contractor shall be decided as per the gravity of the breach of agreement.

23. If any pecuniary or otherwise loss accrues to the Nigam/ Government due to the said work of contract the same shall be recovered as arrears of land revenue.

24. All disputes arising out of this agreement shall be subject to the jurisdiction of the courts in the State of M.P.
25. NOTWITHSTANDING anything contained herein, the Governor shall not be liable for any delay or failure to carry out this agreement whole or in part by reasons of force majeure i.e. act of God, Civil Commotion, Strike, Earthquake, Storm, Flood, Lightning, Explosion, Fire or any other happening which the Governor could not reasonably prevent or control.

    In witness whereof the parties hereto have duly executed these presents the day and year first herein above written;

The common seal of Madhya Pradesh Rajya Van Vikas Nigam Ltd. was here unto affixed in pursuance of resolution of the Board of Directors of the said company passed on .... by Shri .... Directors of the Company who havc in token thereof put their signature hereto in the presence of witnesses:

For and on behalf of the
Governor of Madhya Pradesh

Managing Director
M.P. Rajya Van Vikas Nigam Ltd.,
Bhopal

Witness Signature

1. .... 1. ....
2. .... 2. ....

The common seal of .... was here unto affixed in pursuance of resolution of the Board of Directors of the said Company passed on .... by Shri .... and Shri............................. Directors of the Company who have in token thereof put their signatures here to in the presence of witness :

Witness Signature of Conractor Signature

1. .... 1. ....
2. .... 2. ....

ANNEXTURE—V
Brief Guidelines for Preparation of Plantation Project

(See condition No. 10 of General/Other Terms and Conditions)

Madhya Pradesh Rajya Van Vikas Nigam Limited, Bhopal

PART—I
Technical Bid

PROJECT COMPONENTS

1. ***Objective :*** Objective should include

(a) Purpose of raising the plantation, and end use of the produce.

(b) Whether the tenderer would be using the produce within M.P. or outside the state.

(c) Whether the tenderer would be using the produce himself or would be supplying it to some other industries or would be selling it in the open market.

2. ***Description of project area:*** Description should include the location and extent of the area. The observations/experiments/ analysis etc. carried out to assess the productivity of the area and the social and environmental impacts should also be clearly spelt out.

3. ***Strategy :*** Should include

(a) *Choice of Species :* Basket of species proposed to be planted, suitability of which should adequately be supported by technical documents. (Publications etc.)

(b) *Technical Parameters :* Spacing and patterns of planting including floor management i.e. planting herbs, tubers etc.

(c) *Site Preparation :* Method of Soil Working, size of the pits, pit digging with or without moisture conservation measures, whether any mechanical implements are to be used, time schedule of each activity, etc.

(d) *Planting Technique :* Source and type of the planting stock to be used such as seed sowing, root-shoot technique, polypots, tissue culture, tall planting, etc. and time schedule of every activity. Whether plants are to be bought from Forest Corporation/Forest Department etc. Also it should include the number and pattern of weedings, and use of weedicides if any (please specify names).

(e) *Soil and Moisture Conservation Methods* : Adopted according to the site requirements.

(f) *Irrigation Methods* : Source, intensity, frequency and type of irrigation to be used. Also it should be specified whether any permanent asset would be created.

(g) *Manuring* : Technique, type and doses.

(h) *Insecticides* : Name, quantity, doses and stages of application.

(i) *Protection Methodology* : Proposed against fire, grazing and illicit felling.

(j) *Other Land Uses Proposed* : Such as plantation of medicinal plants and other species in under storey.

(k) *Tending/Silvicultural Operations* : Schedule of cleaning and type of thinning.

(l) Time schedule of various operations.

(m) *Employment Generation* : Arrangement of manpower for implementation of the project.

4. ***Rotation and Final Havesting :*** Should include

(a) Period of rotation/felling cycle for the species raised.

(b) Methodology for assessment of the final harvest of planted crop and the technique to be used for final harvest.

5. Yield estimations of each of the species proposed to be planted, along with frequency of harvest during the contract period.

6. Financial forecast giving details of Cost Benefit Analysis, IRR and *environmental impact analysis* of the project Assessment.

7. *Project organisation, Management and Supervision* : Should include the organisational set up for implementing, managing and supervising the project.

8. *Infrastructure Development* : Approach road, labour/choukidar huts, tube wells, stop dams, electric transformers and any other permanent assets to be developed for the implementation of project should be mentioned.

APPENDIX VI

RESOULUTION FOR COMMUNITY PARTICIPATION IN CONSERVATION AND DEVELOPMENT OF FORESTS

Forest and Culture Department
Mantralaya, Dau Kalyan Singh Bhavan, Raipur

No. F-7-36/2001/F.C./10 Raipur, Dated 22 Oct. 2001

Sub : Resolution for Community Participation in Conservation and Development of Forests

1. The National Forest Policy 1988 envisages participation of local people is necessary for conservation and development of forests. Accordingly Ministry of Environment and Forests, Government of India (on 1st June 1990), has directed all the State Governments that tribals and other communities living in an around forests have their first rights on the forest produce. In consonance with this principles the local people have been involved in the management of forests through *Joint Forest Management.*
2. In acordance with the direction provided by the Government of India, the State Government herewith passes a resolution for people's participation in forest protection and management:

   2.1 Forest Protection Committees will be constituted in the villages situated within 15 km. radius of *dense forest Area.*

   2.2 Village Forest committees will be constituted in the villages, excluding the villages mentioned under section 2.1, situated within 5 km. radius of *degraded forest Area.*

3. Procedure for the Constitution of Committee

3.1 Local staff of forest department will convene meetings of villagers in villages of the State of introduce the concept of joint forest management and if villagers voluntarily decide to join in protection and management of forest then on intimation, a formal meeting will be organised in the presence of a forest officer not below the rank of Forester with the help of local public representatives. If 50% or more villagers, who have the voting rights in the Gramsabha, are present and unanimously pass the resolution for constitution of a committee then a relevant committee as per zonation mentioned above shall be constituted. In case both degraded and dense forest areas are present around the village then, either Forest Protection Committee or Village Forest Committee may be constituted, depending upon the forest area in majority.

3.2 With the aim of constitution of in committees as per Panchayat (Extension to Scheduled Areas) Act, 1966 section 4(b) those hamlets or group of hamlets or small village or its group in which such communities are included which manage their activities as per their traditions and beliefs shall be treated as village whether that village falls in or outside scheduled area (Schedule of Constitution). Divisional Forest Officer (Territorial) concerned shall register the committee as per para-2 within one month after passing of such resolution. Every villager having voting rights, shall be the member of general body of these committees. These members shall elect President and Vice President, whose tenure, will be of two years, out of the villagers in the first meeting of committee, presided over by Sarpanch of concerned village Panchayat. It is obligatory to have at least one female member as President/Vice-President.

3.3 If any forest committee already exists in the village, it will be recognized as either of the mentioned joint forest management committees.

3.4 A confederation of Presidents of the committees shall be constituted at Forest Division level.

4. Executive Committee

The Executive Committee shall be constituted comprising minimum 11 and maximum 21 members out of members of the committee (excluding nominated members) as per following procedure :

1. Executive committee members will be nominated by organising a meeting under chairmanship of the elected President of the committee.
2. President/Vice President of Forest Committee will act as *ex-officio* President/Vice President of the executive committee.
3. Representation of scheduled tribes, scheduled castes and other backward classes in the executive committee will be in accordance with ratio of their population in village.
4. At least the executive committee will have 33% of women members and one member must be from the women savings group, if such group is existing in village.
5. At least two members (one male and one female) of the landless families, if existing and one member from the self help group, if existing in village, must be included in the executive committee.
6. All Panch and Sarpanch residing in village will be *ex-officio* members of the executive committee.
7. At least one beneficiary from different beneficiary groups of Watershed Development Committee under Rajiv Gandhi Watershed Mission will be member of the committee.
8. At least one member from each of various user groups of village resources will be elected as members of working committee for remaining vacancies.
9. Forest Guard or Forester of that forest area will be *ex-officio* Secretary of the executive committee.
10. Members of executive committee, excluding *ex-officio* members, will have maximum 2 years as "tenure of service".

5. Selection of Area

After constitution of the committee Divisional Forest Officer (Territorial) shall select forest area for various committees on the advice of executive committee. The Divisional Forest Officer can authorise an officer, not below the rank of Range Forest Officer, for this purpose. While selecting forest area for the committee the distance of the said forest area from the village and traditional nistar rights of villagers over that forest area will be taken in to consideration. Technically the suitability of the area will be considered on the basis of feasibility report of Assistant Conservator of Forests. In case of any dispute in selection of area, the decision of Divisional Forest Officer (Territorial) shall be final.

6. Micro Plan

6.1 After constitution of the committee, the villagers will prepare a microplan with the help of Forest Department. Area of the village and Forest area selected under committee will be included in this plan. Plan will have provision to include both forest management and village resource development programme. The works included in the micro plan will be based on expected potential of the resource and other works will be added in appendix of the plan on priority basis. Executive agency and possible source of funding will be shown against works. The committee will sent the micro plan to district level forest officer for its approval. The micro plan will be accorded approval after the technical and legal scrutiny.

6.2 Management principles are laid down for forest/wild life management in working plan/management plan. Works proposed in forest area under committee will be in accordance with above principles. It will be ensured that forest/wild life acts or rules are not violated in the micro plan.

6.3 The financial resources for executing the works in forest as per the micro plan will be provided by State Government besides works, which reduces dependence of villagers on forests and are related to better management of forest resources. The funds will be provided, as far as possible, by Forest Department, government funds at the disposal of committee. District Rural Development Agency, other Government departments, Panchayats and other sources Execution of works will comprise, as far as possible, 25% of labour component as "Shram Dan" by the committee members. Amount equivalent to "Shram Dan" will be deposited in the account of committee so that committee can utilise these funds in village resource development programme.

6.4 Forest Department and committee will prepare the micro plan with help of other developmental agencies. The village resource development works included in the micro plan will get technical and financial assistance from other development departments of the State.

6.5 The activities/works for economic development, which are ecologically appropriate and feasible shall be included and prioritized in the micro plan.

6.6 The State Government will constitute a Coordination Committee in each district with President of Forest Sub-committee of District Panchayat as chairman to coordinate the village development works to be executed through micro plan.

This committee will organize at least one meeting in every 4 months, in which President Janpad Panchayat and district level officers of all the departments will be members. A District level forest officer nominated by Conservator of Forests will be its Member Secretary.

7. Meetings

Secretary will call meetings of executive committee with the permission of President. Minimum one meeting in every three months is to be organised. Compulsorily General Committee's President/Vice President will preside over the meeting, but in their absence any member can be selected to preside over the meeting with the consensus of rest of the members. Minimum one general body meeting will be held in every six months. Time and venue of meeting will be decided after consulting the President. Proceedings of the meeting will be recorded in the concerned register by the Secretary. During the terminal year of executive committee's tenure succeeding executive committee will be elected in the existing general body meeting. This meething has to be called one month before the end of tenure of executive committee.

8. Quorum

A quorum will be complete with the presence of 50% members for executive committee and 30% of members for the general body meeting.

9. Right and Duties of Committee

9.1 Rights

On the satisfaction of the District level forest officer those committees which have done satisfactory work under joint forest management, will be eligible to get the following benefits.

1. Families of all committees will be eligible to get annual royalty, free nistar, after deducting expenditure on harvesting as per availability.
2. All the forest committees shall be eligible to get 100% of the produce obtained from time to time, on thinning of timber coupes and cleaning of bamboo clumps in degraded bamboo forest as per the prescriptions of microplan/working plan, on payment of expenditure incurred on harvesting.
3. Forest produce equivalent to the 10% value, of the amount calculated by deducting the expenditure incurred on exploitation (of timber/bamboo) from the total value of timber/bamboo obtained on final felling in coupes in forest area allotted to F.P.C. (as per prescriptions of Working plan) shall be given to the F.P.C. If a loss due to illicit felling in the forest area allocated to committee is detected then concerned forest committee will be accountable to the 50% value of the loss and the equivalent forest produce to that amount will be deducted from the forest produce to be distributed to that committee.
4. Forest produce equivalent to 30% value, calculated by deducting the expenditure incurred on harvesting (of timber/bamboo) from the total value of timber/bamboo obtained and final felling in plantation/rehabilitation of degraded forest/pasture development works etc., shall be given to Village Forest Committee. Computation of value shall be based on the approved Malik Maqbuja rates of timber/bamboo of that year in the

concerned Forest Circle. If a loss due to illicit felling in the forest area allocated to committee is recorded then concerned village forest committee will be accountable to the 50% value of the loss and equivalent forest produce of that amount will be deducted out of forest produce to be distributed to that committee.

5. In case of minor forest produce, rights of committee will be in accordance with the Panchayat (Extension to Schedule Areas) Act, 1996 amended as and when by the Government of Chhattisgarh.
6. If the committee helps in apprehending the offender against the offence occurred under the jurisdiction of the committee, then 50% of fine/fees deposited by offender in lieu of compounding the case or decision of court, shall be deposited in the account of committee, which shall be used for development of the village.
7. If any member of committee does not cooperate in committee's work or does not follow committee's decision or indulges in forest offence, the committee will cancel his membership duly restricting his Nistar Rights. But before this the committee will provide an opportunity to the concerned member to present his case after taking a decision in general body meeting. Action against forest offence will be additionality to the above action.

The committee can delegate its all or specific powers to the executive committee by taking decision in general body meeting.

9.2 Duties

1. Committee will elect Executive body.
2. Committee members will protect forests from fire, illicit grazing, illicit felling, illicit transportation, illicit mining, encroachment and poaching. To achieve this objective the committee will take necessary steps for forest protection with the help of its members. The Committee will be responsible for 50% of the loss due to illicit felling in its jurisdiction.
3. Committee will give information to Forest Department about the unlawful entry/illegal activities of person or persons who cause damage to forest and wild life.

4. The committee will ensure the safety of wandering wild life who incidentally came out of the forests and will inform the nearest forest officer acordingly.
5. Committee with cooperation of Forest Department will prepare Micro plan and annual action plan. Plan will comprise community beneficiary oriented, and site specific need based activities. Priority shall be given to works, which are directly or indirectly related to protection of forest and wild life. Forest Range Officer, on behalf of Forest Department and President of Working Committee, on behalf of the committee will sign on micro plan. Annual Action Plan for each financial year will be prepared on basis of Micro plan. In order to implement annual action plan, 10% of amount will be released in single instalment as an advance to committee. Committee shall execute the approved plan. If execution of works by the committee is not satisfactory or the committee is not interested in execution of works then, the execution of works will be done departmentally.
6. Members of committee, after receiving information of forest offence in their jurisdiction or other forest area, shall immediately inform to beat guard/game guard besides helping forest personnel in apprehending the forest offenders. The forest produce and offenders shall be brought before the concerning forest officers.
7. A Memorandum of Understanding shall be signed by President of the committee and District level forest officer or officer authorised by him.
8. Committee will ensure "Shramdan" as part of the labour component from committee members, as per the para 6.3 of resolution, in execution of the micro plan.
9. Members of committee will help Forest Department staff during the enquiry of forest offence.
10. Committee shall maintain account of the fund received from different sources and audit of expenditure will be done by agency authorised by forest officer.
11. Committee shall maintain a register of its members' list along with other registers and records prescribed by the forest offficer.
12. With reference to forest protection, members of the forest committee shall be treated as public servant at par with forest

personnel under various acts during the forest patrolling in their area and they shall get legal protection as in the case of public servant for the works done in good faith of public and government interest. In the same way if, member of committee gets injured or die while preventing/informing forest offence, he will get all benefits at par with forest personnel.

13. If committee assists in apprehending the forest offender against the offence occurred in its jurisdiction, 50% of the amount recovered as fees or fine after compounding or on the basis of court's decision shall be deposited in the committee's account.

10. Powers and Duties of Forest Officers

District level forest officer shall have following powers and duties. If it is not mentioned otherwise in the Resolution, he can delegate the powers to an officer not below the rank of Forest Range Officer.

10.1 Powers

1. Allocation of Area of committee according to para 5.1.
2. Approval of the micro plan.
3. Inspection of committees account and procedures for sharing of forest produce and other benefits among committee members.
4. If committee does not discharge their duties satisfactorily according to para 9.2 and if there is no improvement after being warned in writing by forest officer, then forest officer can dissolve the committee and terminate the Memorandum of Understanding. In that condition all benefits to committee as cited in para 9.1 will be ceased.

10.2 Duties

1. To organize meeting to constitute committee in accordance with para 3.1.
2. Registration of committee.
3. Supervision of election of committee and executive body.

4. To provide training and technical assistance to committee members on preparation and execution of micro plan.
5. To provide financial resources for execution of Micro Plan as per para 6.3 and to establish coordination between different departments according to para 6.4.
6. To assist committee in execution and monitoring of their duties and to help in resolving their internal conflicts.
7. Monitoring and evaluation of works carried out by the committee.
8. To authorize an agency to inspect the annual acount of committee and to carry out the annual auditing through it.
9. To distribute forest produce and other benefits from allotted area to the committee in accordance with para 9.1 and sub-para 1 to 4.
10. To ensure the participation of the weaker sections of society especially women in decision making and distribution of benefits.

11. Appeal

1. A Person can appeal to the forest officer in jurisdiction and of the rank of Territorial Forest Range Officer, against the order passed under sub-para 7 of para 9.1 within one month from the date of passing of such order.
2. An appeal can be made against forest officer's order of dissolution of committee to the Union constituted under para 3.4 within one month of date of passing of such order.
3. The orders of above Appellant Authorities shall be final.

APPENDIX VII

संकल्प

वन एवं संस्कृति विभाग मंत्रालय, दाऊ कल्याण सिंह भवन, रायपुर

क्रमांक एफ-7-36/2001/व.सं./10 रायपुर, दिनांक 31.10.2002

विषय : वनों के संरक्षण एवं विकास हेतु जन-सहयोग प्राप्त करने के लिए संकल्प से संबंधित इस विभाग की अधिसूचना क्रमांक एफ-7-36/2001/व.सं./10 दिनांक 22.10.2001 में संशोधन।

इस विभाग की अधिकसूचना क्रमांक एफ-7-36/2001/व.सं./10 दिनांक 22.10.2001 में निम्नानुसार संशोधन किया जाता है-

1. कंडिका 9.1 की उप-कंडिका 1, 2, 3 एवं 4 के स्थान पर निम्नानुसार उप-कंडिकाएं प्रतिस्थापित की जाती हैं-

1. सभी समितियों के परिवार लागू निस्तार नीति के अंतर्गत प्रति वर्ष उपलब्धता अनुसार निस्तारी वनोपज प्राप्त करने के लिए पात्र होंगे।
2. सभी वन समितियों को समय-समय पर माइक्रोप्लान/कार्य आयोजना के प्रावधानों के अनुसार समिति को आवंटित क्षेत्र के काष्ठ कूप/बिगड़े वनों के सुधार हेतु पुनर्वास किये गये क्षेत्र के साफ सफाई कार्य (क्लीनिंग) एवं यांत्रिक विरलन तथा बिगड़े बांस वनों के भिर्रा सफाई से प्राप्त शत-प्रतिशत वनोत्पाद विदोहन व्यय लेते हुए दिया जावेगा।
3. वन सुरक्षा समिति को आवंटित वन क्षेत्र में कार्य आयोजना के प्रावधान के अनुरूप बांस या काष्ठ कूप के मुख्य पातन/वन वर्धनिक विरलन से प्राप्त होने वाले वनोत्पाद की स्थल पर कुल कीमत की 15 प्रतिशत राशि तथा 15

प्रतिशत मूल्य तक का वनोत्पाद का 15 प्रतिशत मूल्य की सीमा तक राशि एवं वनोत्पाद समिति को प्रदाय किया जावेगा। वनोपज प्रदाय की दशा में प्रदाय की जाने वाली वनोपज के विदोहन पर हुए व्यय के तुल्य राशि या वनोपज की मात्रा कम कर दी जायेगी। कूप में वनोत्पाद के मूल्य की गणना गैर वाणिज्यिक दारों के आधार पर होगी। यदि समिति को आवंटित वन क्षेत्र में अवैध कटाई से हानि होती है तो शुद्ध हानि की 50 प्रतिशत राशि के लिए संबंधित वन सुरक्षा समिति को उत्तरदायी माना जायेगा और ऐसी राशि या उसके समतुल्य वनोपज समिति को प्रदाय की जाने वाली राशि या वनोपज में से कम कर ली जायेगी।

4. ग्राम वन समिति को आवंटित बिगड़े वन क्षेत्र के सुधार हेतु पुनर्वास कार्य किये जाने पर प्राप्त होने वाली वनोपज एवं पुनर्वास किये गये क्षेत्र के मुख्य पातन या वन वर्धनिक विरलन से प्राप्त होने वाले वनोत्पाद के स्थल पर कुल कीमत की 30 प्रतिशत राशि अथवा 30 प्रतिशत मूल्य तक का वनोत्पाद या 30 प्रतिशत मूल्य की सीमा तक राशि एवं वनोत्पाद समिति को प्रदाय किया जावेगा। वनोपज प्रदाय की दशा में प्रदाय की जाने वाली वनोपज के विदोहन पर हुए विदोहन व्यय के तुल्य राशि या वनोपज काट कर दी जायेगी। कूल में वनोत्पाद के मूल्य की गणना गैर वाणिज्यिक दरों के आधार पर होगी। यदि समिति को आबंटित वन क्षेत्र में अवैध कटाई से हानि होती है तो शुद्ध हानि की 50 प्रतिशत राशि के लिए संबंधित ग्राम वन समिति को उत्तरदायी माना जायेगा और ऐसी राशि या उसके समतुल्य वनोपज समिति को प्रदाय की जाने वाली राशि या वनोपज में से कम कर ली जायेगी।

2. कंडिका 11 की उपकंडिका 2 के स्थान पर निम्नानुसार उपकंडिका प्रतिस्थापित की जाती है–

वनाधिकारी द्वारा समिति भंग करने के आदेश के विरुद्ध आदेश के दिनांक से 1 माह के अंदर कंडिका 3.4 में गठित संघ अथवा संघ के गठन के अभाव में क्षेत्रीय वन संरक्षक को अपील की जा सकेगी।

छत्तीसगढ़ के राज्यपाल के नाम से
तथा आदेशानुसार

(जय सिंह म्हस्के)
उप सचिव
छत्तीसगढ़ शासन
वन एवं संस्कृति विभाग

APPENDEX VIII

STRENGTHENING OF JOINT FOREST MANAGEMENT PROGRAMME—MOEF, GOI

No. 22-8/2000/JFM (FPD)
Government of India
Ministry of Environment and Forest
(JFM Cell)

Paryavaran Bhawan, CGO Complex,
Lodi Road, New Delhi
Dated: 24th December, 2002

To

1. The Secretaries Forest Departments (All States/UTs)
2. The Principal Chief Conservators of Forests (All States/UTs)

Subject : Strengthening of Joint Forest Management (JFM) Programme

Sir,

Attention is invited to the guidelines issued by the Central Government on JFM to the States *vide* No. 22-8/2000-JFM (FPD) dated 21st February, 2000. The Government of India has received considerable feedback from the States or various issues contained in these guidelines. To incorporate these suggestions, following guidelines are issued to further strengthen the JFM programme in the field.

1. Memorandum of Understanding (MOU)

To ensure smooth working relationship between the forest department and the JFM Committees and also to bring in a sense of *empowerment and accountability,* a Memorandum of Understanding (MOU) should be signed between the forest department and the JFM committees outlining the *short term and long term roles and responsibilities, implementation of work programme, pattern of sharing of usufructs and conflict resolution.* In the MOU, JFM Committees should form the basic Forest Management Units to provide them a feeling of empowerment and enable them to effectively protect and conserve the forest resources. The MOU for each committee shall have *location specific work programme based on site-vegetation profile and mutual understanding.* The MOU should reflect the *consumption and livelihood needs of the forest dependent communities, plan for restoration of vegetation and clearly spell out the roles, responsibilities and powers.* The MOU should define the procefure for *necessary transparent accounting of all types of forest produce (seasonal, annual and periodical) accrued from the forests as per the working plans and micro plan prescription, financial accountability and distribution of sharing mechanism including ploughing back of revenue for the (regeneration in terms of Government of India letter No. 22-8/2000-JFM (FPII) dated 21st February, 2000.* All JFM Committees should be assigned specific roles for boundary demarcation, fire prevention and control of grazing, encroachments and illicit felling as well as ensure sustainable non-destructive harvesting of NTFPs including medicinal plants and for this, the Committees should be given authority to act, monetary and other incentives as genuine stakeholders. A provision to assist the JFM Committees has been mae under the Centrally Sponsored Plan Scheme 'Integrated Forest Protection Scheme'. Similar provisions should also be made in other State sector schemes. A GIS based map of the JFM area of the States along with village boundaries may be prepared every two years to monitor the performance and to ascertain the status of vegetation. Similarly, the socio-economic changes brought about by JFM should also be monitored by obtaining regular feedback from the committee members. The action programme should also be linked with the *State Forestry Plan activities in order to make JFM integral to the overall forestry activities including afforestation on all types of lands.* The MOU should also include planning and development of grasslands, other common lands, agro-forestry and water bodies as a reflection of peoples voluntary action for holistic *land use planning and management.* Agroforestry models developed in various states under different conditions should bc studied and adopted. A case in points is thc agro-forestry model developed by Nagaland under the Indo-Canadian Environment Project to control shifting cultivation.

Relationship with Panchayats

The relationship between Panchayats and JFM Committees should be such that the JFM Committees take advantage of the administrative and financial position and organisational capacity of the Pachayats for the management of the forest resources. However, the unique and separate non-political identity of the JFM committees as 'guardian of forests' should be maintained and ensured. The benefits accrued from NTFP sales should be shared with all the members of the gaon sabha including the JFM committees.

In order to achieve a better coordination with Panchayat Raj institutions, a committee may be constituted at the district level under the chairmanship of President, Zila Parishad and under Collector in those districts where the Zila Parishad is non-functional for the time being with the DFO acting as Convener and other district level officers as members.

Capacity Building for the Management of Non-Timber Forest Products (NTFPs)

The success of JFM in good forest areas would depend upon the sustainable development and harvesting of Non-Timber Forest Products. The sustanability of JFM would also depend upon the remunerative prices for the gatherers of NTFPs. This requires a well organised setup and plan of action for the collection, transportation, storage, processing and marketing of Non-Timber Forest Products. For better return to the gatherer and the Committees. State Governments may initiate non-destructive harvesting (in accordance and within the overall prescriptions of the working plans), equity in sharing, institutional reforms and also strengthen the set-up of NTFP management based on the experience in different States. The guiding principles should be to first ensure sustainability of the resources and then maximum benefit to the gatherers and value addition. Panchayats and State Forest Corporations should assist the JFM Committees for developing skills for handling the NTFP collection, manage, coverage, marketing etc.

This issue with the approval of the Minister for Environment and Forests.

Your Faithfully,

Sd/-
(M.K. Sharma)

APPENDIX IX

नीलगिरी से समृद्धि

बिगड़े वनों में निजी पूंजी निवेश

(ओरिएण्ट पेपर मिल, अमलई तथा मरवाही वनमंडल का संयुक्त प्रोजेक्ट)

मरवाही वनमंडल के लगभग 40% वन बिगड़े वनों की श्रेणी में आते हैं। कार्यआयोजना के प्रावधानों के अनुसार इन बिगड़े वनों का सुधार कार्य समितियों की सूक्ष्म प्रबंध योजना बनाकर किया जा रहा है। सर्वाधिक बिगड़े वन मरवाही, पेंड्रा एवं खोडरी परिक्षेत्र में हैं। सुधार कार्य के तहत इन वनों में नीलगिरी, बांस, सागौन, आंवला प्रजाति के पौधों का रोपण किया जा रहा है।

विभागीय रूप से बिगड़े वनों के सुधार कार्य के अतिरिक्त, वन विभाग द्वारा राज्य के विभिन्न जिलों के बिगड़े वनों के सुधार हेतु निजी पूंजी निवेश की संभावनाओं का पता लगाने की पहल माह फरवरी-मार्च 2003 में भी की गई। इसके तहत राष्ट्रीय स्तर पर विज्ञापन प्रकाशित कर निजी निवेशकों से ऑफर आमंत्रित किये। ऐसा करने पर विभाग को दो राष्ट्रीय स्तर के पेपर मिल जे.के. पेपर मिल, रायगढ़, उड़ीसा तथा ओरिएण्ट पेपर मिल, अमलई, म.प्र., द्वारा निजी पूंजी निवेश की इच्छा कुछ शर्तों के तहत की गई। जे.के. पेपर मिल उससे लगे महासमुंद वनमंडल तथा ओरिएण्ट पेपर मिल के द्वारा निकट के मरवाही वनमंडल में बिगड़े वनों के सुधार हेतु निजी पूंजी निवेश की इच्छा व्यक्त की गई।

मरवाही वनमंडल में ओरिएण्ट पेपर मिल द्वारा निजी पूंजी निवेश के कार्य को मूर्तरूप देने के उद्देश्य से प्रधान मुख्य वनसंरक्षक श्री आर.सी. शर्मा की अध्यक्षता में रायपुर में दिनांक 28.04.2003 को एक बैठक आयोजित की गई। बैठक में जनरल

मैनेजर प्लांटेशन ओरिएण्ट पेपर मिल, श्री एम.पी. पाण्डेय, श्री बजाज तथा वन विभाग की ओर से प्रधान मुख्य वनसंरक्षक के अतिरिक्त श्री आर.के. शर्मा अपर प्रधान मुख्य वन संरक्षक (विकास), श्री धीरेन्द्र शर्मा मुख्य वन संरक्षक (उत्पादन/वित्त एवं बजट), श्री के.सी. यादव वन संरक्षक (विकास), श्री कौशलेन्द्र सिंह, वन संरक्षक, बिलासपुर वृत तथा श्री अरुण पांडेय, वनमंडलाधिकारी, मरवाही वनमंडल ने भाग लिया। चर्चा के दौरान प्रधान मुख्य वन संरक्षक ने पेपर मिल के अधिकारियों को गिबड़े वनों के सुधार तथा वन समितियों के बारे में जानकारी दी। श्री पांडेय, जी.एम.ओ.पी. मिल के द्वारा बताया गया कि उनके द्वारा नीलगिरी के क्लोन पौधों को तैयार किया जा रहा है। ये पौधे तेज वृद्धि वाले हैं तथा इन्हें रोपण के छठवें वर्ष में काटा जा सकता है। इनके रोपण में जहां एक ओर वनों की उत्पादकता में वृद्धि होगी वहीं दूसरी ओर वन समितियों को अच्छा आर्थिक लाभ होगा। यूकेलिप्टस का क्लोनल पौधा पांच वर्षों के असिंचित रोपण के बाद औसत 55-60 सेमी. गोलाई तथा 20 मी. उंचाई का हो जाता है तथा लगभग 80 टन प्रति हेक्टेयर का बायोमास उत्पादन अनुमानित है। बिलासपुर जिले हेतु ओरिएण्ट पेपर मिल, अमलई द्वारा रु. 1325 प्रति टन सपोर्ट प्राइज तय किया गया है। पायलट प्रोजेक्ट के तहत मरवाही परिक्षेत्र तथा पेंड्रा परिक्षेत्र के ग्रामों का चयन यूकेलिप्टस के क्लोनल पौधों के वृक्षारोपण हेतु किया जावेगा ताकि उन्हें पेपर मिल से कम से कम लीड पर पौधे मिल सके।

उक्त चर्चा उपरांत 1-2 मई 2003 को वनमंडलाधिकारी मरवाही द्वारा ओरिएण्ट पेपर मिल के वानिकी प्रभाग द्वारा तैयार किये जा रहे क्लोनल पौधों की नर्सरी तथा रोपण के परफारमेंस की जानकारी लेने हेतु अमलई दौरा किया गया एवं मिल के जी.एम. तथा अन्य अधिकारियों से बातचीत की गई। क्षेत्र भ्रमण में यह पाया गया कि ओरिएण्ट क्लोन–15 के पौधों की आठ माह रोपण आयु से औसत 15 सेमी. कालर गर्थ तथा 12 फीट उंचाई है। ओरिएण्ट क्लोन–47 के पौधों की 22 माह की रोपण आयु में औसत छाती गोलाई 30 से.मी. तथा औसत उंचाई 9 मी. है। फील्ड में ओरिएण्ट क्लोन–47 का परफारमेंस अच्छा पाया गया। मिल द्वारा तैयार क्लोनल मल्टीप्लीकेशन एरिया एवं क्लोनल नर्सरी/मिस्ट चेम्बर में तैयार किये जा रहे पौधों का भी निरीक्षण किया गया। मिल के उच्चाधिकारियों से चर्चा उपरांत निम्न विषयों पर सहमति भी व्यक्त की गई–

1. ओरिएण्ट पेपर मिल द्वारा वर्ष 2003 जुलाई में वनमंडल के बिगड़े वन क्षेत्रों में रोपण हेतु 1.00 लाख क्लोनल पौधे नीलगिरी के उपलब्ध कराये जावेंगे।

पौधों की कीमत 6 रुपये प्रति पौधा है तथा रोपण स्थल/परिक्षेत्र नर्सरी तक मिल द्वारा स्वय के व्यय पर पौधों का परिवहन किया जावेगा। अर्थात् प्रथम चरण में मिल द्वारा लगभग 6.5 लाख का पूंजी निवेश किया जावेगा।

2. इन एक लाख क्लोनल पौधों को तैयार की गई अधिकतम 10 रोपण स्थलों पर प्रति स्थल 10 हजार पौधों की दर से लगाया जावेगा। रोपण पूर्व इन रोपण स्थलों से संबद्ध ग्राम वन समिति के सदस्यों को पेपर मिल अमलई का दौरा कराकर उन्हें मिल के क्लोनल रोपण के परिणामों से अवगत कराया जावेगा।

3. इन वृक्षों की छठवें वर्ष पर जब कटाई की जावेगी तब उस समय प्रवृत्त बाजार दर से समितियों द्वारा मिल को बेचा जावेगा। इस आशय का एक बाईबैक एग्रीमेंट मिल के द्वारा किया जावेगा। किन्तु समितियां केवल ओरिएण्ट मिल को अपना माल बेचने हेतु बाध्य नहीं होगी। ज्यादा दाम प्राप्ति की स्थिति में अन्य को भी इसे बेचा जा सकता है किन्तु उक्त दोनों पतिस्थितियों में मिल द्वारा वर्तमान पूंजी निवेश कुछ नोशनल ब्याज के साथ मिल को वापस किया जावेगा। यह ब्याज की राशि विभाग तथा मिल के बीच चर्चा के दौरान तय की जावेगी।

उल्लेखनीय है कि ओरिएण्ट पेपर मिल अमलई का प्रबंधन मिल स्थापना के समय से ही यह कोशिश करता रहा है कि मरवाही, गौरेला, पेंड्रा क्षेत्रों के ग्रामीण अपनी निजी भूमि पर अधिक से अधिक नीलगिरी के पौधे लगावें। इस हेतु उनके द्वारा सीड सूटेड नीलगिरी के पौधों को ग्रामीणों को लगातार सस्ते दाम पर बेचा जाता है। आज गौरेला, पेंड्रा एवं मरवाही के किसानों पर निजी भूमि तथा खेत की मेढ़ों पर बड़ी मात्रा में नीलगिरी के पौधे हैं। इन कृषकों की सूची मिल से प्राप्त कर ऐसी व्यवस्था की जा रही है कि किसान अपनी इच्छानुसार पौधे काटकर मिल को बेच सकें।

ओरिएण्ट मिल ने गत 2 वर्षों से ही क्लोनल पौधे तैयार करने का काम प्रारंभ किया है तथा वन क्षेत्रों में उन्हें समितियों के माध्यम से अधिक से अधिक इन पौधों को लगाये जाने की इच्छा मिल ने व्यक्त की है। भविष्य में इस योजना के तहत मिल प्रबंधन के पास अधिकाधिक क्लोनल पौधों का रोपण वनभूमि की उत्पादकता बढ़ाने हेतु किया जावेगा।

APPENDEIX X

वन समिति एवं वन विभाग के बीच किये जाने वाला (MOU) द्विपक्षीय पारस्परिक समझौते का प्रारूप

(भारत शासन, पर्यावरण एवं वन मंत्रालय के निर्देश दिनांक 24.12.2002 के अनुपालन में)

यह द्विपक्षीय पारस्परिक समझौता आज दिनांक .... को .... में निम्नानुसार दो पक्षकारों के बीच संपादित हुआ।

पक्ष एक ग्राम वन समिति/वन सुरक्षा समिति .... ग्राम पंचायत .... परिक्षेत्र .... वन मंडल .... जिला .... (प्रथम पक्ष को पारस्परिक समझौता में आगे ''समिति'' कहा जायेगा)

पक्ष दो वन विभाग, छत्तीसगढ़ शासन, मंत्रालय, दाऊ कल्याण सिंह भवन रायपुर, छत्तीसगढ़।

(द्वितीय पक्ष का प्रतिनिधित्व क्षेत्रीय स्तर पर स्थानीय क्षेत्रीय वनमंडलाधिकारी करेंगे)

उपरोक्त समिति को प्रबंधन हेतु कक्ष क्रमांक .... संरक्षित/आरक्षित वन कुल क्षेत्रफल .... हेक्टेयर आबंटित किया गया है। समिति को आबंटित वन भूमि का स्वामित्व राज्य शासन का है। वनों की सुरक्षा एवं संवर्धन में जनभागीदारी को सुनिश्चित करने के लिए उक्त वन भूमि समिति को केवल प्रबंधन हेतु आबंटित की गई है।

भारत शासन, पर्यावरण एवं वन मंत्रालय के निर्देश दिनांक 24.12.2002 के अनुपालन तथा छत्तीसगढ़ शासन के संकल्प दिनांक 22 अक्टूबर 2001 एवं संशोधित

संकल्प दिनांक 31 अक्टूबर 2002 के तारतम्य में उपरोक्त दोनों पक्ष यथा समिति तथा राज्य शासन द्वारा समिति को आबंटित बिगड़े वन क्षेत्र में स्थल की उत्पादकता बढ़ाने हेतु तथा समिति को आर्थिक रूप से अधिक सशक्त बनाने के लिए वनाधिकारी द्वारा अनुमोदित सूक्ष्म प्रबंध योजना के तहत कार्य निष्पादित किया जायेगा। इस एम.ओ.यू. के अंतर्गत वनाधिकारी से तात्पर्य ''क्षेत्रीय वनमंडलाधिकारी अथवा उसके द्वारा अधिकृत कोई अन्य अधिकारी'' माना जावेगा।

निम्न बिन्दुओं पर दोनों पक्षकारों के बीच आपस में सहमति हुई।

अनुच्छेद-1

समिति के दायित्व

1.1 समिति आबंटित वन क्षेत्र हेतु वनाधिकारी के तकनीकी मार्गदर्शन में कार्य आयोजना के प्रावधानों के अनुसार सूक्ष्म प्रबंध योजना तैयार करेगी। उक्त योजना तैयार करते समय शासन द्वारा नियत मार्गदर्शी बिन्दुओं, प्रचलित वैधानिक प्रावधानों और तकनीकी सिद्धांतों का ध्यान रखा जायेगा।

1.2 सूक्ष्म प्रबंध योजना में यह सुनिश्चित किया जायेगा कि वन/वन्यप्राणी प्रबंधन हेतु प्रभावशील अधिनियमों/नियमों का उल्लंघन नहीं हो रहा है।

1.3 सूक्ष्म प्रबंध योजना के क्रियान्यवन हेतु आवश्यक धन राशि की व्यवस्था यथा संभव समिति स्वयं के संसाधन अथवा अन्य स्रोतों से संसाधन जुटाकर करेगी।

1.4 समिति आबंटित वनखंड एवं उसमें उपस्थित जलखंड एवं वन को अतिक्रमण, अग्नि, चराई, चोरी इत्यादि से सुरक्षा करेगी और समय-समय पर क्षेत्र की आवश्यकतानुसार उपचार करेगी। समिति सीमाओं के रख-रखाव में विभाग को पूर्ण सहयोग देगी। समिति द्वारा अकाष्ठीय वनोपज के विनाश विहीन विदोहन को सुनिश्चित किया जायेगा।

1.5 समिति समय-समय पर अकाष्ठीय वनोपज जो कि प्राथमिक वनोपज सहकारी समिति द्वारा संग्रहित नहीं की जाती है उसका संग्रहण करेगी एवं उसका लेखा-जोखा संधारण करेगी।

1.6 समय-समय पर वन विभाग या उनके प्रतिनिधियों द्वारा समिति क्षेत्र में कराए गए वृक्षारोपण एवं अन्य कार्यों के निरीक्षण उपरांत दी गई तकनीकी सलाह के अनुसार आवश्यक उपचार/सुधार करेगी।

1.7 समिति स्वीकृत सूक्ष्म प्रबंध योजना के प्रावधानों अनुसार निश्चित समय पर वृक्षों की कटाई का कार्य वन विभाग की अनुमति के बाद करेगी।

1.8 राज्य शासन के संकल्प दिनांक 22 अक्टूबर 2001 एवं संशोधित संकल्प दिनांक 31 अक्टूबर 2002 के अनुसार समिति को आबंटित वनखंड से सूक्ष्म प्रबंध योजना के अनुसार समिति द्वारा रोपित वृक्षों के विदोहन से प्राप्त वनोपज को समिति अधिक से अधिक आय प्राप्त करने के लिए किसी व्यक्ति अथवा संस्थान को बेच सकती है। राष्ट्रीकृत वनोपज को राज्य शासन द्वारा निवर्तित किया जायेगा तथा इसका मूल्यांकन कर समिति को संकल्प में प्रावधानित लाभांश दिया जायेगा।

1.9 समिति द्वारा सूक्ष्म प्रबंध योजना के अंतर्गत बताये गये कार्यों के संपादन हेतु विभिन्न स्त्रोतों से प्राप्त नगद/सामग्री के रूप में प्राप्त देनदारी की अदायगी नियत समय पर की जायेगी। देनदारी की अदायगी का पूर्ण उत्तरदायित्व समिति का होगा।

1.10 समिति किये जा रहे सूक्ष्म प्रबंध योजना में दर्शाये गये विभिन्न कार्यों हेतु वन विभाग के मार्ग दर्शन में निर्धारित अभिलेख एवं पूंजी का संधारण किया जायेगा।

1.11 समिति अपने साधनों से सूक्ष्म प्रबंध योजना में वर्णित किये गये संयुक्त वन प्रबंधन संकल्प के अनुरूप प्राप्त होने वाले वनोपज की अग्नि, चोरी और प्राकृतिक विपदाओं से होने वाली हानि के लिए अपने पक्ष में बीमा कराएगी।

1.12 समिति प्रत्येक सदस्य की सक्रिय भागीदारी सुनिश्चित करेगी ताकि यह समझौता एक समर्थित, सहभागित एवं समन्वित पारस्परिक समझौता हो सके।

1.13 समिति अपने आबंटित क्षेत्र को चिरकालीन सतत् उत्पादन स्त्रोत के रूप में प्रबंधित करेगी।

1.14 समिति आबंटित वन क्षेत्र का कोई भाग किसी भी व्यक्ति अथवा संस्थान को किसी भी प्रयोजन के लिए पट्टे (लीज) पर नहीं दे सकेगी।

अनुच्छेद-2

समिति के अधिकार

2.1 जिला स्तरीय वन अधिकारी क्षरा संतुष्ट होने पर कि समिति द्वारा सूक्ष्म प्रबंध योजना में वर्णित कार्य संतोषप्रद रूप से किया गया है, समिति को निम्नानुसार लाभ प्राप्त हो सकेंगे।

2.2 सभी समितियों के परिवार लागू निस्तार नीति के अन्तर्गत प्रति वर्ष उपलब्धता अनुसार निस्तारी वनोपज प्राप्त करने के लिए पात्र होंगे।

2.3 समिति को समय-समय पर माइक्रोप्लान/कार्य आयोजना के प्रावधानों के अनुसार आवंटित वन क्षेत्र के काष्ठ कूप/बिगड़े वनों के सुधार हेतु पुनर्वास किए गये क्षेत्र के साफ सफाई कार्य (क्लीनिंग) एवं यांत्रिक विरलन तथा बिगड़े बांस वनों के भिर्रा सफाई से शत-प्रतिशत वनोत्पाद विदोहन व्यय लेते हुए दिया जावेगा।

2.4 वन सुरक्षा समिति को आवंटित वन क्षेत्र में कार्य आयोजना के प्रावधान के अनुरूप बांस या काष्ठ कूप के मुख्य पातन/वन वर्धनिक विरलन से प्राप्त होने वाले वनोत्पादत की स्थल पर कुल कीमत की 15 प्रतिशत राशि अथवा 15 प्रतिशत मूल्य तक का वनोत्पाद या 15 प्रतिशत मूल्य की सीमा तक राशि एवं वनोत्पाद समिति को प्रदाय किया जावेगा। वनोपज प्रदाय की दशा में प्रदाय की जाने वाली वनोपज के विदोहन पर हुए व्यय के तुल्य राशि या वनोपज की मात्रा कम कर दी जायेगी। कूप में वनोत्पाद के मूल्य की गणना गैर वाणिज्यिक दरों के आधार पर होगी। यदि समिति को आवंटित वन क्षेत्र में अवैध कटाई से हानि होती है तो शुद्ध हानि की 50 प्रतिशत राशि के लिए संबंधित वन सुरक्षा समिति को उत्तरदायी माना जायेगा और ऐसी राशि या उसके समतुल्य वनोपज समिति को प्रदाय की जाने वाी राशि या वनोपज में से कम कर ली जायेगी।

2.5 ग्राम वन समिति को आवंटित बिगड़े वन क्षेत्र के सुधार हेतु पुनर्वास कार्य किए जाने पर प्राप्त होने वाली वनोपज एवं पुनर्वास किए गए क्षेत्र के मुख्य पातन या वन वर्धनिक विरलन से प्राप्त होने वाले वनोत्पाद के स्थल पर कुल कीमत की 30 प्रतिशत राशि अथवा 30 प्रतिशत मूल्य तक का वनोत्पाद या 30 प्रतिशत मूल्य की सीमा तक राशि एवं वनोत्पाद समिति को प्रदाय किया जावेगा। वनोपज प्रदाय की दशा में प्रदाय की जाने वाली वनोपज के विदोहन पर हुए विदोहन व्यय के तुल्य राशि या वनोपज की मात्रा कम कर दी जायेगी। कूप में वनोपज के मूल्य की गणना गैर वाणिज्यिक दरों के आधार पर होगी। यदि समिति को आवंटित वन क्षेत्र में अवैध कटाई से हानि होती है तो शुद्ध हानि की 50 प्रतिशत राशि के लिए संबंधित ग्राम वन समिति को उत्तरदायी माना जायेगा और ऐसी राशि या उसके समतुल्य वनोपज समिति को प्रदाय की जाने वाली राशि या वनोपज में से कम कर ली जायेगी।

2.6 जिन समितियों द्वारा अपने स्तर पर वित्तीय संस्थानों/निजी संस्थानों/बैंकों से नगद या सामग्री के रूप में ऋण प्राप्त कर प्राप्त राशि में से बिगड़े वनों के सुधार कार्यक्रम के अंतर्गत रोपण हेतु पूंजी लगाई जावेगी उन समितियों को आबंटित वन क्षेत्र के सूक्ष्म प्रबंध योजना के अनुसार वृक्षारोपण/बिगड़े वनों के सुधार कार्य करने एवं उनकी सुरक्षा हेतु प्रोत्साहित करने की दृष्टि से बिगड़े वनों के सुधार कार्य/वृक्षारोपण से तैयार वृक्षों की परिपक्वता पर कटाई से प्राप्त होने वाली सम्पूर्ण वनोपज दे दी जायेगी जिसे समिति निवर्तित कर सकेगी एवं निवर्तन से प्राप्त राशि में से रोपण हेतु लगाई गई पूंजी, पूंजी पर ब्याज तथा अन्य कोई देनदारी हो, को घटाने के उपरान्त शेष बची राशि की 50 प्रतिशत राशि ग्राम वन समिति की आमसभा के निर्णय के अनुसार उपयोग की जायेगी तथा 30 प्रतिशत राशि ग्राम वन समिति क्षेत्र में अधोसंरचना विकास हेतु तथा 20 प्रतिशत राशि वन संसाधनों के विकास हेतु ग्राम वन समिति के सूक्ष्म प्रबंध योजना के अनुसार उपयोग की जावेगी।

अनुच्छेद-3

वन विभाग का दायित्व

3.1 वन विभाग बिगड़े वनों को संयुक्त वन प्रबंध समितियों की सक्रिय भागीदारी से संयुक्त वन प्रबंधन के संकल्प के अनुरूप सूक्ष्म प्रबंध योजना तैयार करवा के समिति को आबंटित वनखंडों को उत्तम श्रेणी के मल्टीटायर उत्पादक वनों में परिवर्तित करने हेतु आवश्यक उपाय करेगा।

3.2 समिति के गठन हेतु ग्राम स्तर पर बैठक का आयोजन, समिति का पंजीयन, समिति एवं कार्यकारिणी के चुनाव का पर्यवेक्षण।

3.3 समिति के सदस्यों को सूक्ष्म प्रबंध योजना बनाने एवं इसके क्रियान्वयन में प्रशिक्षण देना तथा तकनीकी सहयोग प्रदान करना।

3.4 सूक्ष्म प्रबंध योजना में सम्मिलित कार्यों हेतु यथासंभव वित्तीय संसाधन उपलब्ध ा कराना एवं दर्शाये कार्यों हेतु अन्य विभागों/संस्थानों से समन्वय स्थापित करना।

3.5 समिति को उसके कर्तव्यों के निष्पादन एवं अनुश्रवण में सहयोग करना एवं उनके आंतरिक मतभेद समाधान में सहयोग करना।

3.6 समिति के वार्षिक लेखों का परीक्षण करने हेतु एजेंसी का निर्धारण एवं उसके माध्यम से लेखा परीक्षण कराना।

3.7 वनाधिकारी स्वीकृत कार्य आयोजना के प्रावधानों के अंतर्गत समिति को सूक्ष्म प्रबंध योजना बनाने में आवश्यक तकनीकी सहायता उपलब्ध करायेंगे। सूक्ष्म प्रबंध योजना में अन्य तकनीकी जानकारी के साथ-साथ क्षेत्र में प्रस्तावित समस्त कार्य का विस्तृत विवरण मय नक्शे के होना चाहिये। वनाधिकारी द्वारा नियमानुसार सूक्ष्म प्रबंध योजना की स्वीकृति जारी की जायेगी। सूक्ष्म प्रबंध योजना की नियमानुसार स्वीकृति के पश्चात् ही कार्य कराया जायेगा।

3.8 समिति को आबंटित क्षेत्रों से सूक्ष्म प्रबंध योजना में दर्शित प्रावधान के अनुरूप वनोपज एवं अन्य लाभ का प्रदाय करेगा।

3.9 समिति के सदस्यों की क्षमता विकास हेतु उन्हें प्रशिक्षित कर लेखा एवं क्षेत्रीय कार्यों में प्रारंभ में तकनीकी सहायता प्रदान कर आत्मनिर्भर बनाकर सशक्त करने की दिशा में कार्य करेगा।

3.10 समिति की आय में निरंतरता बनाये रखते हुए क्षेत्र के संरक्षण/रोपण कार्य में सहभागिता प्राप्त कर वनों एवं पर्यावरण के प्रति ग्रामीणों की जागरूकता बढ़ायेगा।

3.11 वन विभाग आबंटित वनखंड में वृक्षारोपण हेतु आवश्यक धन राशि की व्यवस्था के लिए समिति को यथासंभव सहायता करेगा।

3.12 समिति को आबंटित वनक्षेत्र का समय-समय पर वन विभाग या उनके प्रतिनिधियों द्वारा निरीक्षण कर समिति को क्षेत्र की उत्पादकता बढ़ाने के लिए तकनीकी सलाह देगा।

3.13 समिति के द्वारा किये जा रहे संयुक्त वन प्रबंधन के फलस्वरूप उस क्षेत्र में जो भी सामाजिक एवं आर्थिक परिवर्तन हो रहा है उसका अनुश्रवण करेगा।

अनुच्छेद-4

वनाधिकारी का अधिकार

4.1 वन प्रबंधन हेतु समितियों को आबंटित किए जाने वाले क्षेत्रों का निर्धारण।

4.2 सूक्ष्म प्रबंध योजना का अनुमोदन।

4.3 समिति का लेखा एवं सदस्यों के मध्य वनोपज एवं अन्य लाभ के वितरण हेतु बनाये गये नियमों का परीक्षण।

4.4 यदि समिति द्वारा सूक्ष्म प्रबंध योजना में वर्णित निर्देशों/प्रतिबद्धता का निष्पादन सही-सही नहीं किया जाता है तथा वनाधिकारी द्वारा लिखित चेतावनी के

उपरांत भी सुधार नहीं किया जाता है तब वनाधिकारी द्वारा समिति को भंग करते हुये मेमोरेंडम ऑफ अन्डरस्टेंडिंग को समाप्त किया जा सकेगा। ऐसी स्थिति में समिति के सदस्यों को दर्शाये लाभों की पात्रता समाप्त हो जायेगी।

अनुच्छेद-5
विवादों का निराकरण

5.1 अनुच्छेद-1 की कंडिका-1.9 के अनुसार समिति द्वारा देनदारी का भुगतान विभाग/संस्थान को निर्धारित समयावधि में नहीं करने बाबत शिकायत प्राप्त होने पर संबंधित वनाधिकारी समिति को नोटिस जारी कर ऋण के भुगतान हेतु, नोटिस की दिनांक से एक माह का समय दिया जायेगा।

5.2 वनाधिकारी द्वारा जारी नोटिस में दी गई अवधि के भीतर समिति द्वारा ऋण/देनदारी का भुगतान न करने की स्थिति में वनाधिकारी समिति को भंग कर समस्त कार्यवाही संचालन हेतु प्रशासक नियुक्त कर सकेंगे।

5.3 कंडिका-5.2 के तहत नियुक्त प्रशासक समिति की समस्त देनदारियों का निपटारा करने के उपरांत अनुच्छेद-2 की कंडिका-2.6 में दर्शाये अनुसार वनोपज/समतुल्य कीमत का बंटवारा या वितरण करेगा।

5.4 प्राकृतिक विपदा के कारण वनोपज की हानि की स्थिति में बीमे की राशि पर प्रथम अधिकार उन संस्थानों का रहेगा जिनसे समिति द्वारा नगद/सामग्री के रूप में ऋण प्राप्त किया गया।

5.5 वनाधिकारी के निर्णय के विरुद्ध आदेश के दिनांक से 1 माह के अंदर छत्तीसगढ़ शासन के संकल्प दिनांक 22 अक्टूबर 2001 के कंडिका 3.4 में गठित संघ अथवा संघ के गठन के अभाव में क्षेत्रीय वन संरक्षक को अपील की जा सकेगी। अपील का निराकरण संबंधित वन संरक्षक एक माह के अंदर करेंगे।

APPENDEX XI

MADHYA PRADESH LOK-VANIKI RULES-2002

Bhopal, the 16th December 2002

No. F-25-46-98-X-2.—In exercise of the powers conferred by Section II of the Madhya Pradesh Lok Vaniki Adhiniyam, 2001 (No. 10 of 2001), the State Government hereby makes the following rules, namely:—

1. Short title, commencement and application : (1) These rules may be called the Madhya Pradesh Lok Vaniki Rules, 2002.

(2) They shall come into force with effect from the date of their publication in the Madhya Pradesh Gazette.

(3) These rules shall apply to such private and revenue areas, which the Bhumiswami, the Gram Panchayat or the Gram Sabha, as the case may be, voluntarily intends to mannage as tree-clad area.

2. Definitions : In these rules, unless the context otherwise requires :

(a) 'Act' means the Madhya Pradesh Lok Vaniki Adhiniyam, 2001 (No. 10 of 2001);

(b) 'Code' means the Madhya Pradesh Land Revenue Code, 1959 (No. 20 of 1959);

(c) 'DFO' means Divisional Forest Officer having territorial jurisdiction;

(d) 'Enrolment Officer' means the Divisional Forest Officer having territorial jurisdiction.

(e) 'Forest Ranger' means the Forest Range Officer having Territorial jurisdiction.

(f) 'Gram Sabha' and 'Gram Panchayat' shall have the same meaning as assigned to them in Madhya Pradesh Panchayat Raj Avam Gram Swaraj Adhiniyam, 1993 (No. 1 of 1994);

(g) 'Lok Van' means a piece of revenue land handed over to a Gram Panchayat or Gram Sabha as tree clad area for the purpose of scientific management and for which a management plan has been prepared under the provisions of these Rules;

(h) 'Management Plan' means a scientific plan prepared for a revenue or private tree-clad area under these rules.

3. Application for Management under Lok Vaniki : (1) A Bhumiswami, who wants to undertake management of a tree-clad area shall submit an application to the Forest Ranger or to any Forest Officer authorised by the DFO.

(2) In a Gram Panchayat or a Gram Sabha wants to unertake the management of any tree-clad revenue land located within its jurisdiction, the concerned Gram Sabha or Gram Panchayat, as the case may be, will apply to the Sub-Divisional Officer (Revenue) to demarcate and hand-over the area to them to be managed as 'Lok Van'. The Sub-Divisional Officer (Revenue) shall decide the application within a period of 30 days. If the Sub-Divisional Officer (Revenue) decides in favour of the applicant, then he shall get the land demarcated and hand over the land to the concerned Gram Sabha or Gram Panchayat within a period of 15 days from the date of taking decision. After taking the possession of the land, the Gram Panchayat or the Gram Sabha, as the case may be, shall submit an application to the Forest Ranger or any Forest Officer authorised by the DFO for scientific management as tree-clad area.

(3) The application under sub-rule (1) and (2) shall be accompanied by a declaration about the land ownership, or possession of land as the case may be with relevant record.

(4) The Forest Range Officer shall send a copy of the declaration to the Patwari within 15 days of the receipt of the application who shall verify the claim from revenue records and give the certificate within 15 days. If the Patwari fails to give the certificate with in the prescribed time, the declaration given by the Bhumiswami shall be

deemed as correct. Subsequently if may discrepancy is detected in such records, the Patwari and the person who has given the declaration shall be held responsible for the same and punishable under the relevant law.

(5) If the boundary is contiguous with the forest land the Forest Range Officer shall satisfy himself about the boundary.

(6) The Forest Ranger shall intimate the Bhumiswami or the Gram Panchayat or the Gram Sabha as the case may be, within a period of 45 days from the receipt of application as per sub-rule (3) about the certification of records. If no such intimation is received within a period of 60 days from the date of submission of application as per sub-rule (3), the records so submitted shall be deemed to be accepted as correct.

4. Preparation of Management Plan : (1) After compliance of the procedure laid down in rule 3, the Bhumiswami, Gram Panchayat or the Gram Sabha as the case may be, shall have to get a management plan prepared for the same area.

(2) The management plan for a tree-clad area shall be prepared by a Chartered Forester as enrolled under rule-8.

(3) The management plan shall be prepared keeping in view the social, economic and environmental roles.

(4) The management plan shall specify the following issues, namely:

(a) Ensuring sustainable production of timber and/or other forest products;

(b) Encouraging and protecting natural regeneration and/or planting of suitable species;

(c) Felling of mature, over mature, dry and diseased trees and removal of wind fallen trees;

(d) Thinning and pruning;

(e) Improving the health and vitality of the crop; and

(f) Ensuring soil and moisture conservation.

5. Sanction of the Management Plan: (1)The application for sanction of a management plan prepared for private areas shall be submitted to the DFO in Form-1 along with 5 copies of management plan.

(2) The application for sanction of a management plan prepared for revenue areas proposed to be managed as 'Lok Van' shall be submitted in Form-2 along with 5 copies of management plan. The application in such cases shall be submitted to the DFO by an authorised representative of the concerned Gram Panchayat or Gram Sabha along with a resolution of the Gram Panchayat or the Gram Sabha, as the case may be, for the sanction of the plan.

(3) The Competent Authority shall have powers to inspect the plan area himself or through his authorised representative, to verify the validity of any prescriptions made in the management plan. Based on such action, the Competent Authority may suggest certain amendments in the proposed management plan. In such case, the applicant/chartered forester shall submit the revised plan incorporating the amendments suggested by the competent authority.

(4) The Competent Authority for sanctioning the management plan shall be the DFO. In case where the management plan area exceeds 10 hectares, the Competent Authority shall submit the management plan with his opinion of the Ministry of Environment and Forest, Government of India for approval through the State Government within 30 days after the receipt of management plan.

(5) The Competent Authority shall take decision regarding the sanction of management plan within 30 days if the management plan area is upto 10 hectares and in case where the management plan area exceeds 10 hectare, the comptent authority after receiving the approval as required in sub-rule (4) shall issue the sanction order of the management plan within 7 days.

(6) The Competent Authority shall pass an order of sanction for a management plan of private area in Form 3 and for a Lok Van in Form 4. Conditions for the implementation of the management plan may be specified in schedule-II/III of the sanction order.

(7) After having sanctioned the management plan, the Competent Authority shall send a copy of the sanctioned plan along with sanction order to the concerned Bhumiswami, Gram Panchayat or Gram Sabha as the case may be, with intimation to the concerned chartered forester. A copy of the sanction order along with a copy of the sanctioned management plan shall also be endorsed to the concerned Sub-Divisional Officer (Revenue) for intimation and for the purpose

of entry into the land record under sub-section (2) of Section 114-A of the Code, as provided under Section-4 of the Act.

(8) In case the Competent Authority denies the sanction of the management plan, he shall record the reasons of denial and such order shall be communicated to the applicant.

(9) If the management plan area is upto 10 hectares, an appeal against the order under sub-rule (8), shall lie before the Conservator of Forests having territorial jurisdiction. Such appeal against the order of the competent authority can be preferred within 30 days of the receipt of the denial order. The Conservator of Forests, after hearing the chartered forester and the concerned bhumiswami/ representative of Gram Panchayat or Gram Sabha, shall decide the appeal within 60 days. The decision of the Conservator of Forests shall be final and binding. This decision shall be communicated to the applicant in writing and a copy shall be endorsed to the Competent Authority.

6. Implementation of the Management Plan : (1) Every Bhumiswami, Gram Panchayat or Gram Sabha, as the case may be, after having received the sanctioned management plan from the competent authority, shall implement the management plan as per the precriptions and conditions.

(2) The Gram Panchayat or the Gram Sabha may authorise Sarvajanic Sampada Samiti of the Gram Sabha for implementation of the plan prescriptions for 'Lok Van'.

(3) The Bhumiswami, the Gram Panchayat or Gram Sabha as the case may be, shall give an intimation regarding the proposed date of felling of trees in the plan area to the concerned Forest Ranger and Tehsildar. This intimation shall be given at least 7 days before the proposed date of felling of trees.

(4) The person implementing the management plan shall maintain a felling register in format as prescribed in the management plan.

(5) Transport of forest produce obtained from felling operation in accordance with the approved management plan shall be subject to the provisions of the Madhya Pradesh Transit (Forest Produce) Rules, 2000.

(6) Disposal of a forest produce declared as a specified forest produce under the Madhya Pradesh Van Upaj (Vyapar Viniyaman) Adhiniyam, 1969 shall be subject to the rules framed in this behalf.

(7) All operations prescribed in the management plan shall be completed within the specified time. If any operation prescribed in the plan is not executed due to some unforeseen reasons, further implementation of the plan shall remain suspended till such time the operations prescribed for the previous year are completed.

7. Monitoring of the Management Plan Implementation: (1) For each development block or a part thereof the Implementation of approved management plans shall be monitored by a committee constituted by the Competent Authority under the chairmanship of Territorial Range Officer and will comprise a Non-Government Individual or Organization, a representative each from the Revenue Department and a Gram Panchayat or a Gram Sabha as the case may be. The committee shall report its observations and recommendations to the Competent Authority. The State Government, wherever necessary, may authorise any official, body or agency to monitor the implementation of such plan for a specified area or period.

(2) The DFO shall take cognizance of the contravention, if reported. On receiving the report, the DFO or the officer authorised by him, shall refer the matter to the Sub-Divisional Officer (Revenue) for further action as provided in rule 10.

8. Enrolment of Chartered Forester : (1) The following persons/category of persons shall be eligible to be considered for enrolment as Chartered Forester :

(a) Persons with bachelor's degree in forestry from a recognised university/institution, or post graduate diploma in forest management from the Indian Institute of Forest Management, having field experience of at least three years in planning and management of forests.

(b) Retired Forest Officers, not below the rank of Forest Ranger, or equivalent officers of Madhya Pradesh Forest Development Corporation.

(c) A registered society or a non-Government organisation consisting of members having qualifications as mentioned in sub-rules (a) and (b) above.

(2) Any serving Forest Officer of the Government of Madhya

Pradesh not below the rank of Forest Ranger shall be deemed to be a chartered forester.

(3) Any person or group of persons fulfilling criteria as per sub-rule (1) above may apply for enrolment as a Chartered Forester in Form 5 with an Affidavit annexed to these rules along with non-refundable application fee of Rs. 1000.

(4) The Enrolment Officer shall enrol the person fulfilling the criteria laid down in sub-rule (1) above within a period of 15 days from the date of receipt of the application in his office. He shall issue an Enrolment Certificate in Form 6 and maintain an Enrolment Register in Form 7 annexed to these rules. A copy of enrolment certificate issued shall be sent to the Officer in charge Lok Vaniki at the State Level every quarter who shall get the updated list displayed at the website of the Forest Department.

(5) In case the applicant is denied enrolment, the Enrolment Officer shall communicate in writing the grounds of denial to the applicant. The appeal against the denial of enrolment shall lie before the Officer in charge Lok Vaniki at the State Level whose decision shall be final and binding.

(6) The enrolment of a Chartered Forester shall be valid until it is terminated by a specific order issued by the Enrolment Officer.

(7) The enrolment of a Chartered Forester shall be liable for termination by the Enrolment Officer on grounds of: (a) gross professional misconduct, or (b) a self written request by the enrolled Chartered Forester.

(8) The Enrolment Officer, before terminating any enrolment, on ground of misconduct, shall serve the concerned Chartered Forester with a show cause notice and provide him full opportunity for explaining his/her stand.

(9) An appeal against such termination shall lie before the Officer in charge Lok Vaniki at the State level. Such an appeal can be preferred within a period of 45 days from the date of order of the enrolment officer and the decision of the Officer in charge Lok Vaniki at the State level shall be final and binding.

9. Duties and Remuneration of the Chartered Forester : (1) The chartered forester shall be responsible for the following :—

(a) Preparation of the management plan; and

(b) Obtain its sanction from the competent authority.

(2) The Chartered forester shall receive remuneration of amount of Rs. 1000 or 6% of the amount received during the 1st year of sale of forest produce by the Bhumiswami/Gram Panchayat or Gram Sabha, whichever is more. However, if the management plan is prepared by the Chartered Forester authorised as per sub-rule (2) of Rule 8, the remuneration payable shall be deposited by the Bhumiswami/Gram Panchayat or the Gram Sabha as the case may be, to the State Government.

10. Punishment for Contravention : (1) The Sub-Divisional Officer (Revenue) on receiving the information about the contravention of an approved management plan from the concerned Forest Ranger/ Competent Authority or the Officer authorised by him, shall issue a show cause notice to the concerned Bhumiswami, or the Gram Panchayat or the Gram Sabha, as the case may be and give a reasonable time for filing the reply to the notice.

(2) If the concerned Bhumiswami or the Gram Panchayat or Gram Sabha as the case may be, fails to submit the reply of the show cause notice, within the specified time limit or after due consideration of the reply to show-cause notice, the Sub-Divisional Officer (Revenue) may decide the case within a period of 30 days as per the provisions of Section 8 of the Act.

11. Appeal : (1) Appellate Authority, for considering the appeal against an order passed by the Sub-Divisional Officer (Revenue) under Section 8 of the Act shall be the District Collector.

(2) The application for appeal shall be received by the Reader of the Collector and will be processed as per procedure laid down in the Code.

(3) Every appeal shall be accompanied by relevant documents of the case along with the order of the Sub-Divisional Officer (Revenue) against which the appeal is preferred and a non-refundable fee of Rs. 100 payable through a Treasury challan or Demand draft.

(4) The Appellate Authority shall hear the parties of the appeal in person or through any agent duly authorised in writing by the applicant and shall decide the appeal within 60 days from the date of receipt of the application.

(5) Copies of the order passed by the Appellate Authority shall be sent to the concerned Sub-divisional Officer (Revenue) for

compliance, or for passing such further order, as may be directed by the Appellae Authority.

SCHEDULE-I

[*See* rule 4(5)]

STANDARD FORMAT OF MANAGEMENT PLAN FOR A PRIVATE TREE CLAD AREA/PRIVATE FOREST AND LOK VAN

PART-ONE

Objectives, The Tract Dealt with and the Crop

Chapter 1: Objectives of the Management Plan

Chapter 2: Introduction to the Village

Chapter 3: Introduction to the private tree clad area/private forest/Lok Van

(Based on information furnished by the Bhumiswami/ Authorised Representative of Gram Panchayat/Gram Sabha)

3.1 General

3.2 Name/Fathers/Husband's Name of the Bhumiswami/ authorised representative

3.3 Category (General/S.C./S.T./O.B.C.)

3.4 Full Postal Address of the Bhumiswami

3.5 Economic status of the Bhumiswami (in case of private areas only) (Details of agriculutre, cattle, income etc.)

3.6 Patwari Halka No./Khasra No.

3.7 Revenue Inspector Circle/Tehsil

3.8 Location of the Khasra

3.9 Area (Hectares)

3.10 Boundaries and Demarcation status

3.11 Legal status of ownership

(According to Khasra Panchasaala of Revenue Department)

3.12 Details of ownership of Khasra last 10 years

3.13 Whether acquired as parental property or purchased

3.14 Topography (specially areas having slopes more than 25 degree)

3.15 Soil Type and depth

3.16 Distance of nearest Government Forest from the Khasra and its details (Type, compartment no., beat etc.) (Whether boundary of holding of contiguous and to Government forest/other private forests/tree clad areas)

3.17 Availability of water in the tract

Chapter 4: Description of the crop

4.1 Type of the crop

4.2 Site quality

4.3 Crop Density

4.4 Age class of the crop

4.5 The following stock

(a) Top canopy (b) Middle story

(c) Ground flora (d) Climbers

(e) Grasses (f) Medicinal plants

(g) Weeds (h) Non-timber forest produce (NTFP)

(i) Bamboo

4.6 Stock Map (On Patwan Map)

(Vegetation type, site quality, density, age class banks etc.)

4.7 Natural Forests/Plantation

(General description and current status)

4.8 History of the tract

(a) Previous harvest/plantations

(b) Outcome of the previous operations

4.9 Availability of bio-diversity

4.10 Injuries to which the crop is liable

Chapter 5: Analysis and valuation of the crop

5.1 Specieswise No. of Trees, Volume and Value thereof

(Abstract of Resource assessment record)

5.2 Result of Regeneration Survey

5.3 Availability of bamboo and grasses

5.4 Availability of NTFP/Medical Plants

5.5 Form Factor

5.6 Growth Rate

5.7 Marketing outlines

5.8 Market prices

PART-TWO

Summary of Proposals

Chapter 1: Scheme for future management

1.1 Management Proposals

1.2 Treatment prescriptions

1.3 Felling cycle/Rotation

1.4 Selection Girth

1.5 Determination of treatment class and treatment map

1.6 Felling rules/rules for other treatment

(a) Trees, (b) Green flora, (c) Natural bamboos

1.7 Calculation of annual yield

Chapter 2: Harvesting

2.1 Harvesting Plan

(a) Timber, (b) Fuel wood, (c) Bamboo, (d) Medicinal plants, (e) Grasses, (g) NTFP, (g) others.

2.2 Seasoning, processing and preservation of forest produce

2.3 Proposed method for disposal of forest produce

(a) Timber, (b) Fuel wood, (c) Bamboo, (d) Grasses, (e) Medicinal plants, (g) NTFP, (g) others.

2.4 Subsidiary silvicultural operations

2.5 Thunning

2.6 Pruning

2.7 Harvest of dead, wind fallen trees

Chapter 3: Plantations

3.1 Gap planting

3.2 Teak plantations

3.3 Bamboo plantations

3.4 Plantation of other species

Chapter 4: Protection and rehabilitation measures

4.1 Protection from grazing

4.2 Fire protection

4.3 Others

4.4 Bio-diversity conservation

Chapter 5: Anticipated average annual expenditure and income

Chapter 6: Format of marking record, felling register (See rule 6(4) of the Lok Vaniki Rules, 2002)

Chapter 7: Monitoring and evaluation of implementation of management plan (See rule 7 of Lok Vaniki Rules, 2002)

Chapter 8: Safeguards suggested by the competent authority if the plan area is adjacent to a Government forest.

DOCUMENT TO BE ENCLOSED WITH
THE MANAGEMENT PLAN

1. Patwari Map of the Khasra in which adjoining survey numbers are also known
2. Land ownership documents
3. Demarcation certificate, if required
4. Stock Map
5. Bhumiswami's written consent (consent of all the partners in case of joint holding)
6. Results of Survey/assessment Reports, if any.

(Signature of the applicant)

SCHEDULE-II

[*See* rule 5(6)]

(Read with From No. 3)

(A) Conditions/restrictions subject to which, this plan is being approved

1. The implementation of the management plan shall be done in accordance with rule 6 of Lok Vaniki Rules, 2002 and any guidelines issued by the State Government in this regard from time to time. The Bhumiswami is advised to read and clearly understand these provisions before starting implementation of the approved plan.

2. Nothing shall be done, during the implementation of the plan, in contravention of any existing Act, Rules, guidelines and executive instructions issued by the Government from time to time.

3. The area shall be managed as per the plan prescriptions and the nature of use of land shall not be changed during the plan period.

(B) Conditions under which this sanction may be cancelled

1. Violation of any conditions/restrictions mentioned in (A) above.

2. (Any other point as may be necessary in the opinion of the Competent Authority)

3. ....

(C) Safeguards for protecting Government forests from illicit felling

(To be provided necessarily if the proposed plan area is adjacent to any type of forest belonging to Government)

1. The burden of proof for establishing ownership of forest produce obtained from implementation of this management plan shall be on the Bhumiswami. (He should take ncessary precautions and always be in position to establish beyond doubt that the forest produce being lifted from his khasra actually comes from trees legitimately felled as per the prescriptions of this plan and not from the adjoining Government forests).

2. If the area has a part of its boundary contiguous to a Government Forest, the Bhumiswami shall be vigilant about the status of the boundary marks separating his holding from the Government Forest. He shall intimate immediately to the concerned forest range office if any damage or deteriorations in such boundary marks, (including boundary pillars or demarcation lines or coal-tar bands on trees) is observed by him. He shall also maintain in proper condition any such marks fixed by revenue authorities on his holding for the purpose of demarcation.

3. Special attention is attracted towards sub-rule (7) of rule 6 of the Madhya Pradesh Lok Vaniki Rules, 2002 which reads: "All operations prescribed in the management plan shall be completed within the specified time. If any operation prescribed in the plan is not executed due to some unforeseen reasons, further implementation of the plan shall remain suspended till such

time the operations prescribed for the previous year are completed."

Endorsement No. .... Dated ....

Copy forwarded to :

1. The District Collector .... district for compliance of Section 1(8) of the Lok Vaniki Adhiniyam, 2001.
2. To Shri .... (Name of the Bhumiswami)
3. To office incharge Lok Vaniki Forest Department Bhopal.

Signature of the Competent Authority

SCHEDULE-III

[*See* rule 5(6)]

(Read with Form No. 4)

(A) Conditions/Restrictions subject to which, this plan is being approved

1. The implementation of the management plan for the Lok Van shall be done in accordance with rule 6 of Lok Vaniki Rules, 2002 and any guidelines issued by the State Government in this regard, from time to time. The authorized representative of the Gram Panchayat/Gram Sabha is advised to read and clearly understand these provisions before starting implementation of the approved plan.
2. Nothing shall be done, during the implementation of the plan, in contravention of any existing Act, Rules, guidelines and executive instructions issued by the Government from time to time.
3. The area shall be managed as per the plan prescriptions and the nature of use of land shall not be changed during the plan period.

(B) Conditions under which this approval may be cancelled

1. Violation of any conditions/restrictions mentioned in (A) above.
2. (Any other point as may be necessary in the opinion of the Competent Authority)

(C) Safeguards for protecting Government forest from illicit felling

(To be provided necessarily if the proposed plan area is adjacent to any type of forest belonging to Government)

1. The burden of proof for establishing ownership of forest produce obtained from implementation of this management plan shall be on the Gram Panchayat/Gram Sabha. The concerned Gram Panchayat/Gram Sabha should take necessary precautions and always be in position to establish beyond doubt that the forest produce being lifted from the Lok Van actually comes from trees legitimately felled as per the prescriptions of this plan and not from the adjoining Government forests.
2. If the area has a part of its boundary contiguous to a Government Forest, the Gram Panchayat/Gram Sabha shall be vigilant about the status of the boundary marks separating his holding from the Government Forest the authorized representative shall intimate immediately to the concerned forest range office if any damage or deteriorations in such boundary marks, (including boundary pillars or demarcation lines or coal-tar bands on trees) is observed by him. The concerned Gram Panchayat/Gram Sabha shall also maintain in proper condition any such marks fixed by revenue authorities on the holding for the purpose of demarcation.
3. Special attention is attracted towards sub-rule (7) of rule 6 of the Madhya Pradesh Lok Vaniki Rules, 2001 which reads: "All operations prescribed in the management plan shall be completed within the specified time. If any operation prescribed in the plan in not executed due to some unforeseen reasons, further implementation of the plan shall remain suspended till such time the operations prescribed for the previous year are completed."

Endorsement No. .... Dated ....

Copy forwarded to :

1. The District Collector .... district for compliance of Section 4(8) of the Lok Vaniki Adhiniyam, 2001.
2. To Shri .... (Name of the Authorized Representative of the Gram Panchayat/Gram Sabha)
3. To office incharge Lok Vaniki Forest Department Bhopal.

Signature of the Competent Authority

FORM NO. 1

[*See* Rules 5(1)]

Application for Sanction of a Management Plan for a Bhumiswami area

(To be filled in triplicate)

1. Applicant's name ....
2. Father's/Husband's name ....
3. Current Postal Address ....
4. Tehsil .... Revenue Sub-Division ....
5. Police Station ....
6. District ....
7. Details of khasra for which plan has been prepared

Sl. No.	Village/Patwari Halka No./ Khasra No. and other location details	Area (Hectares)	No. of trees on Khasra	Owner(s) of the Khasra	Detail of adjoining survey number
(1)	(2)	(3)	(4)	(5)	(6)
					North = East = West = South =

8. Name of the Gram Sabha in which Khasra is located ....
   ....
9. Name of the Village Panchayat in which Khasra is located
   ....
10. Details of the Chartered Forester who has prepared the Plan
   ....

Name and Address of the Chartered Forester	Enrolment Office	Enrolment No.
(1)	(2)	(3)

11. Dates of plan preparation (a) started on .... (b) completed on ....

I ....S/o/D/o/W/o Shri ....resident of .... hereby submit my request for sanctioning of management plan prepared for my said khasra (indicated in para 7) as per the provisions of Lok Vaniki Adhiniyam, 2001 and rules thereunder.

I certify that the management plan being submitted for Sanction is based on true information and documents furnished by me to the Chartered Forester whose name has been indicated in para 8. I also certify that documents appended to the plan are true copies of the original documents of my legal ownership of land and trees standing thereon. I have personally satisfied myself that, the proposed management plan applies only to the area mentioned in para 7 of this application and it does not contain anything, which may have any impact, on adjoining areas outside my legal possession.

Through this application, I also give an undertaking that I shall abide by all rules and regulations subject to which the management plan shall be sanctioned.

Witness : 1.
(with full name 2.
and address)

Name and Signature of the Bhumiswami

Date :
Place :

N.B.: This form shall be filled in triplicate.

1. A copy of this application shall be returned back to the applicant while admitting the application in the office of the competent authority, after putting the seal of the office and date of receipt.
2. Two copies of this application shall be retained in the office of the competent authority. One out of these two copies shall be annexed to the intimation letter sent to the District Collector under Section 4 of Lok Vaniki Adhiniyam 2001 and the Rules thereunder.

FORM NO. 2

[*See* Rules 5(2)]

Application for Sanction of a Management Plan prepared for a *Lok Van*

(To be filled in triplicate)

1. Name of the Gram Panchayat/Gram Sabha on whose behalf the application is being submitted .... located in Janpad Panchayat ....
2. Applicant's name and Post..
3. Father's/Husband's name ....
4. Current Postal Address ....
5. Tehsil .... Revenue Sub Division ....
6. Police Station .... District ....
7. Details of revenue land for which plan has been prepared

Sl. No.	Village/Patwari Halka No./ Khasra No. and other location details	Area (Hectares)	No. of trees on Khasra	Detail of adjoining survey number	Any other details
(1)	(2)	(3)	(4)	(5)	(6)
				North = East = West = South =	

8. No. and date of District Collector's Order *vide* which the area has been handed over to the Gram Panchayat/Gram Sabha for management as Lok Van ....
9. Details of Resolution passed by the Gram Panchayat/Gram Sabha *vide* which the applicant has been authorised to act on their behalf..
10. Details of the Chartered Forester who has prepared the plan ....

Name and Address of the Chartered Forester	Enrolment Place and Office	Enrolment No.
(1)	(2)	(3)

11. Dates of plan preparation (a) started on .... (b) completed on ....

I ....S/o/D/o/W/o Shri ....resident of .... being the authorized person nomiated by the Gram Panchayat/Gram Sabha to act on their behalf for submitting this application hereby submit my request for approval of scientific management plan prepared for the revenue area proposed to be managed as Lok Van (indicated in para 6) as per the provisions of Lok Vaniki Adhiniyam, 2001 and rule thereunder.

I certify, on behalf of the Gram Panchayat/Gram Sabha .... that the area indicated in para 6 of this application has been fully demarcated and handed over to the Gram Panchayat/Gram Sabha by the District Collector to be managed as Lok Van.

I also certify, on behalf of the Gram Panchayat/Gram Sabha .... that the management plan being submitted for approval is based on true information and documents furnished by the Gram Panchayat/Gram Sabha to the Chartered Forester whose name has been indicated in para 9, I also certify that documents appended to the plan are true copies of the original documents of legal ownership of land and trees standing thereon. I have personally satisfied myself that, the proposed management plan applies only to the area mentioned in para 6 of this application and it does not contain anything which may have any impact, on adjoining areas.

Through this application, I also give an undertaking that while acting on behalf of the Gram Panchayat/Gram Sabha .... I shall abide by all rules and regulations subject to which the management plan shall be approved.

Witness : 1.
(with full name 2.
and address)

Name and Signature of the Authorized Representative
of Gram Panchayat/Gram Sabha ....

Seal of the
Office of the
Gram Sabha

Date :

Place :

N.B.: This form shall be filled in triplicate.

1. A copy of this application shall be returned back to the applicant while admitting the application in the office of the competent authority, after putting the seal of the office and date of receipt.
2. Two copies of this application shall be retained in the office of the competent authority. One out of these two copies shall be annexed to the intimation letter sent to the District Collector under Section 4 of Lok Vaniki Adhiniyam, 2001 and the Rules thereunder.

FORM NO. 3

[*See* Rules 5(6)]

Order of the Competent Authority

(For Bhumiswami area)

Order No. .... Date ....

Seal of the
Office

With reference to the application dated .... of Shri .... S/o/D/o/W/o Shri .... resident of .... regarding Sanction of a management plan for his/her holding following decision is taken:

The management plan for Khasra No. .... of Village .... of Tehsil .... of Revenue Sub-Division

.... of .... District, belonging to the applicant is hereby Sanction for a period of .... year commencing from .... and terminating on .... subject to the conditions and restrictions enumerated in SCHEDULE-II annexed to this order.

The implementation of his Sanctioned management plan shall be monitored by ....

Name and Signature of the Competent Authority

Date :

Place :

FORM NO. 4

[*See* Rules 5(6)]

Order of the Competent Authority for Lok Van

(For Bhumiswami area)

Order No. .... Date ....

Seal of the Office

With reference to the application dated .... submitted on behalf of Shri .... Gram Panchayat/ Gram Sabha, by Shri ....S/o/D/o/W/o Shri .... resident of .... regarding Sanction a management plan for a revenue area to be managed as Lok Van, following decision is taken :

The management plan for survey No. .... of Village .... of Tehsil .... of Revenue Sub-Division .... of .... District, belonging to the applicant is hereby sanction for a period of .... year commencing from .... and terminating on .... subject to the conditions and restrictions enumerated in SCHEDULE-III annexed to this order.

The implementation of this sanctioned management plan shall be monitored by ....

Name and Signature of the Competent Authority

Date :
Place :

FORM NO. 5
[*See* Rules 8(3)]

Application for Enrolment as Chartered Forester

1. Applicant's Name ....
2. Father's/Husband's Name ....
3. Date of Birth ....
4. Current Postal Address ....
5. Permanent Address ....

....

Phone No. ....

Fax No. ....

E-mail ....

6. Detail of Educational Qualification*

Sl. No.	Degree	Year	Institution
(1)	(2)	(3)	(4)

(*Enclose copies of certificates)

7. Details of Professional Qualification and Work Experience

Sl. No.	Nature of Qualification/Expereince	Period Details
(1)	(2)	(3)

8. The category to which the applicant belongs [see rule 8(i) (put × mark in the appropriate box)]

(a) Persons with bachelor's degree in forestry from a recognised university/institution, or post graduate diploma in forest management from the Indian Institute of Forest Management, having field experience of at least three years in Planning and Management of Forests.

(b) Retired Forest Officers, not below the rank of Ranger, or equivalent Officers of Madhya Pradesh Forest Development Corporation.

(c) A registered society or a Non-Government Organisation consisting of members having qualification as mentioned in sub-rules (a) and (b) above.

(d) Any serving Forest Officer authorised by the Principal Chief Conservator of Forests, or Officer in charge of Lok Vaniki at the State level.

DECLARATION

I/We .... hereby declare that I/we have read and understood the provisions of the Madhya Pradesh Lok Vaniki Adhiniyam, 2001 and the rules thereunder pertaining to the chartered foresters and agree to abide them. I/We also declare that I/we fulfil the eligibility criteria as laid down in sub-rule (1)/(2) of Rule 8 of the Madhya Pradesh Lok Vaniki Rules, 2002.

I/We also declare that information furnished in this form is correct and true the best of my/our knowledge and I/We have not concealed or suppressed any material fact pertaining to this proposal, I/We agree to abide by any rules and guidelines governing profession and conduct of chartered foresters and hereby request for my/our registration as a chartered forester. I/We hereby agree to the Code of Conduct for Chartered Foresters and enclose the same on an affidavit dated .... with a request for my enrolment as Chartered Forester.

Encl.: (1) Affidavit containing details enumerated in attachment-I ....

(2) Details of payment of application fees ....

(3) Certified copies of the document related to paras 6 and 7 ....

Place :

Date : Signature of the Applicant

AFFIDAVIT

Acceptance of Code of Conduct for Chartered Foresters

[Read with form No. 5 under Rule 8(2 and 3)]

I/We, Shri/Smt./Ms. .... S/o/D/o W/o ....
Aged .... yrs, Resident of ....
....
do hereby, under oath, state that while delivering my/our services as an enrolled chartered forester to an individual/firm/institution/ government of society I shall :

1. Abide by the provisions of the Madhya Pradesh Lok Vaniki Adhiniyam, 2001, and the Rules made thereunder and the relevant guidelines.
2. Keep my integrity beyond doubt and commitment to the nation and forestry profession unquestionable.
3. Use my knowledge, skill and experience of forestry only for good cause and for creating a sensibility in public towards forestry, forest policies and forest dwelling communities.
4. Certify or attest only those maps, document and management plans, which will be correct to the best of my knowledge or prepared under my personal supervision and verified personally with respect to the ground.
5. Not misguide or abet my client to do anything that is ecologically unsustainable or technically not feasible.
6. Accept to undertake only those technical assignments for which I possess the adequate technical knowledge and skills. I shall not undertake any forestry assignment beyond my capability by exaggerating my competence.
7. Intimate the concerned forest officers, if any nefarious activities or anything in contravention of national forest policy, forestry laws and guidelines, being done by any other chartered forester or any other person or institution, comes to my notice.

8. Not enter into an unfair competition with other chartered foresters, I shall also not undertake an assignment being handled by another chartered forester unless either, he/she is unablc to handle it or has left the task incomplete.
9. Not do anything, which undermines the credibility or reputation of forestry profession or chartered forester.

Signature of the Applicant

(Name ....)

Place :

Date :

FORM NO. 6

[*See* Rules 8 (4)]

Enrolment Certificate of a Chartered Forester

Seal of GoMP

Shri/Ms./Smt. .... S/o/D/o/W/o ....
Resident of ....
(permanent address) ....
is hereby enrolled as a Chartered Forester under Rule 8 of M.P. Lok Vaniki Rules, 2002. His/Her Enrolment No. is ....

This enrolment is subject to the terms and conditions of Rule 8 of Lok Vaniki Rules, 2002.

Place ....

Date ....

Signature of Enrolment Officer

(Name ....)

Seal of the Office

FORM NO. 7

[*See* Rules 8(4)]

Enrolment Register of Chartered Forester

Sl. No.	Name/Father's Name of the Enrolled Chartered Forester	M.R. No./Challan No. of Payment of enrolment fee	Qualification
(1)	(2)	(3)	(4)

Date of enrolment	Current Status of Enrolment (to be updated annually)	Remarks	Signature of the Enrolment Officer
(5)	(6)	(7)	(8)

By order and in the name of the
Governor of Madhya Pradesh
RATAN PURWAR, Addl. Secy.

BIBLIOGRAPHY

Acharjee, S. K., Ghosh U and Biswas, A (2004). A study on the perceived benefits of farm forestry in terms of some socio-economic and motivational Correlates in Cooch Behar district of West Bengal. *The Indian Forester* 130(3), 333-339.

ADB: Asian Development Bank (1999). *Annual Report-1999.*

Alverson, W.S., Kuhlman, D. and Wallen, M. (1994). *Wild Forests: Conservation Biology and Public Policy* (Washington D.C.: Island Press).

Anderson, R. and Walter, H. (1988). *The Hour of the Fox :Tropical Forests, The World Bank and Indigenous People in Central India*, Vistaar Publications, New Delhi.

Anon (1993). 'Forestry policies of selected countries in Asia and the Pacific: India', FAO Forestry paper no.115.

Anon (1997). 'ITC Bhadrachalam Paperboards Limited - Report and Accounts 1996-97'.

Anon (1998). *State of the World's Forests.* FAO, Rome.

Anon (1999). *A Millennium Poll on CSR*, Environics International Ltd., Canada.

Anon (2001). *CSR: Perceptions of Indian business Centre for Social Markets,* London/Calcutta.

Anon (2002). "The Front Line: Why business leaders must be good citizens too?" *Financial Times,* U.K.

Anon (2002). 'National Workshop on Promoting Corporate Citizenship in India: Challenges and Opportunities, 8-9 April 2002', New Delhi.

Athreya (1989). *Pilot Study on Experiences in Farm Forestry in Haryana,* Athreya Management Consultants, New Delhi.

Australian Greenhouse Office (1999). National Emissions Trading-Crediting the Carbon-discussion paper-3; Commonwealth of Australia.

Babu, C.R. and Gopinathan, M.C. (1986). The scenario of plant genetic resources; *Bull. BSI,* 28: 159-168.

Banergee, U.K. (1996). Development of degraded forests and industry and the proposed forest act. *The Indian Forester* 122(9), 800-803.

Bayon, R.J., Lovink, S. and Veening, J.W. (2000). *Financing Biodiversity Conservation,* Inter-American Development Bank, Sustainable Development Department, Environment Division.

Bebarta, K.C. (1999 a).On-farm tree cultivation peoples response to the dynamic change in the rural economy. *Journal of Tropical Forestry* 15(2), 81-92.

Bebarta, K.C. (1999 b). *Working Plan for Rajanandgaon forest division.* Forest Department, Madhya Pradesh.

Binkley, C.S. (1998). Forestry in a Post-modern World or what was John Muirdoing running a sawmill in Yosemite Valley? *Policy Sciences Journal,* 31, 133-144.

Brahmi, M.K., Kanswal, B.D., Singh, H.P. and Brahmi, L. (1997). Women's participation in joint forest management—governing factors and a development model. *Advances in Forestry Research in India,* 16, 217-225.

Center For Environment Law and World Wildlife Fund, India (1995). Leasing Forest Lands to Industries.

Chakrabarti, M., Mukherjee, R. and Sarkar, A. (2003). Joint Forest Management in Sub-Himalayan West Bengal, India: Conditions for Institutional Sustainability, Paper accepted for presentation to the XII World Forestry Congress to be held in Quebec, during September 2003.

Chambers, R., Saxena, N.C. and Shah, T. (1989). *To the hands of the poor: water and trees.* Oxford and IBH, New Delhi and Intermediate Technology, London.

Chandra, J.P. (1986). Poplar - A Cash Crop for North Indian Farmers. *The Indian Forester* 112 (8).

Chatha, I.S., Singh, J. and Sidhu, D.S. (1991). An Economic Analysis of Eucalyptus Plantation in Punjab, Paper presented at the workshop on Socio-Economic Aspects of Tree Growing by Farmers in South Asia, held at the Institute of Rural Management Anand, March 11-15, 1991.

Chaturvedi, A.N. (2000). Viability of joint forest management projects. *The Indian Forester,* 126(5), 499-504.

Chaubey, O.P. and Pandey, A. (2001). Eco-silvicultural options in the management of JFM areas in M.P. *Journal of Tropical Forestry* 17(2), 1-17.

Choudhury, S.D. (1993). A Challenging Task in Shyampura, *Wastelands News* Feb.-April, Society for Promotion of Wastelands Development, New Delhi.

Chipeta, M.E. (2000). Indian Forestry on the International Stage: India's Quest for its Own Future. Proceedings of the Workshop 'India's Forests Beyond 2000', April 19-21, 2000, New Delhi, Commonwealth Forestry Association (India Chapter) and Indian Council of Forestry Research and Education, Dehradun, pp. 33-59.

Chitwadgi, S.S. (1999).Wood for Domestic fuel—A crisis. *The Indian Forester* 125(1), 67-71.

Cote, M.A. and Bouthillier, L. (1999). Analysis of the relationship among stakeholders affected by sustainable forest management and forest certification, *Forestry Chronicle* 75(6), 961-965.

CREATE (2002). 'Forest Structure, Resource Use and Institutions: Site Report from Jayanti, Buxa Tiger Reserve, Jalpaiguri, India'.

CSE: Centre for Science and Environment (1985). The State of India's Environment 1984-85: The Second Citizen's Report, Centre for Science and Environment, New Delhi.

Dagar, J.C. (1987). *Species composition and plant biomass of an ungrazed and a grazed grassland at Ujjain, India*; Tropical Ecology 28:208-215.

Dasgupta, S. (2000). People's involvement in resource management in Madhya Pradesh. *The Indian Forester,* 126(5), 525-536.

Datta, R.K. and Aggarwal, M. (2002). Effect of tree plantations on the soil characteristics and microbial activities of coal mine spoil land. *Tropical Forestry* 43(2), 315-324.

De, H.K. and Babu, A.R. (1998). *Constraints in the implementation of joint forest management.* Advances in Forestry Research in India, 18, 1-20.

Dhanagare, D.N. (2000). Joint forest management in UP: people, panchayats and women. *Economic and Political Weekly,* 35(37), 3315-3324.

Dhillon, N. and Malcolm, S. (2001). EU-India initiative for CSR, *Financial Express*, India. (Dated 21-11-2001).

Dixit, A.M. (1997). Ecological evaluation of dry tropical forest vegetation: an approach to environmental impact assessment; *Tropical Ecology* 38(1): 87-99.

Dury, S.J., Polglase, P.J. and Vercoe, T. (2002). *Green House Resource Kit for Private Forest Growers*; Department of Agriculture, Fisheries and Forestry, Australia.

Dutta, A.C. (1985). *Studies in Botany*; Moulik Press, 18-B, Shyamacharan Dc Street, Kolkata-73.

FAO: Food and Agricultural Organisation (1988). *Case Studies of Farm Forestry and Wasteland Development in Gujarat* (mimeo), Bangkok.

FAO: Food and Agricultural Organization (2000). *Global Forest Resource Assessment 2000*—Main Report, Forestry Paper 140.

FIPPI: Federation on Indian Plywood and Panels Industries (2000), *Annual Circular,* 2000, New Delhi.

Foy, T.J. and Pitcher, M.J. (1999). Private sector/community partnerships for commercial forestry development: recent experiences from South Africa. *International Forestry Review* 1(4), 222-226.

French, D. (1985). "The economics of bioenergy in developing countries", in Egneus, H. and Ellegaard, A. (eds.) *Bioenergy 84: Proceedings of an International Conference*, 5 London: Elsevier pp. 161-170.

FSI: Forest Survey of India (2001). *State of Forest Report* 2001, MoEF, Dehradun, India.

Furer-Haimendorf, C.V. (1985). *Tribes of India: The Struggle for Survival,* Oxford University Press, New Delhi.

Ganguli, B.N. (1995). *Breakthroughs in Forestry Development—Experience of the Asian Development Bank* (Asian Development Bank Publication).

Ganguli, B.N. (2000). Forest Management Options Beyond 2000. Issues and Opportunities for India. Proceedings of the Workshop 'India's Forests Beyond 2000', April 19-21, 2000, New Delhi, Commonwealth Forestry Association (India Chapter) and Indian Council of Forestry Research and Education, Dehradun, pp. 33-59.

Ganguli, B.N. (2003). Private sector participation on public forestlands: challenges and policy issues. In: Forest policy for private forestry: Global and regional challenges, CAB International.

Ghosh, S. (2000). *Bipannaya Paribesh; Ganatantrik Mancha,* Kolkata.

Goel, A.K. and Mamgain, S.K. (2001). Ex-situ conservation of some endemic and threatened plant species of India; *Indian Forester* 127(5), 552-562.

GOI: Govt. of India (1990). Involvement of Village Communities and Vas in the Regeneration of Degraded Forest Lands.

GOI: Government of India (1999). National Forestry Action Programme – Ministry of Environment and Forests, New Delhi, India.

GOI: Govt. of India (2000). Guidelines for strengthening of JFM programme in India. Ministry of Environment and Forests, Govt. of India. *The Indian Forester,* 126(5), 583-586.

GOMP: Govt. of Madhya Pradesh (2002a): Lok Vaniki Act, 2001. Madhya Pradesh Government, Bhopal, MP, India.

GOMP: Govt. of Madhya Pradesh (2002b): Lok Vaniki Rule, 2002. Madhya Pradesh Government, Bhopal, MP, India.

Guha, R.C. (1983). Forestry in British and Post-British India, *Economic and Political Weekly,* October 29.

Gupta, R. and Srivastava, S.K. (2004). Impact of participatory Forest Management on socio-economic development of rural people: A case study in Kodsi and Talaichittor villages of Dehradun district. *The Indian Forester* 130(3), 243-252.

Helms, A.J. (1998). Enhancing the Focus on Science in the Profession. *Journal of Forestry* 96(4), 1-7.

Hiremath, S.R., Kanwalli, S. and Kulkarni, S. (1994). *All About Draft Forest Bill and Forest Lands,* Samaj Parivartan Samudaya, Dharwad.

ICFRE: Indian Council of Forestry Research and Education (2001). *Forestry Statistics 2001.* Dehradun, India.

IIM: Indian Institute of Management (1991). *Study of Major and Minor Forest Products in West Bengal,* Prepared for the World Bank. February 4, Indian Institute of Management, Calcutta.

IMRB: Indian Marketing Research Bureau (1989). *Marketing of Farm Forestry Product in Tamil Nadu.* New Delhi: Indian Marketing Research Bureau.

IPCC: Intergovernmental Panel on Climate Change (1998). *The Regional Impacts of Climate Change: An Assessment of Vulnerability* (New York: Cambridge University Press).

IPW India Ply Web (2002) India's forest resources growing rapidly, available at *http://www.indiaplyweb.com/news/news.asp?newsid=68* (accessed on April 2003).

ITTO: International Tropical Timber Organization (2001) Annual Review and Assessment of World Tropical Timber Sitaution, 2001., Yokahama, available at *http://www.itto.or.jp/inside/review2001/download/AnnualReview2001.pdf* (accessed on June 2003).

Jacobson, M.G. and Abt, R.C. (2000). Attitude toward Joint Forest Planning among private landowners. *Journal of Sustainable Forestry,* 34 (2), 95-112.

Jha, M. (2000). Policies on joint forest management and its implementation in Maharashtra. *The Indian Forester,* 126(9), 917-922.

Julitz, T. and Thomas, D. (2001). Simplifying the management of farm forests. *Tropical Forest Update* 11(1), 10-13.

Kadavul, K. and Parthasarathy, N. (1999). Structure and composition of woody species in tropical semi-evergreen forest of Kalyan hills, Eastern Ghats, India; *Tropical Ecology* 40(2):247-260.

Kameswari, V.L.V. (2002). Gendered communication and access to social space: issues in forest management. *Economic and Political Weekly,* 37(8), 797-867.

Kant, S. and Cooke, R. (1999). Jabalpur district, Madhya Pradesh, India: minimizing conflict in joint forest management. Cultivating peace: conflict and collaboration in natural resource management, 19(9), 81-97

Kapoor, R., Ghose, A. and Upadhyaya, A.K. (1998). Status of Joint Forest Management in Rajasthan, *My Forest* 34(3), 825-831.

Kapp, G.B. (1998). Financial evaluation tools for small holder forestry: A methodological comparison of two forms of CBA analysis and optimisation. *International Tree Crops Journal* 9, 233-246.

Khare, A. and Rao, A.V.R. (1991). "Products of Social Forestry - Issues, Strategies and Priorities", *Wastelands News* 6(4): 7-17.

Kumar, A. (2000). A decade of JFM in India: looking back for better foresight. *The Indian Forester,* 126, (5), 579-582.

Kumar, S. (2002). Does "participation" in common pool resource management help the poor? A social cost-benefit analysis of joint forest management in Jharkhand, India. *World Development (Oxford),* 30(5), 763-782.

Kumar, R., Murphy, D.F. and Balsari, V. (2001). Altered Images - The state of corporate responsibility in India poll, *TERI,* India.

Locke, C. (1999). Constructing a gender policy for joint forest management in India. *Development and Change,* 30(2), 265-285.

Lal, P. (1991). Industrial Plantations. In: *Biotechnology in Agriculture.* Ed. M.S. Swaminathan. Macmillan India Limited, Madras, pp. 145-154.

Lal, P. (1992). Economics of Mass Clonal Multiplication of Forest Trees. *The Indian Forester* 120(2), 85-95.

Lal, P. (1994). Technology for Improving Productivity of Plantations. *IPPTA* 6(4), 1-8.

Lal, P. (1995). Role of Support Services in Agroforestry Development. In *Summary Report of "International Workshop on Agroforestry Investment, Production and Marketing".* Eds. James J. French and Ulla Blicher Mathiesen—APAN Report No.20, Asia-Pacific Agroforestry Network, Bogor, Indonesia, pp. 53-58.

Lal, P. (1993). Farm Forestry in India - Problems and Potential. In. *Proceedings of the National Workshop on "Farm Forestry Management",* Eds. Prof.(Dr.) T.H. Babu, *IIFM,* Bhopal. pp. 178-190.

Lal, P. (1995). High Productivity through Clonal Eucalyptus Plantations in India. In: *Tree Breeding and Propagation News.* Eds. K.Vivekanandan and R. Finkeldey. 4(2), p 7.

Lal, P. (1996). Pulp and Paper Industry in India - Captive Plantations for Survival and Growth. *IPPTA Convention Issue,* pp. 119-127.

Lal, P. (2000). National forest policy and raw material supplies for wood based industries in India. *The Indian Forester* 126(4), 351-366.

Lal, P. and Kulkarni, H.D. (1993) Sustainable Development Through Application of Clonal Technology in Forestry/Agroforestry. In: *Biodiversity Implications for Global Food Security.* Ed. M.S. Swaminathan, Macmillan India Limited, Madras : 205-213.

Lal, P., Kulkarni, H.D. and Rao, S.N. (1993). Improving Land Productivity and Returns from Agroforestry Plantations. *The Indian Forester* 119 (6), 431-440.

Lal, P., Kulkarni, H.D. and Srinivas, K. (1996). Clonal Planting Stock of Casuarina—ITC Bhadrachalam Experience. In: *Recent Casuarina Research and Development,* Proceedings of the Third International Casuarina Workshop, Da Nang, Vietnam. Eds. K. Pinyopusarerk, J.W. Turnbull, S.J. Midgley, CSIRO, Canberra, pp. 162-166.

Lal, P., Kulkarni, H.D., Srinivas, K., Venkatesh, K.R. and Santha-Kumar, P. (1997). Genetically Improved Clonal Planting Stock of Eucalyptus—A Success Story From India. *The Indian Forester* 123 (12), 1117-1138.

Le-Net, E. (2000). *Forest resources held by industrial interests: some example from abroad.* Informations-Foret, Afocel 609, p. 6.

Lynch, O.J. (1992). Securing Community-based Tenurial Rights in the Tropical Forests of Asia, Issues in Development, November, *WRI,* Washington

Maithani, G.P. (1988). "Survey of Shrubs for hastening the process of reclamation of ecologically vulnerable areas of Central Himalayas", *Indian Forester* 114 (5).

MPFD: Madhya Pradesh Forest Department (1999). *Madhya Pradesh Forestry Action Plan.* Forest Department, Government of MP, Bhopal, India.

Malhotra, K.C. and Poffenherger, M. (1989). *Forest regeneration through community protection: the West Bengal experience.* Calcutta, West Bengal Forestry Department.

Mayers, J. and Vermeulen, S. (2002). *Company-community forestry partnerships: from raw deals to mutual benefits.* London, UK, International Institute for Environment and Development (IIED).

Mills, N.L. and Ford, J. (1999). *Privatising Sustainable Forestry. A Global Review of Trends and Challenges.* International Institute of Environment and Development (London).

Misra, K.C., Rao, D.N., Ambasht, R.S., Dwivedi, R.S., Mukherjee, K.L. and Ananthakrishnan, P.V. (1970). *Ecology (study of Ecosystem)*; A.H. Wheeler and Co. (P) Ltd., Allahabad, India.

MoEF: Ministry of Environment and Forests (1997), *Report of the expert committee on 'Review of afforestation policies and rehabilitation of wastelands'*, Ministry of Environment and Forests, Government of India, New Delhi.

Mohan, A. (2001). Corporate Citizenship: Perspectives from India, *Journal of Corporate Citizenship*, Greenleaf Publishing Ltd.

Murthy, K.N. (2003). International timber trade – Indian imperatives, *Wood News* 13(3), 8-11.

Nair, C.T.S. (1985). "Crisis in forest resource management", in Bandyopadhyaya, J. *et al.* (eds), *India's Environment: Crisis and Responses*, Natraj Publishers, Dehradun.

Nair, K.S.S., Chundamannil, M., Chacko, K.C., Ganapathy, P.M. and Chidambara Iyer, S. (1996) *Private Sector Involvement in Forestry Research in India;* Kerala Forest Research Institute, Peechi.

National Bureau of Soil Survey and Land Use Planning. (NBSS&LUP) Nagpur. (A Unit of ICARM New Delhi). Soil Resource Map of M.P. 1997.

NBSS and LUP: National Bureau of Soil Survey and Land Use Planning (1997). *Soil resource map of Madhya Pradesh.*

NCAER: National Council of Applied Economic Research (1987). *Haryana Wood Balance Study* (mimeo), National Council of Applied Economic Research, New Delhi.

Narayanswamy, T. and Raghupati, D. (2003). Developing an eco-development strategy for a degraded forest villages of Andhra Pradesh. *My Forest* 39(2), 117-123.

Nayak, P.K. (2003). Community based forest management in India: the significance of tenure. *Forests, trees and livelihoods* 13 (2), 135-160.

Negi, M.S., Ansari, M.Y. and Singh, H.P. (2001). Market channels in Agro-forestry products—A case study of Yamuna Nagar. *The Indian Forester* 127(5), 519-525.

Ngantou, D., Ndikumagenge, C., Wehiong, S.M. and Angu, K.A. (2003). Paths to partnerships in Central Africa. *Tropical Forest Update* 13 (4), 11-13.

Nimbkar, B.V. (1988). "The Greening of Maharashtra", *Economic and Political Weekly*, 20th February.

ODA : Overseas Development Agency (1989). *Evaluation of the Karnataka Social Forestry Project.* London: ODA.

Odum, E.P. (1966). *Ecology*; Holt, Rinehart and Winston (Modern Biology Series), U.S.A.

Odum, E.P. (1971). *Fundamentals of Ecology*; W.B.Saunders Company, Philadelphia.

Pal, M. (1993). Advancements made in clonal propagation in India and their potential large-scale implementation. Proceedings of the Regional Symposium on Recent Advances in Mass Clonal Multiplication of Forest Trees for Plantation Programmes held between 1-8 December, 1992, at Cisarua, Bogor, Indonesia. UNDP/ FAO Field Document No. 4, RAS/91/004. Ed. John Davidson : 196-203.

Pathan, R.S. (1998). Role of industries in eco-development through afforestation—A case of Dadar and Nagar Haveli. *The Indian Forester* 128(1), 31-36.

Planning Commission (1998). Leasing of degraded forestlands. Working Group's report on the prospects of making degraded forests available to private entrepreneurs, Planning Commission, Government of India, New Delhi.

Poffenberger, M., Bhatia, K. and McGean, B. (1990). *Forest management partnerships: regenerating India's forests*. New Delhi. The Ford Foundation and the Indian Environmental Society.

Poffenberger, M. and McGean, B. (1995). Village Voices, Forest Choices, *OUP,* New Delhi.

Pratima and Jattan, S.S. (1999). Participatory Forest Management to meet basic needs of forest dependent communities. *The Indian Forester,* 125(9), 895-901.

Pratima and Jattan, S.S. (2002). Emerging issues in participatory forest management. *The Indian Forester,* 128(8), 863-867.

Prasad, B. (2000). Monitoring of joint forest management in India: issues and methods. *The Indian Forester,* 126(5), 516-524.

Rao, S.K. (2002) Current trends of wood use: an Indian perspective, *Wood News* 12 (2), 24-28.

Rathore, B.M.S. and Campbell, J. (1995). Silvicultural Needs for Joint Forest Management, *Wastelands News,* Society for Promotion of Wastelands Development, New Delhi.

Rawat, J.K. (1994). Role of industry in the development of degraded forests: A need for change in policy. *The Indian Forester* 120(5), 390-394.

Ray, P.N. (2000). Status of joint forest management in Tripura. *The Indian Forester,* 126(5), 483-492.

Reddi, P.R. and Reddy, P.S. (1995). *Commentaries on Land Reforms Laws in Andhra Pradesh.* 3rd Edn. Asia Law House, Hyderabad, p. 100.

Rishi, P. (2002). Joint forest management at village level: a cognitive analysis. *The Indian Forester,* 128(5), 539-544.

Rukmantara (2003). Repairing the Road to SFM. ITTO *Tropical Forest Update* 13 (4), 6-7.

Saigal, S., Arora, H. and Rizvi, S.S. (2002). *The new foresters: the role of private enterprise in the Indian forestry sector.* Instruments for sustainable private sector forestry series. Ecotech Services, New Delhi and International Institute for Environment and Development, London.

Sarin, M., Singh, N.M., Sundar, N. and Bhogal, R.K. (2003). Devaluation as a Threat to Democratic Decision-making in Forestry? Findings from Three States in India, Working Paper 197, Overseas Development Institute, London.

Saxena, N.C. (1990). *Farm Forestry in North-west India,* Ford Foundation, New Delhi.

Saxena, N.C. (1991). "Marketing Constraints for Eucalyptus from Farm Forestry in India", *Agroforestry Systems,* 13(1) 73-85.

Saxena, N.C. (1994). *India's Eucalyptus Craze: The God that Failed,* Sage, New Delhi.

Saxena, N.C. (1995). Forest Policy and the Rural Poor in Orissa, *Wasteland News,* Oct.-Dec. Society for Promotion of Wastelands Development, New Delhi.

Saxena, N.C. (2003). From monopoly to deregulation of NTFPs: Policy shifts in Orissa (India)". *The International Forestry Review* 5(2), 168-176.

Saxena, N.C. and Ballabh, V. (1995). Farm forestry and the context of farming systems, In Saxena, N.C. and Ballabh, V. (eds.), *Farm Forestry in South Asia,* Sage Publications, New Delhi.

Sedjo, R.A. (1999). Potential for Carbon Forest Plantations in Marginal Timber Forests: The Case of Patagonia, Argentina. Resources For the Future, Discussion Paper 19-27.

Sedjo, R.A. and Botkin, D. (1997). Using forest plantations to spare natural forests. *Environment* 39(10),14-20.

Sedjo, R.A., Goetzl, A. and Morfat, S.O. (1998). Sustainability of Temperate Forests (Resources For the Future, Washington DC).

Sethi, P. and Khan, H. (2001a). Structuring financial empowerment for localized development within Joint Forest Management (JFM): examples from Madhya Pradesh, India. Sustainable Development, 9(2), 87-102.

Sethi, P. and Singh, T.P. (2001b). Village resource development as an incentive to sustain the joint forest management programme. *The Indian Forester,* 127(11), 1215-1222.

Sharma, J. (1998). Joint forest planning and management approach: An analysis of the experience gained. *My Forest* 34(3), 833-837.

Sharma, R.C. (1997). Total Forest Management (TFM) an innovative approach for conservation of natural resources with human face. *The Indian Forester,* 123(6), 447-459.

Sharma, V.P. and Kumar, A. (1999). Demand for forest products in India: Post trend and projections to 2010. *The Indian Forester* 125(1), 111-126.

Shiva, B. (1991). *Ecology and the Politics of Survival*; Sage Publications India Pvt. Ltd., 32, M-Block Market, Greater Kailash-I, New Delhi-110048.

SIDA : Swedish International Development Agency (1990). Report on Pilot Market Studies of Farm Forestry Produce in India.

Singh, A. (1992). Trends in timber and pulp import in Anil Agarwal (ed.). *The Price of Forests,* Centre for Science and Environment, New Delhi.

Singh, D.P. (1998). *Socio-economic Development of Madhya Pradesh through Forestry by the people,. of the people and for the people.* Madhya Pradesh Government, Bhopal, MP, India.

Singh, H.S. (1997). Appropriate silvicultural models under JFM have—Gujarat State. *The Indian Forester* 123(6), 477-483

INDEX